Satellite Data Compression

Bormin Huang

Editor

Satellite Data Compression

 Springer

Editor
Bormin Huang
Space Science and Engineering Center
University of Wisconsin – Madison
Madison, WI, USA
bormin@ssec.wisc.edu

ISBN 978-1-4614-1182-6   e-ISBN 978-1-4614-1183-3
DOI 10.1007/978-1-4614-1183-3
Springer New York Dordrecht Heidelberg London

Library of Congress Control Number: 2011939205

Printed on acid-free paper

Springer is part of Springer Science+Business Media (www.springer.com)

Contents

Contributors

Andrea Abrardo Dip. di Ingegneria dell'Informazione,
Università di Siena, Siena, Italy

Bruno Aiazzi IFAC-CNR, Via Madonna del Piano 10, 50019 Sesto F.no, Italy

Luciano Alparone Department of Electronics & Telecommunications,
University of Florence, Via Santa Marta 3, 50139 Florence, Italy

Mauro Barni Dip. di Ingegneria dell'Informazione, Università di Siena,
Siena, Italy

Stefano Baronti IFAC-CNR, Via Madonna del Piano 10, 50019 Sesto F.no, Italy

Andrea Bertoli Carlo Gavazzi Space S.p.A., Milan, Italy

Ian Blanes Universitat Autònoma de Barcelona, E-08290 Cerdanyola
del Vallès (Barcelona), Spain

Roberto Camarero Ctr. National d'Études Spatiales (CNES), Toulouse, France

Chein-I Chang Remote Sensing Signal and Image Processing Laboratory,
Department of Computer Science and Electrical Engineering,
University of Maryland, Baltimore, MD, USA

Department of Electrical Engineering, National Chung
Hsing University, Taichung, Taiwan

Qian Du Department of Electrical and Computer Engineering,
Mississippi State University, Starkville, USA

Yong Fang College of Information Engineering, Northwest A&F University,
yangling, China

James E. Fowler Department of Electrical and Computer Engineering,
Mississippi State University, Starkville, USA

Enrique García-Berro Institut d'Estudis Espacials de Catalunya,
Barcelona, Spain

Departament de Física Aplicada, Universitat Politècnica de Catalunya,
Castelldefels, Spain

Raoul Grimoldi Carlo Gavazzi Space S.p.A., Milan, Italy

Bormin Huang Space Science and Engineering Center,
University of Wisconsin–Madison, Madison, WI, USA

L.C. Jiao School of Electronic Engineering, Xidian University, xi'an, China

Chulhee Lee Yonsei University, South Korea

Jonghwa Lee Yonsei University, South Korea

Sangwook Lee Yonsei University, South Korea

Jie Lei State Key Laboratory of Integrated Service Networks,
Xidian University, Xi'an, China

Yunsong Li State Key Laboratory of Integrated Service Networks,
Xidian University, Xi'an, China

Kai Liu State Key Laboratory of Integrated Service Networks,
Xidian University, Xi'an, China

Enrico Magli Dip. di Elettronica, Politecnico di Torino, Torino, Italy

Jarno Mielikainen School of Electrical and Electronic Engineering,
Yonsei University, Seoul, South Korea

Antonio J. Plaza Department of Technology of Computers and Communications,
University of Extremadura, Escuela Politecnica de Caceres, Caceres, SPAIN

Jordi Portell Departament d'Astronomia i Meteorologia/ICCUB,
Universitat de Barcelona, Barcelona, Spain

Institut d'Estudis Espacials de Catalunya, Barcelona, Spain

Shen-En Qian Canadian Space Agency, St-Hubert, QC, Canada

Haleh Safavi Remote Sensing Signal and Image Processing Laboratory,
Department of Computer Science and Electrical Engineering,
University of Maryland, Baltimore, MD, USA

Peter Schelkens Department of Electronics and Informatics,
Vrije Universiteit Brussel, Pleinlaan 2, B-1050 Brussels, Belgium

Interdisciplinary Institute for Broadband Technology, Gaston Crommenlaan 8,
b102, B-9050 Ghent, Belgium

Joan Serra-Sagristà Universitat Autònoma de Barcelona, E-08290 Cerdanyola del Vallès (Barcelona), Spain

Juan Song State Key Laboratory of Integrated Service Networks, Xidian University, Xi'an, China

Carole Thiebaut Ctr. National d'Études Spatiales (CNES), Toulouse, France

Alberto G. Villafranca Institut d'Estudis Espacials de Catalunya, Barcelona, Spain

Departament de Física Aplicada, Universitat Politècnica de Catalunya, Castelldefels, Spain

Raffaele Vitulli European Space Agency – ESTEC TEC/EDP, Noordwijk, The Netherlands

Keyan Wang State Key Laboratory of Integrated Service Networks, Xidian University, Xi'an, China

Lei Wang School of Electronic Engineering, Xidian University, xi'an, China

Shih-Chieh Wei Department of Information Management, Tamkang University, Tamsui, Taiwan

Chengke Wu State Key Laboratory of Integrated Service Networks, Xidian University, Xi'an, China

Jiaji Wu School of Electronic Engineering, Xidian University, xi'an, China

Wei Zhu Department of Electrical and Computer Engineering, Mississippi State University, Starkville, USA

Chapter 1
Development of On-Board Data Compression Technology at Canadian Space Agency

Shen-En Qian

Abstract This chapter reviews and summarizes the researches and developments on data compression techniques for satellite sensor data at the Canadian Space Agency in collaboration with its partners in other government departments, academia and Canadian industry. This chapter describes the subject matters in the order of the following sections.

1 Review of R&D of Satellite Data Compression at the Canadian Space Agency

The Canadian Space Agency (CSA) began developing data compression algorithms as an enabling technology for a hyperspectral satellite in the 1990s. Both lossless and lossy data compression techniques have been studied. Compression techniques for operational use have been developed [1–16]. The focus of the development is the vector quantization (VQ) based near lossless data compression techniques. Vector quantization is an efficient coding technique for data compression of hyperspectral imagery because of its simplicity and its vector nature for preservation of the spectral signatures associated with individual ground samples in the scene. Reasonably high compression ratios (>10:1) can be achieved with significantly high compression fidelity for most of remote sensing applications and algorithms.

S.-E. Qian (✉)
Canadian Space Agency, St-Hubert, QC, Canada
e-mail: shen-en.qian@asc-csa.gc.ca

B. Huang (ed.), *Satellite Data Compression*, DOI 10.1007/978-1-4614-1183-3_1,
© Springer Science+Business Media, LLC 2011

1

1.1 Lossless Compression

In the 1990s, a prediction-based approach was adopted. It used a linear or nonlinear predictor in spatial or/and spectral domain to generate prediction and applies DPCM to generate residuals followed by an entropy coding. A total of 99 fixed coefficient predictors covering 1D, 2D and 3D versions were investigated [15]. An adaptive predictor that chose the best predictor from a pre-selected predictor bank was also investigated. The Consultative Committee for Space Data System (CCSDS) recommended lossless algorithm (mainly the Rice algorithm) [17] was selected as the entropy encoder to encode the prediction residuals. In order to evaluate the performance of the CCSDS lossless algorithm, an entropy encoder referred to as Base-bit Plus Overflow-bit Coding (BPOC) [18] was also utilized to compare the entropy coding efficiency.

Three hyperspectral datacubes acquired using both the Airborne Visible/Near Infrared Imaging Spectrometer (AVIRIS) and the Compact Airborne Spectrographic Imager (*casi*) were tested. Two 3D predictors, which use five and seven nearest neighbor pixels in 3D, produced the best residuals. Compression ratios around 2.4 were achieved after encoding the residuals produced by using these two entropy encoders. All other predictors produced compression ratios smaller than 2:1. The BPOC slightly outperformed the CCSDS lossless algorithm.

1.2 Wavelet Transform Lossy Data Compression

We have also developed a compression technique for hyperspectral data using wavelet transform techniques [16]. The method for creating zerotrees as proposed by Shapiro [19] has been modified for use with hyperspectral data. An optimized multi-level lookup table was introduced which improves the performance of an embedded zerotree wavelet algorithm. In order to evaluate the performance of the algorithm, this algorithm was compared to SPIHT [20] and JPEG. The new algorithm was found to perform as well as or to surpass published algorithms which are much more computationally complex and not suitable for compression of hyperspectral data. As in Sect. 1.1 testing was done with both AVIRIS and *casi* data and compression ratios over 32:1 were obtained with fidelities greater than 40.0 dB.

1.3 Vector Quantization Data Compression

We selected VQ technique for data compression of hyperspectral imagery because of its simplicity and because its vector nature can preserve the spectra associated with individual ground samples in the scene. Vector quantization is an efficient coding technique. The generalized Lloyed algorithm (GLA) (sometimes also called LBG) is the most widely used VQ algorithm for image compression [21].

In our application, we associate the vectors used in the VQ algorithm with the full spectral dimension of each ground sample of a hyperspectral datacube that we refer to as a *spectral vector*. Since the number of targets in the scene of a datacube is limited, the number of trained spectral vectors (i.e. codevectors) can be much smaller than the total number of spectral vectors of the datacube. Thus, we can represent all the spectral vectors using a codebook with comparatively few codevectors and achieve good reconstruction fidelity.

VQ compression techniques make good use of the high correlation often found between bands in the spectral domain and achieves a high compression ratio. However, a big challenge to VQ compression techniques for hyperspectral imagery in terms of operational use is that it requires large computational resources, particularly for the codebook generation phase. Since the size of hyperspectral datacubes can be hundreds of times larger than those for traditional remote sensing, the processing time required to train a codebook or to encode a datacube using the codebook could also be tens to hundreds of times larger. In some applications, the problem of training time can largely be avoided by training a codebook only once, and henceforth applying it repeatedly to all subsequent datacubes to be compressed as adopted in the conventional 2D image compression. This works well when the datacube to be compressed is bounded by the training set used to train the codebook.

However, in hyperspectral remote sensing, it is in general very difficult to obtain a so-called "universal" codebook that spans many datacubes to the required degree of fidelity. This is partly because the characteristics of the targets (season, location, illumination, view angle, the needs for accurate atmospheric effects) and instrument configuration (spectral and spatial resolution, spectral range, SNR of the instruments) introduce high variability in the datacubes and partly because of the need for high reconstruction fidelity in its downstream use. For these reasons, it is preferred that a new codebook is generated for every datacube that is to be compressed, and is transmitted to the decoder together with the index map as the compressed data. Thus, the main goal in the development of VQ based techniques to compress hyperspectral imagery is to seek a much faster and more efficient compression algorithm to overcome this challenge, particularly for on-board applications. In this chapter, the compression of a hyperspectral datacube using the conventional GLA algorithm with a new trained codebook for a datacube is referred to as 3DVQ herein.

The CSA developed an efficient method of representing spectral vectors of hyperspectral datacubes referred to as Spectral Feature Based Binary Code (SFBBC) [1]. With SFBBC code, Hamming distance, rather than Euclidean distance, could be used in codebook training and codevector matching (coding) in the process of the VQ compression for hyperspectral imagery. The Hamming distance is a simple sum of logical bit-wise exclusive-or operations. It is much faster than Euclidean distance. The SFBBC based VQ compression technique can speed up the compression process by a factor of 30–40 at a fidelity loss of PSNR <1.5 dB compared to the 3DVQ. This compression technique is referred to as SFBBC based 3DVQ.

Later, the Correlation Vector Quantization (CVQ) was developed [2]. It uses a movable window to cover a block of 2×2 spectral vectors of adjacent ground samples of a hyperspectral datacube and removes both the spectral and spatial correlation of the datacube simultaneously. The coding time (CT) can be improved by a factor of $1/(1 - \beta)$, where β is the probability that a spectral vector in the window can be approximated by one of the three coded spectral vectors in the window. The experimental results showed that the coding time could be improved by a factor of around 2 and the compression ratio was 30% higher than that using 3DVQ. CVQ can be combined with SFBBC to further speed up the coding time of 3DVQ [3].

The codebook generation phase is an iterative process and dominates the overall processing time of the 3DVQ, thus it is critical to reduce the codebook generation time (CGT). Since the CGT is roughly proportional to the size of the training set, it follows that a faster compression system can be obtained simply by reducing the size of the training set. The CSA developed three spectral vector selection schemes to sub-sample spectral vectors in a datacube to be compressed to form a small and yet efficient training set for codebook generation. The numerical analysis showed that a sub-sampling rate of 4% is the optimal for preserving the reconstruction fidelity and reducing the CGT. The experimental results showed that the processing time could be improved by a factor of 15.6–17.4 at a loss of PSNR of 0.6–0.7 dB, when the training set was composed of sub-sampling the datacube at a rate of 2.0% [4].

The CSA further improved 3DVQ using remote sensing knowledge contained in a hyperspectral datacube to be compressed. A spectral index, such as the Normalized Difference Vegetation Index (NDVI), was introduced to benefit the compression algorithm. A novel VQ based compression technique referred to as spectral index based Multiple Sub-Codebook Algorithm (MSCA) was developed [5]. A spectral index map is first created for a datacube to be compressed. Then it is segmented into n (usually 8 or 16) distinct regions (or classes) based on the index values. The datacube is divided into n subsets according to the segmented index map. Each subset corresponds to a region (or class). An independent codebook is trained for each of the subsets, and is applied to compress the corresponding subset. The MSCA can speed up both the CGT and CT by a factor of around n. The experimental results showed that both CGT and CT were improved by a factor of 14.1 and 14.8 when the scene of the test datacube is segmented into 16 regions, while the reconstruction fidelity was almost the same as that using 3DVQ.

Three VQ data compression systems for hyperspectral imagery were created and tested using the combination of the previously developed fast 3DVQ techniques: SFBBC, sub-sampling, and MSCA. The simulation results showed that the CGT could be reduced by over three orders of magnitude, while the quality of the codebooks remained good. The overall processing speed of the 3DVQ could be improved by a factor of around 1,000 at an average PSNR penalty of <1.0 dB [6].

A fast search method for VQ based compression algorithm has been proposed [7]. It makes use of the fact that in the full search of the GLA a training vector does not require a search to find the minimum distance partition if its distance to the partition is improved in the current iteration compared to its distance to the partition in the

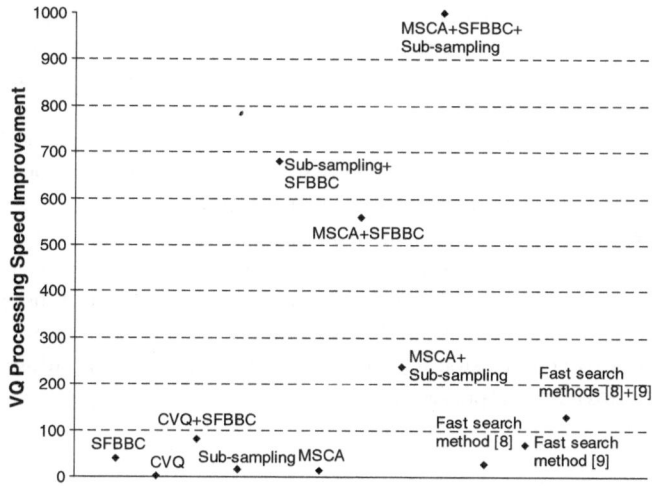

Fig. 1.1 Improvements on processing speed of 3DVQ attained by the CSA fast algorithms and their combination

previous iteration. The proposed method has the advantage of being simple, producing a large computation time saving and yielding compression fidelity as good as the GLA. Four hyperspectral datacubes covering a wide variety of scene types were tested. The experimental results showed that the proposed method improved the compression time by a factor of 3.08–27.35 for the four test datacubes with codebook size from 16 to 2048. The larger the codebook size, the more the time savings. The loss of spectral information due to compression was evaluated using the spectral angle mapper and a remote sensing application.

Following the work in [7], CSA further improved search method for vector quantization compression techniques [8]. It makes use of the fact that in GLA a vector in a training sequence is either placed in the same minimum distance partition (MDP) as in the previous iteration or in a partition within a very small subset of partitions. The proposed method searches for the MDP for a training vector only in this subset of partitions plus the single partition that was the MDP in the previous iteration. As the size of this subset is much smaller than the total number of codevectors, the search process is speeded up significantly. The proposed method generates a codebook identical to that generated using the GLA. The experimental results show that the computation time of codebook training was improved by factors from 6.6 to 50.7 and from 5.8 to 70.4 for two test data sets when codebooks of sizes from 16 to 2048 were trained. The computation time was improved by factors from 7.7 to 58.7 and from 13.0 to 128.7 for two test data sets when it was combined with the fast search method in [7].

Figure 1.1 shows the processing speed improvements of the fast VQ techniques above and their combination compared against to the 3DVQ. With a single fast technique, the factor of processing speed improvement is up to 70. The largest factor of processing speed improvement is around 1,000 when the three fast techniques are combined. The PSNR fidelity loss of the fast techniques is <1.5 dB compared to the 3DVQ when SFBBC technique is used. The fidelity loss of the fast techniques is <1.0 dB when sub-sampling is used. Other fast techniques have almost no loss of fidelity.

CSA has developed and patented two near lossless VQ data compression techniques for on-board processing: Successive Approximation Multi-stage Vector Quantization (SAMVQ) [10, 11] and Hierarchical Self-Organizing Cluster Vector Quantization (HSOCVQ) [12, 13]. Both of them are simple and efficient and have been designed specifically for on-board use [14] with multi-dimensional sensor data. These algorithms also have application to on-ground data compression although they are optimized for on-board hyperspectral data. In the next two sections we will briefly describe these two compression techniques and their features that allow them to be termed as near lossless compression.

2 Near Lossless Compression Technologies: SAMVQ and HSOCVQ

In our data compression development, we restrict the compression error introduced in the lossy compression process to the level of the intrinsic noise of the original data set. The intrinsic noise here refers to the overall noise or error contained in an original data set that is caused by the instrument noise and other error sources of the data set, such as errors or uncertainty introduced in the data processing chain (e.g. detector dark current removal, non-uniformity correction, radiometric calibration and atmospheric correction, etc.). This level of compression error is expected to have small to negligible impact on remote sensing applications of the data set compared to the original data. This kind of lossy data compression is referred to as *near lossless* compression in our practice. It is different from the visual near lossless reported in [22, 23] for medical images and the virtual-near lossless in [24, 25] for multi/hyperspectral imagery.

2.1 Successive Approximation Multi-Stage Vector Quantization (SAMVQ)

The SAMVQ is a multi-stage VQ compression algorithm and compresses a datacube using extremely small codebooks in successive approximations manner. The computational burden present in the conventional VQ methods is no longer a problem, as the codebook size N is over two orders of magnitude smaller. Assume that SAMVQ compresses a datacube using four codebooks in the multi-stage approximation process each containing eight codevectors. The equivalent conventional VQ codebook would need $N = 8^4 = 4,096$ codevectors to achieve the similar reconstruction fidelity, whereas the SAMVQ codebooks contain only $N' = 8 \times 4 = 32$ codevectors between them. Both the codebook training time and the coding time are improved by a factor of approximately $N/N' = 4{,}096/32 = 128$, as they are both proportional to the codebook size. Since the total number of codevectors is much smaller, the compression ratio of SAMVQ is greater than the conventional VQ method for the same fidelity of the reconstructed data. Equivalently, SAMVQ can obtain much higher reconstruction fidelity than the conventional VQ, at the same compression ratio as conventional VQ.

In addition, SAMVQ adaptively classifies/divides a datacube to be compressed into clusters (subsets) based on the similarity of spectrum features and compresses each subset individually. This feature further speeds up the processing time for on-board use, as it allows parallel operation in hardware implementations by assigning each subset to an individual processing unit. The processing time can be further improved by a factor of roughly 8, for example, if a datacube is divided into eight subsets. This feature also improves the reconstruction fidelity, since spectral vectors in each cluster are similar and can be encoded with much smaller coding distortions when the same number of codevectors is used.

The compression ratio and fidelity can be easily controlled by properly selecting the codebook size and the number of stages. The greater the number of stages, the higher the compression fidelity. In the course of compression, the algorithm can adaptively select the size of the codebook at each approximation stage to minimize distortion and maximize the compression ratio. For on-board use, SAMVQ can be set to operate in either Compression Ratio (CR) mode or Fidelity mode. In CR mode, a desired CR can be achieved by setting the parameters prior to compression. The compression fidelity then varies for different datacubes. The root mean squared error (RMSE) is often used as measure of fidelity; in the Fidelity mode, a threshold RMSE is set prior to compression and the algorithm will then ensure that the compression error will be less than or equal to the set threshold. Near lossless compression can be achieved when the threshold is set to the level consistent with the intrinsic noise of the original data. A detailed description of the SAMVQ algorithm can be found in [10, 11, 14].

2.2 Hierarchical Self-Organizing Cluster Vector Quantization (HSOCVQ)

Given a desired fidelity measure (such as RMSE), HSOCVQ compresses clusters of spectral vectors in a datacube until each spectral vector is encoded with an error less than the threshold. This feature allows HSOCVQ to better preserve the spectra of infrequent or small targets in compressed hyperspectral datacubes.

HSOCVQ first trains an extremely small number of codevectors (such as 8) from a datacube to be compressed and uses these codevectors to classify spectral vectors of the datacube into clusters. It then compresses spectral vectors in each of the clusters by training a small number of new codevectors. If all the spectral vectors in the cluster are encoded with an error less than the threshold, it completes to encode the current cluster and goes to next cluster. Otherwise, it splits the cluster into sub-clusters by classifying the spectral vectors in the cluster. A cluster is split into sub-clusters hierarchically until all the spectral vectors in each of the sub-clusters are encoded with an error less than the threshold. In HSOCVQ, the number of sub-clusters of a cluster to be split (i.e. the number of new codevectors) is determined adaptively. If the fidelity of a cluster is far from the threshold, a larger number of sub-clusters are generated. The clusters generated in this way are disjoint and their sizes decrease with splitting going to deep levels. Thus the compression process is fast and efficient, as both the training set (cluster) size and the codebook size are small and the spectral vectors in each cluster or sub-cluster are only trained once.

Thanks to the unique way of clustering and splitting, codevectors trained in HSOCVQ have a well-controlled reconstruction fidelity and there are few codevectors overall. High reconstruction fidelity is attained with a high compression ratio. For on board application, HSOCVQ operates only in Fidelity mode. Similar to SAMVQ, near lossless compression can be achieved when the error threshold is set to a level consistent with the intrinsic noise of an original data. A detailed description of the HSOCVQ algorithm can be found in [12–14].

3   Evaluation of Near Lossless Features of SAMVQ and HSOCVQ

The CSA carried out the evaluation study to examine the near lossless features of SAMVQ and HSOCVQ by comparing the compression errors with the intrinsic noise of the original data to see if the level of compression errors is consistent with that of intrinsic noise of the original data [14].

A data set acquired using the Airborne Visible/Near Infrared Imaging Spectrometer (AVIRIS) at low altitude in the Greater Victoria Watershed District, Canada on August 12, 2002 was used [26]. The ground sample distance (GSD) of the data set is 4×4 m with AVIRIS nominal SNR of 1,000:1 in the visible and near infrared (VNIR) region. A low-resolution datacube was derived by spatially averaging the 4×4 m GSD data set to form a 28×28 m GSD datacube. The nominal SNR of the spatially aggregated datacube is $1,000 \times \sqrt{7 \times 7} = 7,000 : 1$. This datacube is viewed as a noise free datacube in the evaluation, as the noise is too small to have a significant impact. This datacube is referred to as the "reference datacube". A "simulated datacube" with SNR = 600:1 was generated by adding simulated instrument noise and photon noise to the "reference datacube". The simulated datacube is considered to be representative of a real satellite hyperspectral data set, since SNR for such an instrument is likely to be around that level [27]. This simulated datacube was used as an input to the compression techniques for evaluation of the SAMVQ at a compression ratio of 20:1 and for HSOCVQ at a compression ratio of 10:1. The output of the techniques is "compressed data". The amplitude within all datacubes is expressed as a 16-bit digital numbers (DN).

When the "simulated datacube" is compared with the "reference datacube", the calculated error is the "intrinsic noise". When the "reconstructed datacubes" that were produced by decompressing the compressed data is compared with the "simulated datacube", the calculated error is the "compression error". When "reconstructed datacubes" is compared with the "reference datacube" the calculated error is the "intrinsic noise + compression error", which is the overall error or the final noise budget of the datacube, if the reconstructed data is sent to a data user for deriving their remote sensing products.

The standard deviation for each comparison was calculated and plotted as a function of spectral band number (wavelength). These standard deviations were used to evaluate the "compression error", "intrinsic noise" and the "intrinsic noise + compression error" as shown in Fig. 1.2. For SAMVQ, the "compression error" (*solid line*) is generally smaller than "the intrinsic noise" (*dotted line*) for most bands except in the *blue region* where the input data has low signal levels. The "intrinsic noise + compression error" (*thick broken line*) are smaller than the intrinsic noise (*dotted line*) in all bands. For HSOCVQ, the "compression error" is about 5–10 DN larger than the "intrinsic noise" for most bands, but the "compression error" is smaller for the bands with high amplitude. The "compression error + intrinsic noise" is smaller than "the intrinsic noise" between bands 35 and 105.

The evaluation results demonstrated that the compression errors introduced by SAMVQ and HSOCVQ are smaller than or comparable to the "intrinsic noise", which justifies that SAMVQ and HSOCVQ algorithms are considered as nearly lossless for remote sensing applications.

4 Effect of Anomalies on Compression Performance

It is important to evaluate the effect of anomalies in raw hyperspectral imagery on data compression. The evaluation results could help to decide whether or not an on-board data cleaning is required before compression. The CSA carried out the evaluation of the effect of anomalies in the raw hyperspectral data caused by detector and instrument defects on data compression. The anomalies examined were dead detector pixels, frozen detector pixels, spikes (isolated over-responded pixels) and saturation. Two raw hyperspectral datacubes acquired using airborne hyperspectral sensors Short Wave Infrared Full Spectrum Imager II (SFSI-II) and Compact Airborne Spectrographic Imager (*casi*) were tested. Statistics based measures RMSE, SNR and %E were used to evaluate the compression performance. Difference spectra between the original and reconstructed datacubes at spatial locations where anomalies occur were plotted and verified. The compressed SFSI-II datacubes were also evaluated using a remote sensing application – target detection [28].

Dead detector elements (zeros) are of fixed pattern in the same bands of the raw data and have no impact on VQ data compression, since they do not contribute to the codevector training or to the calculation of the compression fidelity. Frozen detector

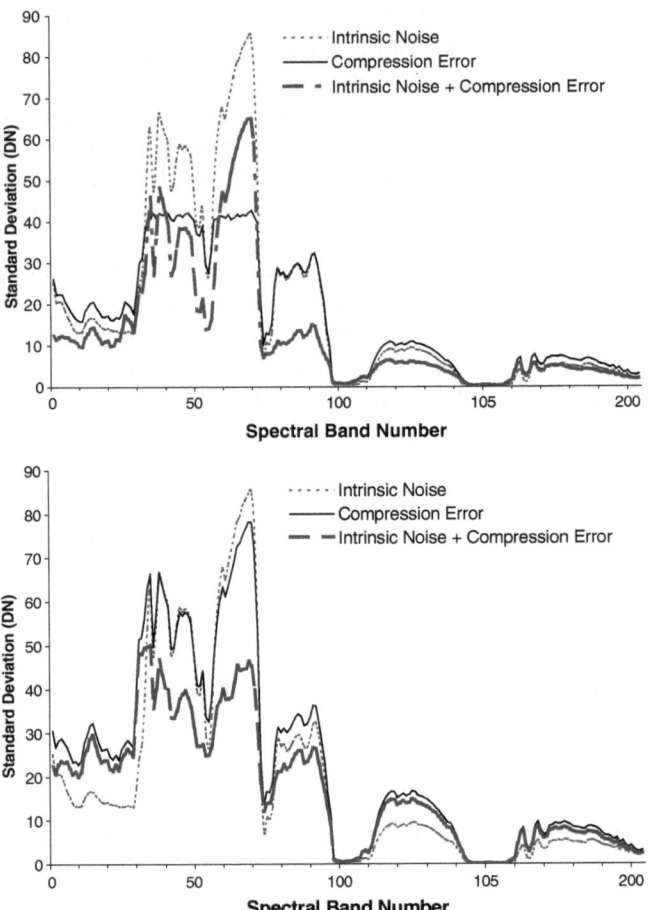

Fig. 1.2 Standard deviations of single band images for "intrinsic noise", "compression error" and overall error ("intrinsic noise + compression error"). *Left*: compressed using SAMVQ at 20:1, *Right*: compressed using HSOCVQ at 10:1

elements have a minor impact on data compression, since their values do not change in the bands of spectra where frozen detector elements occur. Thus, the evaluation was focused on the impact of spikes and saturation in raw hyperspectral data.

The experimental results showed that HSOCVQ is insensitive to both the spikes and saturations when the raw hyperspectral data is compressed. It produced almost the same statistical results, no matter if the spike and saturation anomalies were removed or not before compression.

The experimental results showed that SAMVQ is almost insensitive to spikes removal, since the compression fidelity is only slightly reduced (0.12–0.2 dB of SNR) after spikes were removed from the raw datacube before it was compressed. SAMVQ produced slightly better compression fidelity (from 0.08 to 0.3 dB of SNR) with removing the saturations than without removing the saturations when the raw datacube was compressed at ratios of 10:1 and 20:1. This is because (1) removal of the saturations did not change the dynamic range of the datacube and (2) an entire spectrum was replaced by a unique typical spectrum if a spectrum was found containing saturation in a single spectral band. This approach to removing saturations increases the occurrence frequency of the typical spectrum and ultimately increases the compressibility of the datacube.

For the SFSI-II datacube, target detection was selected as an example of remote sensing applications to assess the anomaly effect. Double blind test approach was adopted in the evaluation. There are five targets in the scene of the test datacube. Each target was assigned a full score of four points if it was perfectly detected. The total score of all the targets is 20 points for the test datacube. Two reconstructed SFSI-II datacubes compressed using HSOCVQ at compression ratio 10:1 without and with removing the spikes before compression were assessed. The reconstructed datacube without removing the spikes before compression received 10 points out of the full score of 20, while the reconstructed datacube with the spikes removed before compression received 11 points. Two reconstructed SFSI-II datacubes compressed using SAMVQ at compression ratio 12:1 without and with removing the spikes before compression were also evaluated. The reconstructed datacube without removing the spikes before compression received 14 points, while the reconstructed datacube with the spikes removed before compression received 15 points. The experimental results showed that there is no impact on the application with respect to removal of spikes, since the evaluation scores are close.

It was concluded that an on-board data cleaning to remove the anomalies before compression is not recommended, since the evaluation results did not show significant gain of the compression performance after the anomalies were removed.

5 Impact of Pre-Processing and Radiometric Conversion on Compression Performance

The CSA carried out studies to evaluate the impact of pre-processing and radiometric conversion to radiance on data compression on board hyperspectral satellites to examine whether or not these processes should be applied onboard before compression [29]. In other words, the compression should be applied on either a raw data or on its radiance version [30]. The pre-processing includes removal of detector's dark current, offset, noise and correction of non-uniformity. Radiometric conversion refers to the conversion of the raw detector digital number data to the at-sensor radiance.

Since the pre-processing and radiometric conversion processes alter the raw data, the evaluation of their impact on data compression could not be performed by comparing the statistical measures, such as RMSE, SNR and percentage error, obtained for the raw data with those for the radiance data. Two remote sensing products were used as metrics to evaluate the impact. The retrieval of leaf area index (LAI) from hyperspectral datacubes in agriculture applications was selected.

The spectral un-mixing based target detection from short wave infrared hyperspectral datacubes in defence applications was also selected. Double blind test approach was adopted in the evaluation.

Three *casi* datacubes for retrieval of LAI in agriculture applications and one SFSI-II datacube acquired for target detection were tested. For the *casi* datacubes, the LAI images derived from the compressed datacubes were assessed using visual inspection as the qualitative measure (Fig. 1.3). The R^2, absolute RMSE and relative RMSE between the LAI derived from a compressed datacube and the measured LAI (ground truth) were used as the quantitative measures. The evaluation results showed that pre-processing and radiometric conversion applied before or after compression had no impact on retrieval of LAI product, since there was no significant difference between the R^2, absolute RMSE and relative RMSE values obtained for the compressed raw datacubes and for the compressed radiance datacubes. The visual inspection did not find difference between the LAI images derived from the compressed datacubes with pre-processing and radiometric conversion applied after and before the compression.

For the SFSI-II datacube, the impact of pre-processing and radiometric conversion on compression was evaluated on a target-by-target basis using four quantitative criteria measured by the total scores per datacube. The evaluation results showed that pre-processing and radiometric-conversion did have impact on the target detection application. Compression on the raw SFSI-II datacube produced lower evaluation scores and poorer user acceptability than the compression on the radiance datacube that had undergone pre-processing and conversion of raw data to radiance units before compression.

The evaluation studies concluded that pre-processing and radiometric conversion applied before or after compression have no impact on retrieval of LAI products of the *casi* datacubes, but have impact on the target detection application of the SFSI-II datacube.

6   Effect of Keystone and Smile on Compression Performance

The effect of spatial distortion (keystone) and spectral distortion (smile) of hyperspectral sensors on compression was also evaluated at CSA to examine whether or not these distortions have any impact on compression performance, thus should be corrected on board before compression [47, 48]. In an imaging spectrometer, the keystone refers to the across-track spatial mis-registration of the ground sample pixels of the various spectral bands of the spectrograph. It is caused by geometric distortion, as can be seen in camera lenses, or chromatic aberration, or a combination of both. The smile, also known as spectral line curvature, refers to the spatial non-linearity of monochromatic image of a straight entrance slit as it appears in the focal plane of a spectrograph. It is caused by dispersion element, prism or grating, or by aberrations in the collimator and imaging optics. These distortions have the potential to affect compression performance.

Fig. 1.3 LAI images derived from the original datacube (IFC-2) and from the compressed datacubes using SAMVQ and HSOCVQ at compression ratio 20:1 with compression applied on the raw and on the radiance (Rad.) datacube

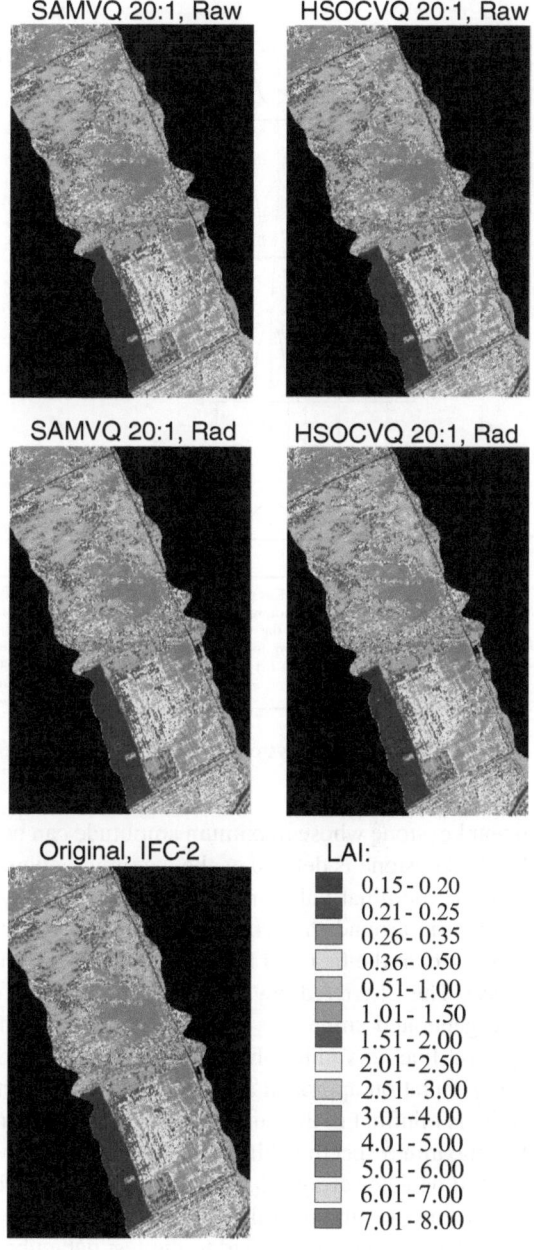

A datacube acquired using an airborne hyperspectral sensor *casi* in an application of boreal forest environment was used as a test datacube. It has 72 spectral bands with a spectral sampling distance of approximately 7.2 nm. The half-bandwidth used is 4.2 nm. Keystone and smile were simulated and then ingested into the test datacube (Fig. 1.4). The generated keystone was to simulate the shift of nominal data due to a

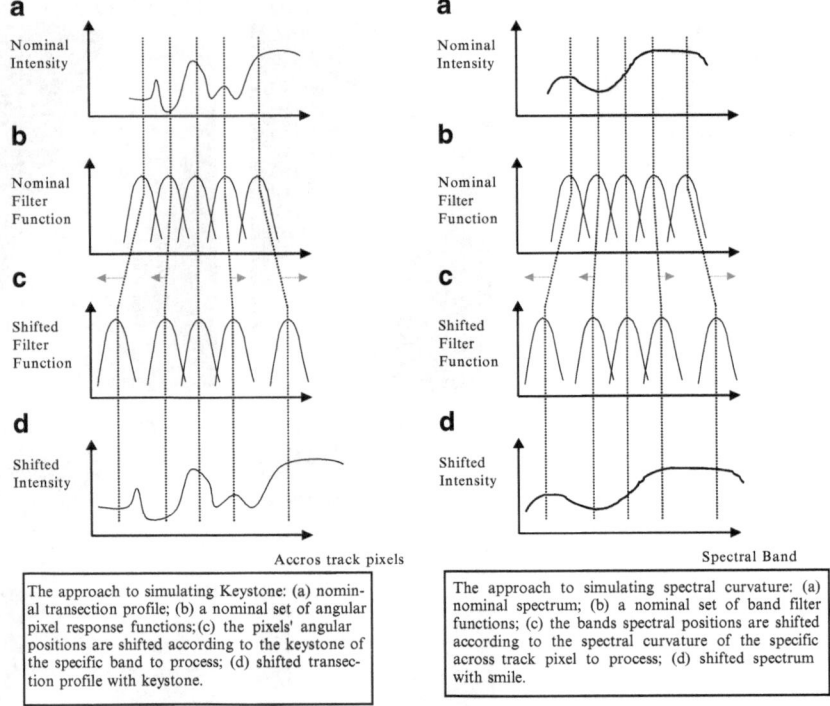

The approach to simulating Keystone: (a) nominal transection profile; (b) a nominal set of angular pixel response functions; (c) the pixels' angular positions are shifted according to the keystone of the specific band to process; (d) shifted transection profile with keystone.

The approach to simulating spectral curvature: (a) nominal spectrum; (b) a nominal set of band filter functions; (c) the bands spectral positions are shifted according to the spectral curvature of the specific across track pixel to process; (d) shifted spectrum with smile.

Fig. 1.4 Simulation of keystone (*left*) and smile (*right*) of hyperspectral datacubes

linear keystone whose maximum amplitude can be specified by a user. The amplitude for the keystone is defined as the maximum angular shift in the pixel center position from the nominal value in the full detector array. Since the keystone is linear and symmetric around the array center, the maximum keystone is located at the array edges (i.e. at the first and last pixel in the array and for the first and last bands in the array). The generated keystone is the same from one focal plane frame to the next. The generated smile was to simulate the shift of nominal data due to a quadratic spectral line curvature whose maximum amplitude can also be specified by a user. The simulation approach assumes that the diffraction slit is curved in order that the smile is minimal in the middle of the array. The amplitude for the smile is defined as the maximum spectral shift in the band center wavelength from the nominal value in the full detector array. Since the smile is quadratic and symmetric around the array center, the maximum smile is located at the array edges.

Compression was applied to the test datacube and the datacubes with simulated keystone and smile. Experimental results showed that keystone has little or no impact on the compression fidelity produced by both SAMVQ and HSOCVQ. The PSNR fidelity loss is <1 dB. Smile has little to some impact on the compression fidelity. The PSNR fidelity loss is typically 2 dB with HSOCVQ and <1 dB with SAMVQ. Figure 1.5 shows the curves of compression fidelity (PSNR) produced using SAMVQ as function of magnitude of keystone and smile.

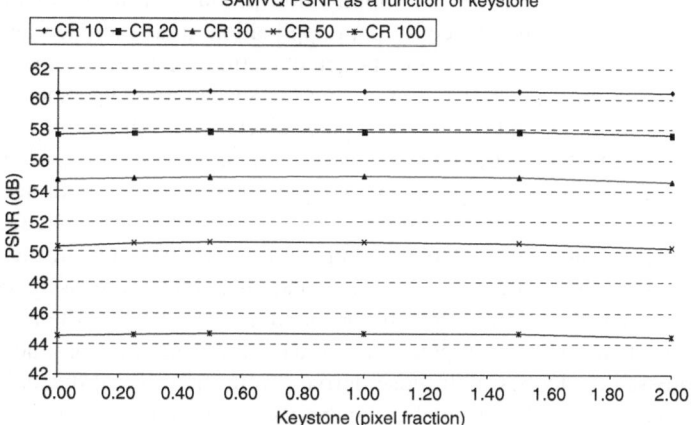

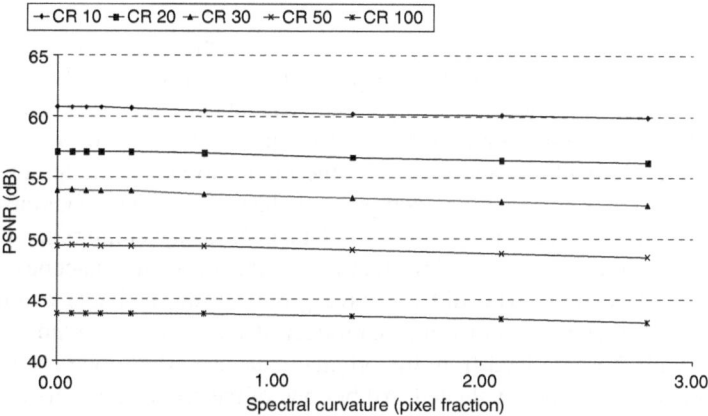

Fig. 1.5 Compression fidelity (PSNR) produced using SAMVQ as function of magnitude of keystone (*left*) and smile (*right*)

7 Multi-Disciplinary User Acceptability Study for Compressed Data

Since the SAMVQ and HSOCVQ are lossy compression techniques, users of hyperspectral data are concerned about the possible loss of information as a result of compression [30]. In order to respond to this concern, a multi-disciplinary user acceptability study has been carried out [31]. Eleven hyperspectral data users covering a wide range of application areas and a variety of hyperspectral sensors assessed the usability of the compressed data qualitatively and quantitatively using their well understood datacubes and application products in terms of predefined evaluation criteria. The hyperspectral data application areas included agriculture, geology, oceanography, forestry and target detection. A total of nine different hyperspectral sensors were covered including the spaceborne hyperspectral sensor Hyperion. These users ranked and accepted/rejected the compressed datacubes according to the impact on the remote sensing applications. The original datacubes were provided by the users prior to the compression using both techniques. Evaluations are made on the original datacube and reconstructed datacubes with a variety of compression ratios. Double blind testing was adopted in the study to eliminate bias in the evaluation. That is, random names were assigned to the original and reconstructed datacubes before sending them back to the users.

The study intentionally attempted to avoid a comparison of the products derived from compressed datacubes with those derived from the original datacube. When making comparisons to the original datacube, users may focus on minute changes whose significance is not well assessed. Since an original datacube is not exempt from intrinsic noise and errors due to calibration or atmospheric correction, these errors can also propagate into the remote sensing products derived from the original data. Whenever possible, the products derived from blind compressed datacubes were assessed and ranked according to their agreement with ground truth.

The 11 users evaluated the compressed datacubes using both SAMVQ and HSOCVQ at compression ratios between 10:1 and 50:1. Four out of the 11 users had ground truth available and used it as the metric to assess the remote sensing products derived from the blind compressed datacubes. They qualitatively and quantitatively accepted all the compressed datacubes, as the compressed datacubes provided the same amount of information as the original datacubes for their applications. Two users who did not have ground truth available evaluated the impact of compression by comparing the products derived from the blind compressed datacubes with those derived from the original datacube. These two users accepted the compressed datacubes evaluated. All but one of the remaining users accepted or marginally accepted the compressed datacubes. These users rejected six datacubes out of the 48 compressed datacubes using SAMVQ, and rejected six datacubes out of the 44 compressed datacubes using HSOCVQ. In general, SAMVQ shows better acceptability than HSOCVQ.

Details of the user acceptability study are provided in [31–33]. Individual studies for the impact of the CSA developed VQ data compression techniques on hyperspectral data applications can also be found in [34–42].

8 Enhancement of Resilience to Bit-Errors of the Compression Techniques

After data compression the compressed data become vulnerable to bit-errors. Compressed data produced by the traditional compression algorithms can be easily corrupted by bit-errors in the downlink channel and the bit-errors are propagated to

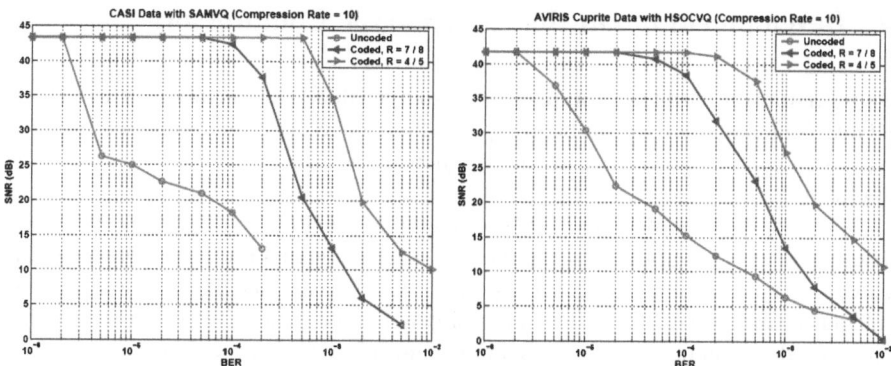

Fig. 1.6 Compression fidelity (SNR) versus bit-error rate before and after the enhancement of resilience of compressed hyperspectral data against bit errors

the whole data set. Experimental results showed that compressed data using SAMVQ or HSOCVQ are much more robust to bit-errors than those compressed using the traditional compression algorithms. There is almost no loss of compression fidelity when bit-error rate (BER) is smaller than 10^{-6}. Although both SAMVQ and HSOCVQ are more bit-error resistant than the traditional compression techniques, when the BER exceeds 10^{-6}, the compression fidelity starts to drop. The level of resilience to bit-error rate of 10^{-6} may not be adequate for certain applications. The CSA explored the benefits of employing forward error correction (FEC) on top of data compression to enhance the resilience to bit-errors of the compressed data to deal with higher BERs [43].

Error control in digital communications is often accomplished using FEC, also known as channel coding. Channel codes add redundancy to the information in a controlled way to give the receiver the opportunity to correct the errors induced by the noise in the channel. The convolutional codes recommended by the CCSDS were employed to protect hyperspectral data compressed by SAMVQ and HSOCVQ against bit errors. Convolutional codes add redundancy to the compressed data in a controlled way before the transmission of the data over noisy channels. At the receiver side, the channel decoder uses the added redundancy to correct the errors that are induced by the noise in the channel. Afterwards, the compressed data is decompressed and the hyperspectral data is reconstructed for evaluating the fidelity loss. The experimental results for three test datacubes showed that while the uncoded compressed data hardly endured bit errors even at BERs as low as 2×10^{-6}, the coded compressed data perfectly tolerates bit errors even at BERs as high as 1×10^{-4} to 5×10^{-4} by allowing redundancy overheads of 12.5–20%, respectively (Fig. 1.6).

It is demonstrated that by proper use of convolutional codes, the resilience of compressed hyperspectral data against bit errors can be improved by close to two orders of magnitude.

9   Development of On-Board Prototype Compressors

Two versions of hardware compressor prototypes that implement the SAMVQ and HOSCVQ techniques for on-board applications have been built. The first version was targeted for real-time application whereas the second was for non-real-time application.

Three top-level topologies were considered to meet the initial design objective. These processing approaches included a digital signal processor (DSP) engine based, a high performance general purpose CPU based and Application Specific Integrated Circuits (ASIC) or Field Programmable Gate Arrays (FPGA). The resulting on-board data compression engines were evaluated for various configurations. After studying the topologies, the hardware and software architectural options and candidate components, an architectural preference was placed on a hardware compressor that would exhibit both modularity and scalability. The performance trade-off studies for these architectures showed that the best performance and scalability could be achieved from a dedicated compression engine (CE) based on an ASIC/FPGA topology. The advantages of the ASIC/FPGA approach include the ability to:

– Apply parallel processing to increase throughput.
– Provide for successive upgrades of compression algorithms and electronic components over a long term.
– Support high speed direct memory access (DMA) transfers for read and write operations.
– Optimize the scale of the design to mission requirements.
– Provide data integrity features throughout the data handling process.

Candidate components were procured and performance simulation was carried out for the candidate architecture by coding the FPGA using Very high-speed integrated circuit Hardware Description Language (VHDL). This also verified that the proposed architecture supports expansion to arrays of CEs. The design of a real-time data compressor, using VHDL tools, benefited from generic functions that provide for rapid re-design or re-sizing. With these infrastructure tools, the CEs can adapt to the scale of different data requirements of a hyperspectral mission. Figure 1.7 shows a block diagram of the real-time compressor. A proof-of-concept prototype compressor has been built. Figure 1.8 shows the prototype compression engine board. It is composed of multiple standalone CEs each with the ability to compress a subset of spectral vectors in parallel. These are autonomous devices and once programmed perform compression in continuous mode, subset by subset. A CE is composed of a FPGA chip. The prototype board also has a Network Switch, a Fast Memory and a PCI bus interface. The Network Switch which is composed of a FPGA chip is used to serve the data flow transfer in and out of each of the CEs serving up to eight CEs in parallel using a high-speed serial link.

In the prototype compressor, the Fast Memory is temporally treated as the continuous data flow source of focal plane images from a hyperspectral sensor. The imagery data are fed into CEs via Wide Bus and the PCI Bus of the controller

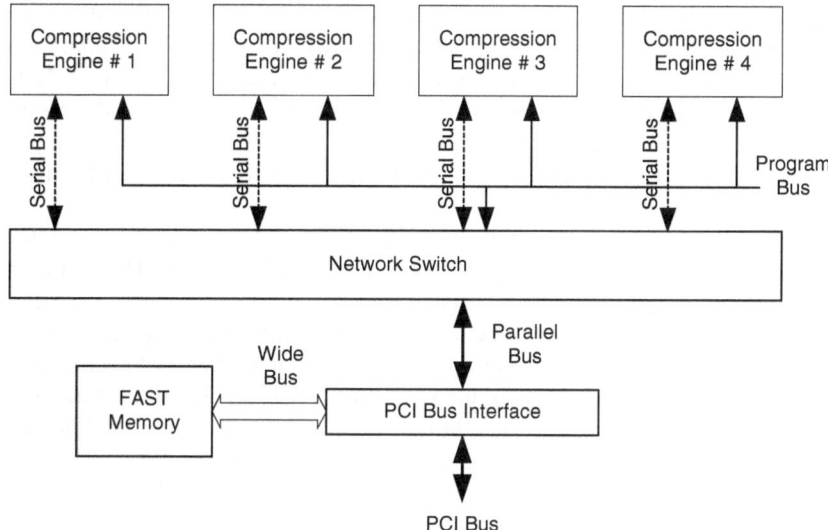

Fig. 1.7 Block diagram of the real-time hardware data compressor

Fig. 1.8 The proof-of-concept prototype compressor board (on the board there are four compression engines and one network switch each of which uses a FPGA chip)

computer and distributed to each CE by the Network Switch. The data rate of transfer from the Fast Memory to the Network Switch may be lower than the real data rate of the focal plane images produced by a hyperspectral sensor. But the throughput of the compressor from the point where data reaches the Network Switch to the point of output of the compressor must be greater than or equal to the real data rate. In the real case, the Fast Memory will be replaced by the data buffer after the A/D or pre-processing of a hyperspectral sensor.

The calculation of distance between a spectral vector and a codevector is the most frequent operation in the VQ based compression algorithms. The architecture of computing the distance between a spectral vector and a codevector dominates the performance of a CE. During the design and breadboarding phases, two CE architectures referred to as "Along Spectral Bands" and "Across Spectral Bands" were developed. They are the two main families of CE architectures. Each family has many variable CE architectures. In the "Along Spectral Bands" architecture, a spectral vector in a data subset to be compressed is compared to a codevector by processing all spectral bands in parallel. The distance between the spectral vector and the codevector is obtained in a few system clock cycles. This architecture uses a large amount of the available hardware resources. In the "Across Spectral Bands" architecture, a matrix of $m \times n$ vector distances is computed for one band in a system clock cycle [where m spectral vectors and n codevectors are processed]. The number of spectral bands determines the number of system clock cycles to obtain the whole $m \times n$ vector distances. This architecture is less resource intensive. A patent has been granted for both architectures that capture the unique design of the hardware compressor and the CEs. For a detailed description of the techniques please refer to [44].

The design of the hardware compressor is capable of accepting varying datacube sizes, numbers of spectral bands, and codebook sizes. The system is truly scalable, as any number of CE components can be used according to the mission requirements. The compression board shown in Fig. 1.8 was developed in 2001. The Xilinx XC2V6000 FPGA chips were used. The size of the fast memory was 64 MB. The bus interface width was 128 bits. The power consumption of each FPGA chip was about 5 W (0.2 W/Msamples/s). It has been benchmarked that each CE can compress data at a throughput of 300 Mbps (25 Msamples/s). Four CEs in parallel on the board can provide a total throughput up to 1.2 Gbps, which met the requirement of the throughput ≥ 1 Gbps.

In the second version of the hardware compressor implementation, a non real-time hardware compressor has been developed based on a commercial-off-the-shelf (COTS) board as shown in Fig. 1.9. The use of a COTS product decreased development cost and provided a shorter design cycle. The board accommodates two Virtex II Pro FPGA chips, a 64 MB memory with each chip and ancillary circuitry. Each FPGA chip has a 160-bit width bus to the PCI interface. There are two sets of 50 differential pairs and two sets of 20 Rocket I/O pairs between the two FPGA chips. The architecture uses a two-stage cascade compression. The first stage CE groups spectral vectors of a hyperspectral datacube into clusters (subsets) and performs coarse compression, the second stage CE performs fine compression on a

Fig. 1.9 Non real-time compressor based on a COTS board (2 Virtex II Pro FPGA chips on the board)

subset-by-subset basis. In the non real-time option, the ratio of time to compress an imagery compared to the time to acquire imagery is assumed to be 12:1 (based on the HERO mission concept [27]). For example, if there are only 7.4 min for imaging within one orbit of duration 96 min, a non real-time compression with process ratio of 12:1 meets the requirement. The throughput of the first stage CE was 500 Mbps, while the throughput of the second stage CE is 120 Mbps. Thus the throughput of the system was 120 Mbps, which met the requirement for non real-time compression.

Recently, a new COTS board, which contains two Virtex-6 LXT FPGA chips, has been procured and is to replace the existing COTS board, which was based on 9 years old Xilinx Virtex II technology. The FPGA chips in the new board have significantly more gates and multipliers, larger internal RAM and flash memory than in the current board. High processing performance is expected with the new board.

10 Participation in Development of International Standards for Satellite Data Systems

The Consultative Committee for Space Data System (CCSDS) [45] is developing new international standards for satellite multispectral and hyperspectral data compression. The CSA's SAMVQ compression technique has been selected as a candidate. The preliminary evaluation results show that the SAMVQ produces competitive rate-distortion performance on the CCSDS test images acquired by the hyperspectral sensors and hyperspectral sounders. There is a constraint to achieve lower bit rates when the SAMVQ is applied to the multispectral images due to their small number of bands. This is because the SAMVQ was designed for compression of hyperspectral imageries, which contain much more spectral bands than the multispectral images. In response to CCSDS' action items, the CSA carried out studies to compare the rate-distortion performance of the SAMVQ with other proposed compression techniques using the CCSDS hyperspectral and hyperspectral sounders test images and to investigate how to enhance the capability of the SAMVQ for compressing multispectral images while maintaining its unique properties for hyperspectral images [46].

The CCSDS test images contain:

- Four sets of hyperspectral images for different applications acquired using four hyperspectral sensors, i.e., AVIRIS, *casi*, SFSI-II and Hyperion.
- Two sets of hyperspectral sounder images acquired using Infrared Atmospheric Sounding Interferometer (IASI) and Atmospheric Infrared Sounder (AIRS).
- Six sets of multispectral images acquired using SPOT5, Landsat, MODIS, MSG (Meteosat Second Generation) and PLEIATES (simulation).

Seven lossy data compression techniques selected by the CCSDS working group were compared. These compression techniques are:

- JPEG 2000 compressor with bit-rate allocation (JPEG 2000 BA)
- JPEG 2000 compressor with spectral decorrelation (JPEG 2000 SD)
- CCSDS Image Data Compressor – Frame Mode – 9/7 wavelet floating – Block 2 (CCSDS-IDC)
- ICER-3D
- Fast lossless/near lossless (FL-NLS)
- Fast lossless/near lossless, updated version in 2009 (FLNLS 2009)
- CSA SAMVQ

The experimental results show that SAMVQ produces the best rate-distortion performance over the other six compression techniques for all the tested hyperspectral images and sounder images when the bit rates are lower (e.g. ≤ 1.0 bits/pixel, an example shown in Fig. 1.10). For the multispectral data sets acquired using MODIS, the SAMVQ still outperforms to other compression techniques.

Due to the compression ratio attained by the SAMVQ proportional to the length of vectors, the short vectors formed in multispectral data compression prevent SAMVQ from achieving lower bit rates (i.e. higher compression ratios). Three schemes to form longer vectors have been investigated. The experimental results showed that for multispectral images with relatively larger number of bands (>10, such as MODIS images) after longer vectors are used, SAMVQ not only produced the lower bit rates, but also improved (up to 4.0 dB) the rate-distortion performance. For the multispectral images whose number of bands is extremely small (e.g. 4), SAMVQ produced the best rate-distortion performance from the high bit rates to a certain low bit rate when vector length is equal to the number of bands. Beyond this point of bit rate, SAMVQ produced better rate-distortion performance only when the longer vectors were used. The detailed description on this study can be found in [46].

11 Summary

This chapter reviewed and summarized the researches and developments of the near lossless data compression techniques for satellite sensor data at the Canadian Space Agency in collaboration with its partners and industry in the last decade. It briefly

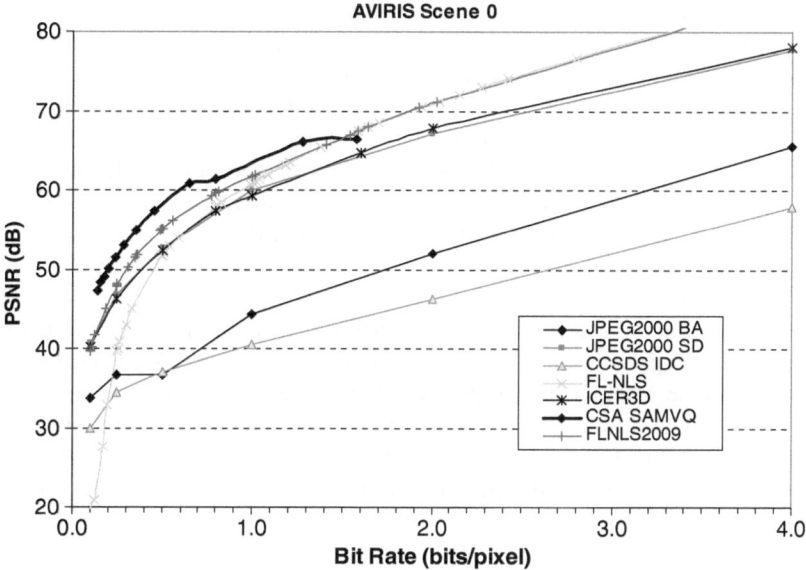

Fig. 1.10 Rate–distortion curves of a test AVIRIS datacube compressed using the SAMVQ and the six compression techniques selected by the CCSDS

described two vector quantization based near lossless data compression techniques for use on-board a hyperspectral satellite: Successive Approximation Multi-stage Vector Quantization (SAMVQ) and Hierarchical Self-Organizing Cluster Vector Quantization (HSOCVQ). It reviewed the evaluation of the features of the compression techniques that allow them to be termed near lossless compression. The evaluation results demonstrated that the compression errors introduced by SAMVQ and HSOCVQ are smaller than or comparable to the intrinsic noise. This level of compression errors has no impact or minor impact on the afterwards application utilization comparing with the original data. This kind of compression is referred to as near lossless compression in our practice.

For on board data compression, it is important to know how the data quality related to the sensor's system characteristics and the data product level impact the compression performance. This chapter reviewed the activities for evaluating these impacts. The study of the impact of the anomalies in the raw hyperspectral data was reviewed. These anomalies, such as dead detector pixels, frozen detector pixels, spikes (isolated over-responded pixels) and saturation, are caused by detector and instrument defects. This study was to help making decision whether or not an on-board data cleaning is required before compression. The evaluation study concluded that an on-board data cleaning to remove the anomalies before compression is not required, since the evaluation results did not show significant gain of the compression performance after the anomalies were removed.

The activities to study the impact of pre-processing and radiometric conversion to radiance on data compression were reviewed. The pre-processing includes removal of dark current, offset, noise and correction of non-uniformity. Radiometric conversion refers to the conversion of the raw detector digital number data to at sensor radiance. These studies were to examine whether or not the pre-processing and radiometric conversion should be applied on-board before compression. In other words, the compression should be applied on either raw data or on its radiance version. The evaluation results did not provide a unanimous conclusion with the two applications studied. The pre-processing and radiometric conversion applied before or after compression had no impact on retrieval of LAI products from the *casi* datacubes, but had impact on the target detection application of the SFSI-II datacube.

The studies on the effect of spatial distortion (keystone) and spectral distortion (smile) of hyperspectral sensors on compression was also reviewed. These studies were to examine whether or not these distortions have an impact on compression performance, thus should be corrected on board before compression. Experimental results showed that keystone has little or no impact on the both SAMVQ and HSOCVQ compression, while smile has minor impact on the compression performance. The PSNR fidelity loss is <1 dB with SAMVQ and typically 2 dB with HSOCVQ.

This chapter also summarized the activities of systematically assessing the impacts of the near lossless compression techniques on remote sensing products and applications in a multi-disciplinary user acceptability study. This study was carried out by 11 users covering a wide range of application areas and a variety of hyperspectral sensors. The study concluded that most of the users qualitatively and quantitatively accepted the compressed datacubes, as the compressed datacubes provided the same amount of information as the original datacubes for their applications.

Although for on board use, both SAMVQ and HSOCVQ are more bit-error resistant than the traditional compression algorithms, the compression fidelity starts to drop when the bit-error rate exceeds 10^{-6}. This chapter reviewed the CSA's effort to explore the benefits of employing forward error correction on top of data compression to enhance the resilience to bit-errors of the compressed data to deal with higher bit-error rates. It is demonstrated that by proper use of convolutional codes, the resilience of compressed hyperspectral data against bit errors can be improved by close to two orders of magnitude.

This chapter summarized the activities of CSA and its industry on hardware implementation of the two compression techniques. Two versions of hardware compressor prototypes, which implement the SAMVQ and HOSCVQ techniques for on-board use, have been built. The first version was targeted for real-time application whereas the second was for non-real-time application. The design of the hardware compressor is capable of accepting varying datacube sizes, numbers of spectral bands, and codebook sizes. The system is scalable, as any number of compression engines can be used according to the mission requirements. The prototype compressor has been benchmarked. A commercial-off-the-shelf (COTS) FPGA board based hardware compressor prototype has been developed for a non real-time application. The use of a COTS product decreased development cost and provided a shorter design cycle.

The Consultative Committee for Space Data System (CCSDS) is developing new international standards for satellite multispectral and hyperspectral data compression. The CSA's SAMVQ compression technique has been selected as a candidate. This chapter reported the CSA's participation in the development of international standards for satellite data systems within the CCSDS organization. The experimental results show that SAMVQ produces the best rate-distortion performance compared to the six compression techniques selected by the CCSDS for the tested hyperspectral images and sounder images when the bit rates are lower (e.g. ≤ 1.0 bits/pixel).

Acknowledgments The author would like to thank his colleagues A. Hollinger, M. Bergeron, I. Cunningham and M. Maszkiewicz; his post-doctor visiting fellows C. Serele, H. Othman and P. Zarrinkhat, and over 30 internship students, for their contributions to the work summarized in this chapter. The author thanks D. Goodenough at the Pacific Forestry Centre, Natural Resources Canada, K. Staenz (now at University of Lethbridge), L. Sun and R. Neville at the Canada Center for Remote Sensing, Natural Resources Canada, J. Levesque and J.-P. Ardouin at the Defence Research and Development Canada, J. Miller and B. Hu at York University, for providing data sets and for actively collaborating on the user acceptability study and the impact assessments. The author thanks the following users for their participation and contribution to the multi-disciplinary user acceptability study: A. Dyk at the Pacific Forestry Centre, B. Ricketts and N. Countway at Satlantic Inc., J. Chen at University of Toronto, H. Zwick, C. Nadeau, G. Jolly, M. Davenport and J. Busler at MacDonald Dettwiler Associates (MDA), M. Peshko at Noranda/Falconbridge, B. Rivard and J. Feng at the University of Alberta, J. Walls and R. McGregor at RJ Burnside International Ltd., M. Carignan and P. Hebert at Tecsult, J. Huntington and M. Quigley at the Commonwealth Scientific and Industrial Research Organization in Australia, R. Hitchcock at the Canada Centre for Remote Sensing. The author thanks L. Gagnon, W. Harvey, B. Barrette, and C. Black at former EMS Technologies Canada Ltd. (Now MDA Space Missions) and the technical teams for the development and fabrication of the hardware compression prototypes. The author thanks V. Szwarc and M. Caron at Communication Research Centre, Canada for discussion on enhancement of the resilience to bit-errors of the compression techniques, and P. Oswald and R. Buckingham for discussion on on-board data compression. The author also thanks the CCSDS MHDC Working Group for providing the test data sets and the members of the working group for providing the compression results. The Work in the chapter was created by a public servant acting in the course of his employment for the Government of Canada and within the scope of his duties in writing the Work, whereas the copyright in the Work vests in Her Majesty the Queen in right of Canada for all intents and purposes under the *Copyright Act* of Canada© Government of Canada 2011.

References

1. S.-E. Qian, A. Hollinger, D. Williams and D. Manak, "Fast 3D data compression of hyperspectral imagery using vector quantization with spectral-feature-based binary coding," Opt. Eng. 35, 3242–3249 (1996) [doi:10.1117/1.601062]
2. S.-E. Qian, A. Hollinger, D. Williams and D. Manak, "A near lossless 3-dimensional data compression system for hyperspectral imagery using correlation vector quantization," *Proc. 47th Inter. Astron. Congress*, Beijing, China (1996).

3. S.-E. Qian, A. Hollinger, D. Williams and D. Manak, "3D data compression system based on vector quantization for reducing the data rate of hyperspectral imagery," in Applications of Photonic Technology II, G. Lampropoulos, Ed., pp. 641–654, Plenum Press, New York (1997)

4. D. Manak, S.-E. Qian, A. Hollinger and D. Williams, "Efficient Hyperspectral Data Compression using vector Quantization and Scene Segmentation," *Canadian J. Remote Sens.*, **24**, 133–143 (1998).

5. S.-E. Qian, A. Hollinger, D. Williams and D. Manak, "3D data compression of hyperspectral imagery using vector quantization with NDVI-based multiple codebooks," *Proc. IEEE Geosci. Remote Sens. Symp.*, 3, 2680–2684 (1998).

6. S.-E. Qian, A. Hollinger, D. Williams and D. Manak, "Vector quantization using spectral index based multiple sub-codebooks for hyperspectral data compression," IEEE Trans. Geosci. Remote Sens., 38(3), 1183–1190 (2000) [doi:10.1109/36.843010]

7. S.-E. Qian, "Hyperspectral data compression using a fast vector quantization algorithm," IEEE Trans. Geosci. Remote Sens., 42(8), 1791–1798 (2004) [doi: 10.1109/TGRS.2004.830126]

8. S.-E. Qian, "Fast vector quantization algorithms based on nearest partition set search," IEEE Trans. Image Processing, 15(8), 2422–2430 (2006) [doi:10.1109/TIP.2006.875217]

9. S.-E. Qian and A. Hollinger, "Current Status of Satellite Data Compression at Canadian Space Agency," Invited chapter in *Proc. SPIE* **6683**, 04.01-12 (2007).

10. S.-E. Qian, and A. Hollinger, "System and method for encoding/decoding multi-dimensional data using Successive Approximation Multi-stage Vector Quantization (SAMVQ)," U. S. Patent No. 6,701,021 B1, issued on March 2, 2004.

11. S.-E. Qian and A. Hollinger, "Method and System for Compressing a Continuous Data Flow in Real-Time Using Cluster Successive Approximation Multi-stage Vector Quantization (SAMVQ)," U.S. Patent No. 7,551,785 B2, issued on June 23, 2009.

12. S.-E. Qian, and A. Hollinger, "System and method for encoding multi-dimensional data using Hierarchical Self-Organizing Cluster Vector Quantization (HSOCVQ)," U. S. Patent No. 6,724,940 B1, issued on April 20, 2004.

13. S.-E. Qian and A. Hollinger, "Method and System for Compressing a Continuous Data Flow in Real-Time Using Recursive Hierarchical Self-Organizing Cluster Vector Quantization (HSOCVQ)," U.S. Patent No. 6,798,360 B1 issued on September 28, 2004.

14. S.-E. Qian, M. Bergeron, I. Cunningham, L. Gagnon and A. Hollinger, "Near Lossless Data Compression On-board a Hyperspectral Satellite," IEEE Trans. Aerosp. Electron. Syst., 42(3), 851-866 (2006) [doi:10.1109/TAES.2006.248183]

15. S.-E. Qian, A. Hollinger and Yann Hamiaux, "Study of real-time lossless data compression for hyperspectral imagery," Proc. IEEE Int. Geosci. Remote Sens. Symp., 4, 2038–2042 (1999)

16. S-E Qian and A Hollinger, "Applications of wavelet data compression using modified zerotrees in remotely sensed data," Proc. IEEE Geosci. Remote Sens. Symp., 6, 2654–2656 (2000)

17. CCSDS, "Lossless Data Compression," CCSDS Recommendation for Space Data System Standards, Blue Book 120.0-B-1, May 1997.

18. S.-E. Qian, "Difference Base-bit Plus Overflow-bit Coding," *Journal of Infrared & Millimeter Waves*, **11**(1), 59–64 (1992).

19. J. M. Shapiro, "Embedded image coding using zerotrees of wavelet coefficients," *IEEE Trans. Signal Processing*, 41, 3445–3462 (1993).

20. A. Said and W.A. Pearlman, "A new, fast and efficient image codec based on set partitioning in hierarchical trees," *IEEE Trans. on Circuits and Systems for Video Technology*, 6, 243–250 (1996).

21. A. Gersho and R.M. Gray, Vector Quantization and Signal Compression, Boston, MA: Kluwer, 1992.

22. K. Chen and T.V. Ramabadran, "Near-lossless compression of medical images through entropy-coded DPCM," IEEE Trans. Med. Imaging 13, 538–548 (1994) [doi:10.1109/42.310885]

23. X. Wu and P. Bao, "Constrained high-fidelity image compression via adaptive context modeling," IEEE Trans. Image Processing, 9, 536–542 (2000) [doi:10.1109/83.841931]
24. B Aiazzi, L. Alparone and S. Baronti, "Near-lossless compression of 3-D optical data," IEEE Trans. Geosci. Remote Sens., 39(11), 2547–2557 (2001) [doi:10.1109/36.964993]
25. E. Magli, G. Olmo and E. Quacchio, "Optimized onboard loeelsess and near-lossless compression of hyperspectral data using CALIC," IEEE Geosci. Remote Sens. Lett., 1, 21–25 (2004) [doi:10.1109/LGRS.2003.822312]
26. Low altitude AVIRIS data of the Greater Victoria Watershed District, http://aviris.jpl.nasa.gov/ql/list02.html
27. A. Hollinger, M. Bergeron, M. Maszkiewicz, S.-E. Qian, K. Staenz, R.A. Neville and D.G. Goodenough, "Recent Developments in the Hyperspectral Environment and Resource Observer (HERO) Mission" Proc. IEEE Geosci. Remote Sens. Symp., 3, 1620–1623 (2006).
28. S.-E. Qian, M. Bergeron, A. Hollinger and J. Lévesque, "Effect of Anomalies on Data Compression Onboard a Hyperspectral Satellite", Proc. SPIE 5889, (02)1–11 (2005)
29. S.-E. Qian, M. Bergeron, J. Lévesque and A. Hollinger, "Impact of Pre-processing and Radiometric Conversion on Data Compression Onboard a Hyperspectral Satellite," Proc. IEEE Geosci. Remote Sens. Symp., 2, 700–703 (2005).
30. S.-E. Qian, B. Hu, M. Bergeron, A. Hollinger and P. Oswald, "Quantitative evaluation of hyperspectral data compressed by near lossless onboard compression techniques," Proc. IEEE Geosci. Remote Sens. Symp., 2, 1425–1427 (2002).
31. S.-E. Qian, A. Hollinger, M. Bergeron, I. Cunningham, C. Nadeau, G. Jolly and H. Zwick, "A Multi-disciplinary User Acceptability Study of Hyperspectral Data Compressed Using Onboard Near Lossless Vector Quantization Algorithm," Inter. J. Remote Sens., 26(10), 2163–2195 (2005) [doi:10.1080/01431160500033500]
32. C. Nadeau, G. Jolly and H. Zwick, "Evaluation of user acceptance of compression of hyperspectral data cubes (Phase I)," final report of Canadian Government Contract No. 9F028-013456/001MTB, Feb. 6, 2003.
33. C. Nadeau, G. Jolly and H. Zwick, "Evaluation of user acceptance of compression of hyperspectral data cubes (Phase II)," final report of Canadian Government Contract No. 9F028-013456/001MTB, July 24, 2003.
34. S.-E. Qian, A. B. Hollinger, M. Dutkiewicz and H. Zwick, "Effect of Lossy Vector Quantization Hyperspectral Data Compression on Retrieval of Red Edge Indices," IEEE Trans. Geosc. Remote Sens., 39, 1459–1470 (2001) [doi:10.1109/36.934077]
35. S.-E. Qian, M. Bergeron, C. Serele and A. Hollinger, "Evaluation and comparison of JPEG 2000 and VQ based on-board data compression algorithm for hyperspectral imagery," Proc. IEEE Geosci. Remote Sens. Symp., 3, 1820–1822 (2003).
36. B. Hu, S.-E. Qian and A. Hollinger, "Impact of lossy data compression using vector quantization on retrieval of surface reflectance from CASI imaging spectrometry data," Canadian J. Remote Sens., 27, 1–19 (2001).
37. B. Hu, S.-E. Qian, D. Haboudane, J.R. Miller, A. Hollinger and N. Tremblay, "Impact of Vector Quantization Compression on Hyperspectral Data in the Retrieval Accuracies of Crop Chlorophyll Content for Precision Agriculture," Proc. IEEE Geosci. Remote Sens. Symp., 3, 1655–1657 (2002).
38. B. Hu, S.-E. Qian, D. Haboudane, J. R. Miller, A. Hollinger and N. Tremblay, "Retrieval of crop chlorophyll content and leaf area index from decompressed hyperspectral data: the effects of data compression," Remote Sens. Environ., 92(2), 139–152 (2004) [doi:10.1016/j.rse.2004.05.009]
39. K. Staenz, R. Hitchcock, S.-E. Qian and R.A. Neville, "Impact of on-board hyperspectral data compression on mineral mapping products," Int. ISPRS Conf. 2002, India (2002).
40. K. Staenz, R. Hitchcock, S.-E. Qian, C. Champagne and R.A. Neville, "Impact of On-Board Hyperspectral Data Compression on Atmospheric Water Vapour and Canopy Liquid Water Retrieval," Int. ISSSR Conf. (2003).

41. C. Serele, S.-E. Qian, M. Bergeron, P. Treitz, A. Hollinger and F. Cavayas, "A Comparative Analysis of two Compression Algorithms for Retaining the Spatial Information in Hyperspectral Data," *Proc. 25th Canadian Remote Sens. Symp.*, Montreal, Canada, 14–16 (2003).

42. A. Dyk, D.G. Goodenough, S. Thompson, C. Nadeau, A. Hollinger and S.-E. Qian, "Compressed Hyperspectral Imagery for Forestry," *Proc. IEEE Geosci. Remote Sens. Symp.*, 1, 294–296 (2003).

43. P. Zarrinkhat and S.-E. Qian, "Enhancement of Resilience to Bit-Errors of Compressed Data On-board a Hyperspectral Satellite using Forward Error Correction" Proc. SPIE 7084, 07.1–9 (2008)

44. S.-E. Qian, A. Hollinger and L. Gagnon, "Data Compression Engines and Real-Time Wideband Compressor for Multi-Dimensional Data" U.S. Patent No. 7,251,376 B2, issued on July 31, 2007.

45. Consultative Committee for Space Data System (CCSDS), http://public.ccsds.org/default.aspx.

46. S.-E. Qian, "Study of hyperspectral and multispectral images compression using vector quantization in development of CCSDS international standards" Proc. SPIE 7477A, 23.1–11 (2009)

47. M. Bergeron, K. Kolmaga and S.-E. Qian "Assessment of Keystone Impact on VQ Compression Fidelity," Canadian Space Agency internal technical report on May 20th, 2003.

48. M. Bergeron, K. Kolmaga and S.-E. Qian "Assessment of Spectral Curvature Impact on VQ Compression Fidelity," Canadian Space Agency internal technical report on May 20th, 2003.

Chapter 2
CNES Studies for On-Board Compression of High-Resolution Satellite Images

Carole Thiebaut and Roberto Camarero

Abstract Future high resolution instruments planned by CNES for space remote sensing missions will lead to higher bit rates because of the increase in resolution and dynamic range. For example, the ground resolution improvement induces a data rate multiplied by 8 from SPOT4 to SPOT5 and by 28 to PLEIADES-HR. Lossy data compression with low complexity algorithms is then needed while compression ratio must considerably rise. New image compression algorithms have been used to increase their compression performance while complying with image quality requirements from the community of users and experts. Thus, DPCM algorithm used on-board SPOT4 was replaced by a DCT-based compressor on-board SPOT5. Recent compression algorithms such as PLEIADES-HR one use wavelet-transforms and bit-plane encoders. But future compressors will have to be more powerful to reach higher compression ratios. New transforms have been studied by CNES to exceed the DWT but other techniques as selective compression are required in order to obtain a significant performance gap. This chapter gives an overview of CNES past, present and future studies of on-board compression algorithms for high-resolution images.

1 Introduction

The French Space Agency (Centre National d'Etudes Spatiales – CNES) is in charge of the conception, development and operation of satellites. For more than 20 years, optical Earth observation missions have been one of its specialities. Indeed, since 1986, CNES has launched several Earth observation satellites with gradual improvement of the spatial resolution. The SPOT family is a good example of this progress: SPOT1/2/3/4 launched from 1986 to 1998 had a spatial resolution of 10 m and

C. Thiebaut (✉) • R. Camarero
Ctr. National d'Études Spatiales (CNES), Toulouse, France
e-mail: Carole.Thiebaut@cnes.fr; Roberto.Camarero@cnes.fr

B. Huang (ed.), *Satellite Data Compression*, DOI 10.1007/978-1-4614-1183-3_2,
© Springer Science+Business Media, LLC 2011

Fig. 2.1 SPOT5 (*left hand side*) and PLEIADES-HR (*right hand side*) images of a famous place in Toulouse. SPOT5 image resolution is 2.5 m and PLEIADES-HR one is 0.7 m

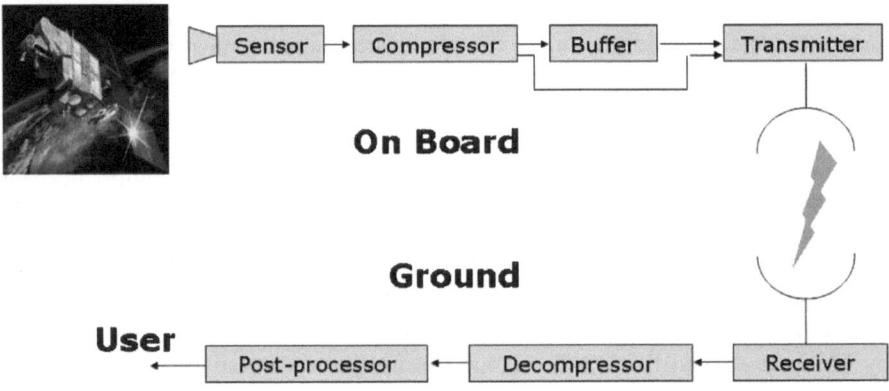

Fig. 2.2 On-board to ground image channel

SPOT5 launched in 2002 had a resolution of 5 m for the panchromatic band (HRG instrument) with an unchanged swath of 60 km [1]. Moreover, the THR mode of SPOT5 could produce images of 2.5 m thanks to a quincunx sampling. With the very agile satellite PLEIADES-HR, CNES is going further with a panchromatic band at 70 cm and a swath reduced to 20 km [2]. This spatial resolution improvement from 10 m to 0.7 m (see Fig. 2.1) induces a natural increase of data rate. Simultaneously, transmission bit rate of telemetry has not increased in the same order of magnitude. For example, for SPOT1-4, only one channel of 50 Mbits/s was used. Two such channels are used for SPOT5.

For PLEIADES-HR, each one of the three telemetry channels will have a capacity of 155 Mbits/s. This limitation combined with the growth of instrument data rate leads to an increasing need in compression. As shown on the on-board to ground image channel depicted in Fig. 2.2, on-board compression is useful to reduce the amount of data stored in the mass-memory and transmitted to the ground. On-board compression algorithms studied and implemented by CNES have been adapted to the user's constraints in terms of image quality while using performing tools available in the image processing domain. High-resolution Earth observation images have very strong constraints in terms of image quality as described in [3]. Requirements such as perfect image quality are asked whatever the landscape viewed by satellite. Furthermore, the on-board implementation issue is also a big challenge to deal with. Fortunately, highly integrated circuits (ASIC technology) make possible the implementation of high bit-rate functions such as image compression. Indeed, high-resolution missions have very high instrument bit-rate (128 Mbits/s for SPOT5, 4.3 Gbits/s for PLEIADES–HR) which makes impossible the implementation of software compression units. The hardware circuits available for space applications have lower performances than ground-based ones which prevent the chosen algorithms to have comparable performances. This chapter gives an overview of CNES studies in terms of development of image compression algorithms for high-resolution satellites. Section 2.2 gives a brief overview of implemented compressors since 1980s up to current developments. Section 2.3 illustrates present and future of on-board compression domain. New spatial decorrelators are also described and the principle of selective compression is introduced. Authors explain why and how this type of techniques could lead to higher compression ratios. Section 2.4 is a conclusion of this overview.

2 On-Board Compression Algorithms: History and Current Status

2.1 First Compressors

In 1986, a 1D-DPCM (Differential Pulse Code Modulation) with fixed length coding was implemented on-board SPOT1. The same algorithm was used up to SPOT4. As shown in Fig. 2.3, it provided a low compression ratio equal to 1.33. Every three pixels, one complete pixel was transmitted (on 8-bits) and the two following values were predicted as the mean value of the previously coded and

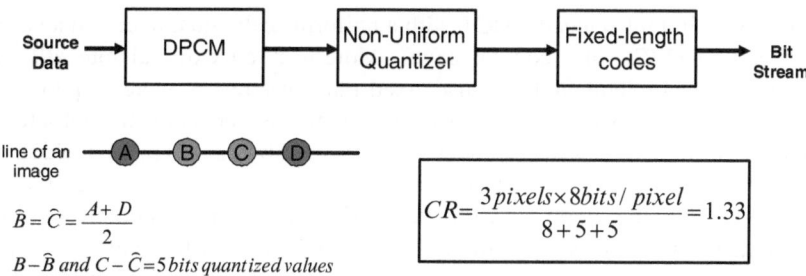

$$\hat{B} = \hat{C} = \frac{A+D}{2}$$

$B - \hat{B} \text{ and } C - \hat{C} = 5 \text{ bits quantized values}$

$$CR = \frac{3 \, pixels \times 8 bits / pixel}{8+5+5} = 1.33$$

Fig. 2.3 SPOT1-4 compression scheme and its resulting compression ratio

Fig. 2.4 SPOT5 satellite and its compression unit

the next one. Errors between prediction and real values for those pixels were non-uniformly quantized and coded on 5 bits. This simple algorithm had a complexity of three operations per pixel which was compatible to the very poor space qualified electronics of that time.

Shortly after, studies on the Discrete Cosine Transform (DCT) started, first with a fixed length coding (used on Phobos missions) and an off-line software implementation. Then, in 1990, with the development of a compression module using a DCT and a variable length encoder with space qualified components [4]. The throughput of this module was 4 Mpixels/s and the compression ratio was adjustable between 4 and 16. This module – called MCI for Module de Compression d'Image (Image Compression Module) – was used for several exploration missions (CLEMENTINE Lunar Mission, Mars 94/96 Probes, Cassini Probe . . .). Since this module did not specifically target Earth observation missions, another algorithm was developed for SPOT5 satellite images.

2.2 DCT-Based Compressor

SPOT5 algorithm introduced a DCT with a uniform scalar quantizer and a variable length coding (JPEG-like coding stage). Moreover, an external rate allocation procedure was implemented because fixed-rate bit-streams were required. The rate regulation loop was adapted to push-broom scanned data. It computed the complexity criteria over a line of blocks (8 lines high by 12,000 pixels) which gave an optimal quantization factor for each line of blocks using rate-prediction parameters [3]. After iteration with the user's community, a compression ratio was decided. This ratio is equal to 2.81 and is associated with a very good image quality both around block boundaries and in the homogeneous regions. The compression unit and SPOT5 satellite are shown in Fig. 2.4.

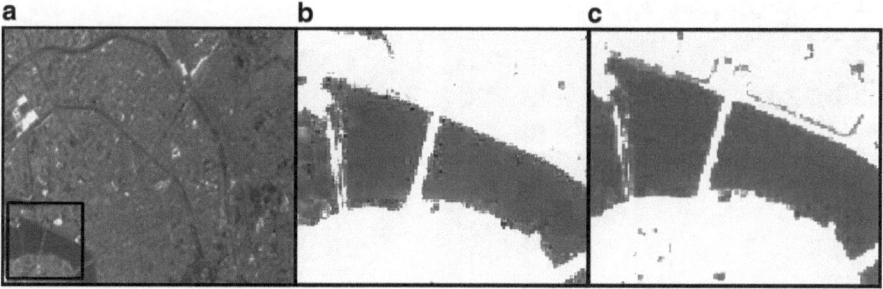

Fig. 2.5 (a) SPOT5 image of Toulouse (5 m). (b) Zoom before exceptional processing. (c) Zoom after exceptional processing

Nevertheless, a so-called "exceptional processing" was performed for some blocks presenting poor image quality compared to the whole line of blocks. These blocks are low-energy areas which are too much quantized compared to the mean complexity of the line of blocks. According to the number of DCT-transformed AC coefficients, the quantization step could be divided by 2 or 4 for those blocks [3]. Due to the very local characteristic of the modification (3.4% of exceptional blocks in average), the algorithm's behavior is not perturbed and the rate regulation remains stable. The proposed modification leads to a slight increase of the quantization factor and a negligible rise in average RMSE (Root Mean Square Error). However, as seen in Fig. 2.5, the obtained image quality is more uniform. In fact, the [signal]/[compression noise] ratio between image blocks is more homogeneous and the image quality of the final product is significantly enhanced.

This algorithm and its exceptional processing were validated on simulated and real SPOT4 images before SPOT5 launch date and then during its commissioning period.

Even with the exceptional processing described above, it was observed that beyond a compression ratio of 3, block artefacts due to the DCT appeared. It is a well-known drawback of this decorrelator at high compression ratios [5]. Accordingly, this algorithm is limited to compression ratios lower than 3, meaning a bit rate larger than 3 bits/pixel for 8-bits images. This was the reason why CNES looked for a new transform for PLEIADES-HR satellite images. Furthermore PLEIADES images are encoded on 12 bits with a targeted compression ratio close to 6.

2.3   Wavelet-Based Compressor

A wavelet-based algorithm was developed for PLEIADES-HR images [5]. This algorithm uses a 9/7 biorthogonal filter and a bit-plane encoder to encode the wavelet coefficients. The PLEIADES-HR panchromatic images are compressed with a bit rate equal to 2.5 bits/pixel and the multispectral bands are compressed at 2.8 bits/pixel. As for SPOT5, user's community, including French army, tuned this

Fig. 2.6 PLEIADES-HR satellite and its compression unit

data rate for preserving the image quality of final products. The in-flight commissioning period will confirm this choice. A module with ASICs including both Mass-Memory and compression module was designed for this mission.

This unit integrates a generic compression module called WICOM (Wavelet Image Compression Module). This high-performance image compression module implements a wavelet image compression algorithm close to JPEG2000 standard. ASIC optimized internal architecture allows efficient lossless and lossy image compression at high data rate up to 25 Mpixels/s. The compression is done at a fixed bit-rate and enforced on each strip or on full images. No compression parameter needs to be adjusted except the compression data rate. Input image dynamic range up to 13-bits can be handled by the module which has a radiation-tolerant design. The compression unit and the satellite are presented in Fig. 2.6.

2.4 A Standard for Space Applications: CCSDS Recommendation

CNES chose a non-standard algorithm for PLEIADES-HR compression because JPEG2000 standard [6] was considered too complex for a rad-tolerant hardware implementation. In the same time, CNES was involved in the Image Data Compression (IDC) Working Group of the CCSDS (Consultative Committee for Space Data System). In 2006, the new recommendation CCSDS-IDC was published [7]. In this algorithm, a wavelet transform and a bit-plane encoding of wavelet coefficients organized in trees (8×8 coefficients) are performed. Even if it was too late to be used for PLEIADES-HR, this recommendation was adapted to Earth observation missions' throughput. An ASIC implementation is currently available, it has been developed at the University of Idaho's Center for Advanced Microelectronics and Biomolecular Research (CAMBR) facility where the Radiation-Hardness-By-Design (RHBD) technique is being applied to produce high-speed space-qualified circuits. The projected throughput is over 20 Mpixels/s.

This implementation separates the DWT and the Bit Plane Encoder into two ASICs. CNES has performed several comparative studies on several reference data sets (PLEIADES simulated images, very-high resolution images from airborne sensors, CCSDS reference data set). Both PLEIADES and CCSDS have very similar performances in terms of quantitative criteria (Mean Error, Maximum Error and Root Mean Squared Error).

2.5 Image Quality Assessment

As explained in [3], image quality is a very strong requirement for the choice of a compression algorithm. Criteria that are usually taken into account in this trade-off are statistical, such as the Root Mean Squared Error, though it has been proved that these quantitative criteria are not enough to specify a compression algorithm. Expert users' analyses are useful to evaluate different algorithms and several compression ratios. These experimentations belong to an iterative process between algorithm refinement and image quality requirement assessment. Lately, better statistical criteria, more related to the experimental results, are being considered. The signal variance to noise variance ratio is a candidate. In fact, a set of criteria should be used to validate and finalize a compression algorithm but users' feedback remains necessary. In addition, high-resolution imagers need a restoration phase including deconvolution and denoising. These steps are performed on-ground after decompression. These techniques are necessary to produce a good-quality image without blurring effect due to the MTF (Modulation Transfer Function) and instrumental noise. Until now, statistical criteria used to evaluate the compression efficiency were computed between original and decompressed images, meaning that restoration functions were not taken into account. In 2009, CNES started a study to optimize both compression and restoration steps. The complete image chain from instrument through on-board compression to ground restoration will be considered.

3 On-Board Compression Algorithms: Present and Future

3.1 Multispectral Compression

Future high resolution instruments planned by CNES will have higher number of spectral channels than current instruments. In the case of so-called multi-spectral or super-spectral missions about ten bands are acquired simultaneously with a narrow swath and a spatial resolution from 10 m to 1 km. In the case of very high resolution instruments, a smaller number of bands are acquired, typically four bands: blue, red, green and near infra-red, sometimes completed with the short-wave infrared or a

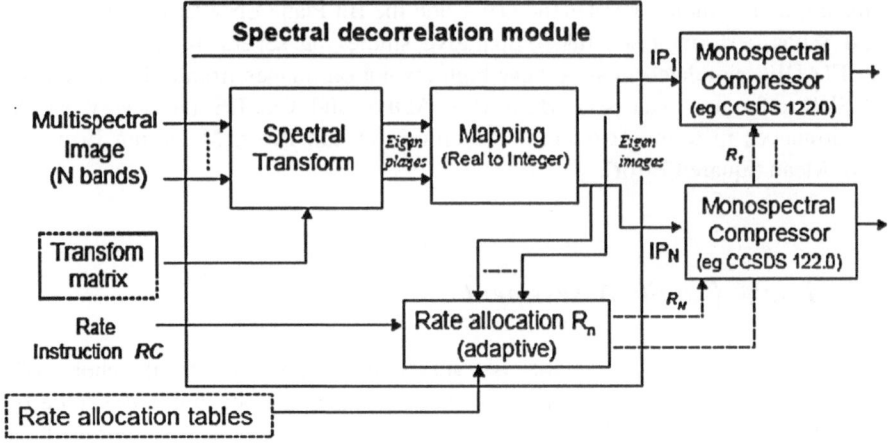

Fig. 2.7 Spectral decorrelation module based on an exogenous KLT

panchromatic band with a better spatial resolution. This is the case for PLEIADES images which have a panchromatic band and four multispectral bands. Up to now, data compression is done independently on each channel, which means on the panchromatic one and on each multispectral channel. In this case, the so-called "monospectral" compressor only exploits the spatial redundancies of the image, ignoring the redundancy between the different images of the same scene taken in different spectral bands. For optimum compression performance of such data, algorithms must take advantage of both spectral and spatial correlation. In the case of multispectral images, CNES studies (in cooperation with Thales Alenia Space) studies have led to an algorithm using a fixed transform to decorrelate the spectral bands, where the CCSDS codec compresses each decorrelated band using a suitable multispectral rate allocation procedure [8].

As shown in Fig. 2.7 this low-complexity decorrelator is adapted to hardware implementation on-board satellite. It is suited to high-resolution instruments for a small number of spectral bands. For higher number of bands (superspectral and hyperspectral images), CNES has also led several studies based on a spectral decorrelator followed by an entropy encoder (CCSDS, SPIHT3D, JEPG2000) [9]. In the framework of the new CCSDS Multispectral and Hyperspectral Data Compression Working Group, CNES is currently studying a hyperspectral compression algorithm suitable for space applications and based on a spectral decorrelator and the CCSDS Image Data Compression recommendation [10].

3.2   Wavelet Limitations

Using the PLEIADES-HR compressor or the CCSDS recommendation, it can be seen that artefacts appear for high compression ratios. These artefacts can be well-known blurring effects, high quantization of low-complexity regions (due to

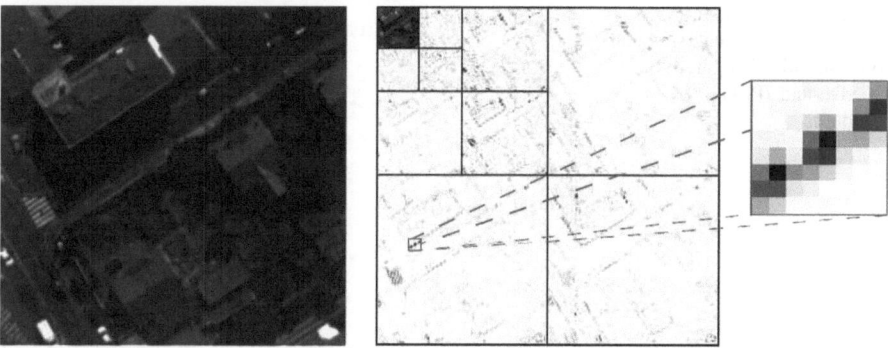

Fig. 2.8 High-resolution image from airborne sensor (*left*) and its wavelet transform (*right*) with a zoom on some coefficients

the rate-allocation procedure over large swath of pixels) but also bad definition of image edges. This last artefact is due to the wavelet transform which has residual directional correlation between wavelet coefficients in a small neighborhood (see Fig. 2.8). In [11], Delaunay has shown that EBCOT is very efficient in capturing these residual values. This context coding makes JPEG2000 among the best existing compressors but its implementation complexity issue has been previously explained in this chapter.

3.3   A New Transform for On-Board Compression

Since 2004, CNES has investigated new spatial decorrelators while considering the on-board satellite implementation constraints. Several promising transforms such as contourlets, curvelets, ridgelets and bandelets have been studied. Finally a post-transform optimization based on wavelet coefficients and very close to the basic idea of the bandelet transform has been achieved [12]. Transform bases are designed based on directional groupings and on Principal Component Analysis (hereafter PCA) on blocks of wavelet coefficients. A Lagrangian rate-distortion optimization process is used to select the best transform for each 4 × 4 wavelet coefficients block. An internal study showed that this size was optimal in terms of performances versus complexity trade-off. In [13], it is proved that bases resulting from PCA on various sets of blocks are better than directional bases. These performances are compared to CCSDS and JPEG2000 compressors on a set of very high resolution images (Fig. 2.9). We observe a gain from 0.5 to 1 dB in the range [0.6, 2.6 bpp] over the CCSDS. Nevertheless, the JPEG2000 performances are never reached and whatever the post-transform, results are around 0.6 dB lower than JPEG2000.

In terms of implementation complexity, the post-transform studied in this case does not mean complex operations. Wavelet coefficients are projected on 12 bases

Fig. 2.9 Post-transforms performances in peak signal to noise ratio compared to CCSDS and JPEG2000 standards

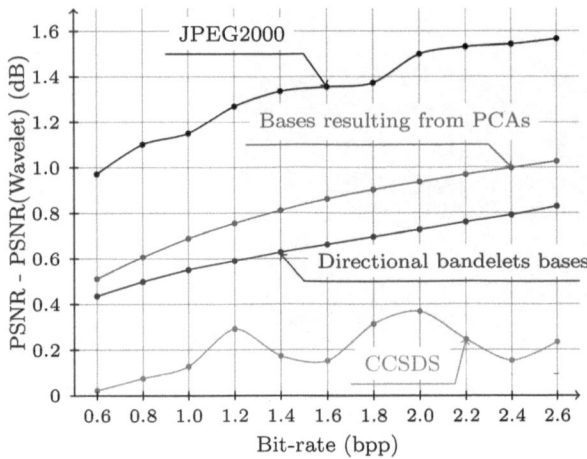

of 16×16 coefficients. Then, a Lagrangian cost is computed and post-transformed coefficients are encoded.

The arithmetic coder is used for experimental testing but a last study of best basis selection was performed to do a complete analysis of optimal quantization steps, Lagrangian multiplier and transform bases. We plan to replace the arithmetic coding stage, which is known to have difficult implementation issues, by a bit-plane encoder allowing bit accurate and progressive encoding. This encoder has to be adapted to the post-transform wavelet coefficients. An efficient bit-plane encoding procedure should provide as good results as the arithmetic coder. The final bit-stream should be fully embedded like the CCSDS recommendation or the PLEIADES-HR compressor allowing progressive transmission.

3.4 COTS for On-Board Compression

CNES, as a member of the JPEG2000 standard committee plan to use this algorithm on-board satellites. Consequently it performed in 2000, an implementation study of the JPEG2000 standard with radiation hardened components. The results were quite disappointing because this algorithm was considered too complex to be implemented on this kind of hardware, principally because of the arithmetic coder and the rate allocation procedure (handling of optimal truncation points). This was the same conclusion of the CCSDS Image Data Compression Working Group when it started to look for a candidate for the CCSDS-IDC recommendation. The published recommendation is finally half complex than the JPEG2000 standard with performances 2 dB lower in a scan-based mode (memory limited). In 2008, CNES started a complete study of the commercial component from Analog Device compliant to the JPEG2000 standard (ADV212). The ADV212 integrated circuit is a System-On-Chip designed to be a JPEG2000 codec and targeted for video and high bandwidth image

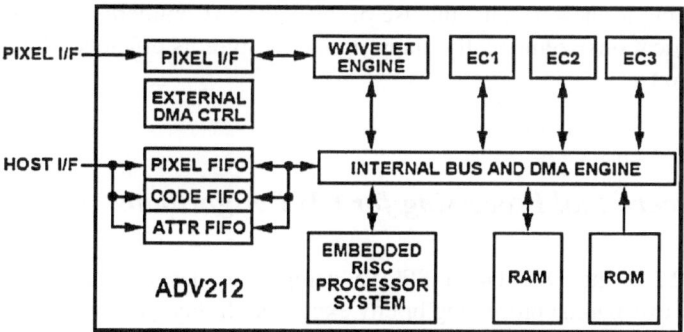

Fig. 2.10 ADV212 internal architecture

Fig. 2.11 ADV212 evaluation board (ADV212 integrated circuit is at the bottom right)

compression applications. The Integrated circuit includes multiple hardware functions such as wavelet engine, RiSC processor, DMA, memories and dedicated interfaces. Figure 2.10 presents the ADV212 architecture. The software executed within the circuit allows the chip to perform compression or decompression. The evaluation board is shown in Fig. 2.11. This study was both a performance analysis and a spatial environment tolerance study (radiation sensitivity). The algorithmic performances were compared to Kakadu software and CNES took a special care about tile partitioning limitations, compression performances and fixed bitrate ability. The radiation campaign was led in Louvain-La-Neuve cyclotron during first quarter of 2010. This circuit was tested against latch-up and SEFI events. Results were not satisfying for latch-up as the Integrated Circuit revealed a high sensitivity (from 2 latch-ups per second under high energy Xenon beam with 100 particle/s, to a bare 1 latch-up per minute under low energy Azote beam with 1,000 particle/s). The last beam setup allowed to perform functional tests that led to timeout error, configuration error… Not a single image was compressed successfully (within the ten tests done) under heavy ions beam.

According to these results, the use of this COTS (Commercial of the Shelf) for space applications seems really inappropriate despite its efficiency for JPEG2000 compression.

3.5 Exceptional Processing for DWT Algorithms

In Sect. 2.2.2, the DCT-based SPOT5 compressor has been presented and the associated exceptional processing briefly explained. In that particular case the occurrence of the defects or artefacts was linked to the choice of a locally unsuitable quantization factor in the rate regulation loop, as this factor was the same for the whole line of blocks. The exceptional processing developed and validated for SPOT5 locally corrected the quantization factor by a factor of 2 or 4, depending on the number of AC coefficients in the block. For both DWT-based PLEIADES compressor described in 2.3 and CCSDS standard described in 2.4, no rate regulation loop is needed, because bit plane encoders hierarchically organize the output bit-stream so that targeted fixed bit rate can be obtained by truncating this bit-stream. However, because of on-board memory limitations, the DWT, bit-plane encoding and truncation are performed over a fixed-size image area. For PLEIADES compressor, this image area is 16 lines (width equal to image width). For CCSDS compressor, this image area is called a segment and its size is defined by the user. In both cases, the quantization induced by truncation of the bit planes description is the same for the studied image area and some defects already observed in SPOT5 compressed images still appear with these DWT-based compressors. In order to locally correct the defected blocks, CNES has studied exceptional processing. The criteria used to decide whether exceptional processing is needed for a block is the [signal variance]/[compression noise variance ratio]. As defined in [7], a block of wavelet coefficients consists of a single coefficient from the LL3 subband, referred to as the DC coefficient, and 63 AC coefficients. Depending on this ratio value, the wavelet coefficients of the block are multiplied by a positive factor before bit-plane encoding. These multiplied coefficients will be treated earlier than in the nominal case (without exceptional processing) by the encoder. This wavelet coefficients processing is similar to what is done in JPEG2000 standard for Region of Interest handling [6]. The image quality improvement brought by the exceptional processing has been confirmed by image analysis. However, its utilization is not needed for PLEIADES images because the targeted bit rate is high enough to prevent such defects.

3.6 Selective Compression

Algorithms presented above are associated with low performance gain while preserving a good image quality whatever the scene and wherever in the scene. However, this gain is not enough compared to the increase of data rate. Unfortunately, any transform has been able to obtain a significant gain on compression ratio

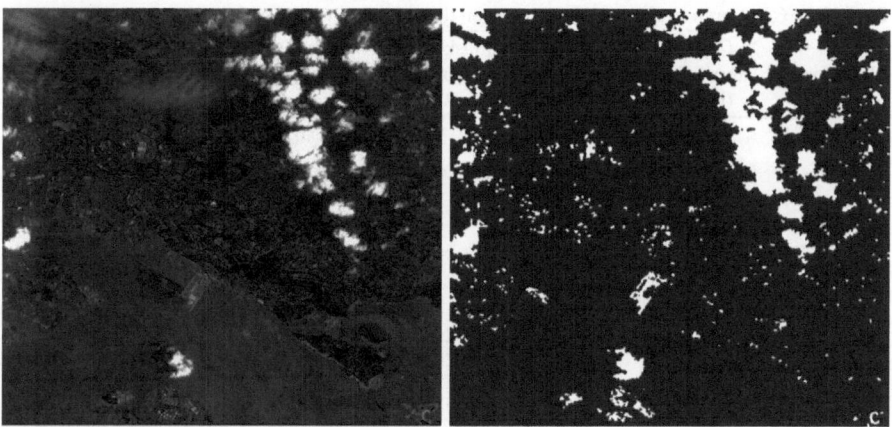

Fig. 2.12 Cloudy SPOT5 image and its associated cloud mask

for the same image degradation. To reach such bigger compression ratios, it is necessary to perform "smart" compression, meaning different compression ratios inside a scene. Thus, on-board detection of useful or non-useful data is required, the non-useful data being more compressed to ensure a compression gain. This kind of compression is called "selective compression" and consists of detecting and then compressing a so-called region-of-interest (ROI) or non-interest. Nevertheless, on-board detection for Earth observation missions must be performed at high data rate and detection algorithms are often too complex. Moreover, selective compression is not so famous because it is hard to describe useful or non-useful data. Fortunately, one type of data can be considered, for almost all CNES High-Resolution applications, as non-useful data: the clouds. In fact, despite the use of weather forecast in satellite scheduling, most of optical satellite images are cloudy; this is the case of SPOT5 for which up to 80% of the images are classified as cloudy by the SPOT-image neural network classifier.

Figure 2.12 gives an example of a cloudy image and its associated binary mask indicating a cloud-pixel when pixel is white (output of a classifier) and a non-cloud-pixel when pixel is black. Considerable mass-memory and transmission gains could be reached by detecting and suppressing or significantly compressing the clouds. Compression algorithms should use this kind of mask during the encoding stage to differently compress both regions of the scene.

3.6.1 On-Board Cloud-Detection Feasibility

During the last years, CNES has studied the implementation of a cloud detection module on-board satellite [14]. The idea was to simplify and optimize for on-board implementation an already existing cloud-detection module used on-ground for PLEIADES-HR album images [15]. This algorithm uses a Support Vector Machine

Table 2.1 Comparative global results between the reference model and fixed-point/VHDL model

Cloud mask coverage	Cloud mask surface in pixels (ref. model)	Cloud mask surface in pixels (fixed-point model)	Common coverage (ref. vs. fixed point) (%)	Different coverage (ref. vs. fixed point) (%)
Acapulco_1	298,570	299,871	100.0	0.4
Bayonne_1	243,113	244,345	100.0	0.5
Dakar_2	475,644	481,180	100.0	1.1
La_Paz_1	167,772	168,542	100.0	0.4
La_Paz_2	749,313	753,932	100.0	0.6
London_1	418,660	416,966	99.5	0.4
Los_Angeles_2	120,225	120,428	100.0	0.1
Marseille_1	153,887	155,251	100.0	0.8
Papeete_1	361,610	362,916	100.0	0.3
Quito_1	724,569	726,453	100.0	0.2
Quito_2	544,185	545,662	100.0	0.2
Seattle	275,310	276,590	100.0	0.4
Toulouse	123,691	123,872	100.0	0.1

classifier on images at low resolution (wavelet transform 3rd decomposition level). The main stages of this algorithm consist in computing top of atmosphere radiance of the image using the absolute calibration factor, computing classification criteria and finally classifying these criteria with the trained SVM configuration.

The on-board implementation study firstly analyzed independently all the phases of this process to propose an on-board simplified model of cloud cover detection algorithm based on SVM technique. The proposed restrictions computed together, via a floating-point software model, showed that equivalent performances could be obtained by the on-board simplified model (<1% of error).

In order to prepare the VHDL description other restrictions were taken into account as the ranges and targeted accuracies of each computing parameter for fixed-point operations.

Finally, an HLS tool was used to obtain fixed-point and VHDL descriptions and to verify the performances if compared with the reference model. Table 2.1 shows some main results for 13 different sites: maximum error complies with the initial specifications with about 1% of error in cloud detection coverage and mask generation. Furthermore, as worst-case errors correspond to cloud pixels considered as ground pixels (common coverage ~100%), almost any additional loss will be introduced in the region of interest by this cloud compression stage.

3.6.2 ROI Coding

The last step of selective compression is the "smart" encoding of the Region Of Interest. In the case of on-board cloud-detection, the background is the cloud-mask and the foreground (ROI) is the rest of the image. ROI coding methods already exist and the principle of the general ROI scaling-based method is to scale (or shift) wavelet coefficients so that the bits associated with the ROI are placed in higher

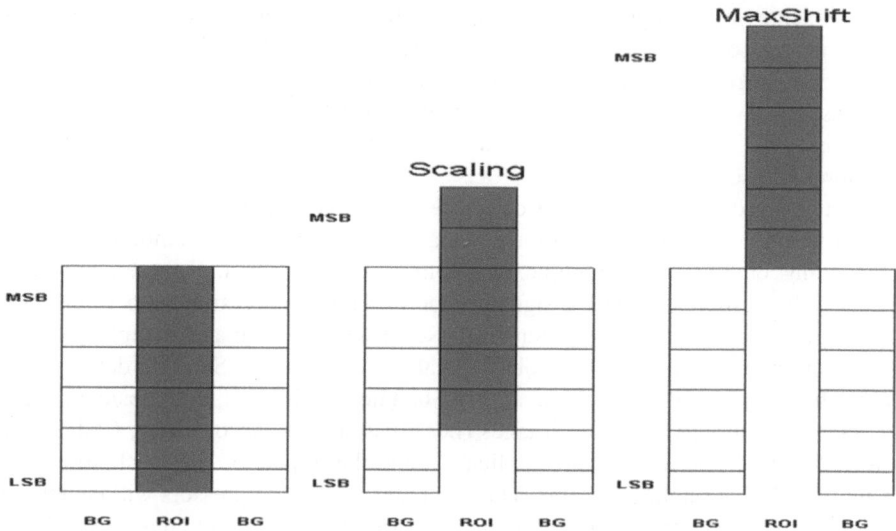

Fig. 2.13 Standard vs. shifted bit plane coding with "Scaling" and "MaxShift" methods

bit-planes than those of the background (see Fig. 2.13). Then, during the embedded coding process, the most significant ROI bit-planes are placed in the bit-stream before background bit-planes of the image.

Two different ROI coding capabilities, Maxshift and Scaling, have been already added by CNES to existing compressors (CCSDS IDC. . .) in order to perform selective compression over all kind of image including cloudy ones.

Of course, both methods have particular advantages and drawbacks considering the application: thus, Maxshift is a better candidate for cloud detection applications as it preserves the best reachable image quality over the ROI and the decompressor can automatically decode the cloud covering mask. Its main drawback is that users cannot control the image quality over the background region (e.g. clouds), or between regions with different degree of interest. The Scaling method will be preferable for this kind of applications but extra over-head must be expected for ROI mask transmission. Other techniques, as Bitplane-By-Bitplane Shift, may also be studied with the aim of allying the advantages of these two methods.

On-going CNES studies will provide rate allocation compatibility for these ROI compression techniques.

3.7   Fixed Quality: Variable Data Rate Compression

Actual on-board compressors assign "almost" the same data rate to every image segment on the scene in order to globally simplify the system data handling by providing highly-predictable volumes and data rates. This is not, however, the best way of achieving the optimum image quality for a scene or a group of scenes for a given amount of data (i.e.: on-board mass memory capacity).

Even if selective compression can offer a significant improvement if compared to classical fixed data rate compression, it is still hard to describe the different regions of interest and even more to handle them in order to obtain efficient compression schemes.

Thus, Fixed Quality-Variable data rate compression algorithms seem to be a better option to optimally compress images with the lowest amount of data: depending on the image complexity (computed by block or segment inside a scene), the compressor will assign the appropriate compression ratio to comply with the selected quality requirements. Nevertheless, as image complexity will vary, variable data rates will be obtained after compression. Satellite data system must be then overvalued to be able to handle the highest volumes/data rates associated with very complex images (i.e. urban areas). Accordingly, CNES will study during the next years the application and the impact of such techniques for High-Resolution satellite systems. The main idea is that compressors will be able to prior compute the image complexity in order to assign the optimum description level (bitplane) for a chosen quality limit. A global limit in terms of data amount should be also imposed for on-board handling and storage issues. Some good approaches, as quality-limit parameters BitPlaneStop and StageStop of CCSDS IDC algorithm, will certainly play an important role in these future compression methods.

4 Conclusions

In this chapter, we have firstly given an overview of past compression algorithms that have been implemented on-board CNES satellites devoted to Earth observation missions. The development of compression algorithms has to deal with several constraints: telemetry budget limiting the data rate after compression, image quality constraints from expert image users and space qualified electronics devices also limiting the implementation and its performances. By using performing techniques such as DCT and then DWT, CNES has developed several modules for Earth observation missions that have been described here. As shown in Fig. 2.14, the used techniques allow a gain in compression ratio while preserving a very good image quality. This gain is more and more necessary to face up to the growing data rate of very high resolution missions. But for coming missions, new compression techniques have to be found. New transforms with better decorrelation power, on-board detection techniques for selective compression or algorithms performing fixed quality compression represent the CNES main fields of study to prepare the future.

Acknowledgments Authors would like to thank all the people who have contributed to CNES on-board compression studies and developments presented in this chapter.

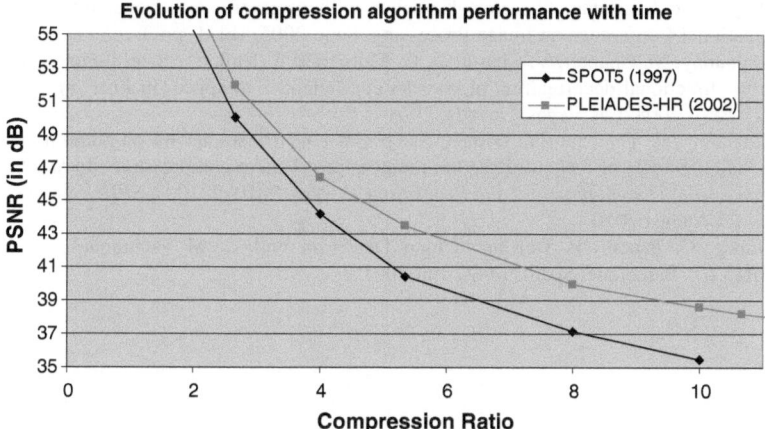

Fig. 2.14 Evolution of compression algorithm performance with time: from SPOT5 to PLEIA-DES-HR

References

1. C. Fratter, M. Moulin, H. Ruiz, P. Charvet, D. Zobler, "The SPOT5 mission", *52nd International Astronautical Congress*, Toulouse, France, 1–5 Oct 2001.
2. P. Kubik, V. Pascal, C. Latry, S. Baillarin, "PLEIADES image quality from users' needs to products definition", *SPIE Europe International Symposium on Remote Sensing, Bruges, Belgium, September 19–22, 2005*.
3. P. Lier, G. Moury, C. Latry and F. Cabot, "Selection of the SPOT-5 Image Compression algorithm", *in Proc. of SPIE 98*, San Diego, vol.3439–70, July 1998.
4. C. Lambert-Nebout, J.E. Blamont, "Clementine: on-board image compression", *26th Lunar and Planetary Science Conference* LPCE, Houston, 1995.
5. C. Lambert-Nebout, G. Moury, J.E. Blamont, "A survey of on-board image compression for CNES space missions", *in Proc. of IGARSS 1999*, Hambourg, June 1999.
6. ISO/IEC 15444–2, "Information technology – JPEG 2000 image coding system: Extensions", 2004
7. CCSDS, Image Data Compression Recommended Standard, CCSDS 122.0-B-1 Blue Book, Nov. 2005
8. C. Thiebaut et al., "On-Board Compression Algorithm for Satellite Multispectral Images", *in Proc. of Data Compression Conference 2006*, Snowbird, March 28–30, 2006.
9. C. Thiebaut, E. Christophe, D. Lebedeff, C. Latry, "CNES Studies of On-Board Compression for Multispectral and Hyperspectral Images", Proc. of SPIE Satellite Data Compression, Communications and Archiving III, San Diego CA, vol. 6683, pp. 668305.1-668305.15, August 2007.
10. C. Thiebaut & R. Camarero, "Multicomponent compression with the latest CCSDS recommendation", Proc. of SPIE Satellite Data Compression, Communications and Processing V, San Diego CA, Vol. 7455, 745503, August 2009.
11. X. Delaunay, M. Chabert, G. Morin and V. Charvillat "Bit-plane analysis and contexts combining of JPEG2000 contexts for on-board satellite image compression", In *Proc. of ICASSP'07*, I-1057–1060, *IEEE*, April 2007, Honolulu, HI.

12. G. Peyre and S. Mallat, "Discrete bandelets with geometric orthogonal filters," in IEEE International Conference on Image Processing, Sept. 2005, vol. 1, pp. I– 65–8.
13. X. Delaunay, M. Chabert, V. Charvillat, G. Morin and R. Ruiloba "Satellite image compression by directional decorrelation of wavelet coefficients", to appear in *Proc. of ICASSP'08, IEEE,* April 2008, Las Vegas, Nevada, USA
14. R. Camarero, C. Thiebaut, Ph. Dejean, A. Speciel « CNES studies for on-board implementation via HLS tools of a cloud-detection module for selective compression", In *Satellite Data Compression, Communication, and Processing VI,* Vol. 7810, 781004, *SPIE,* San Diego, CA, USA, 24 August 2010.
15. C. Latry, C. Panem, P. Dejean, "Cloud Detection with SVM Technique", *in Proc. of IGARSS'07,* Barcelone, Spain, 13–27 Jul. 2007.

Chapter 3
Low-Complexity Approaches for Lossless and Near-Lossless Hyperspectral Image Compression

Andrea Abrardo, Mauro Barni, Andrea Bertoli, Raoul Grimoldi, Enrico Magli, and Raffaele Vitulli

Abstract There has recently been a strong interest towards low-complexity approaches for hyperspectral image compression, also driven by the standardization activities in this area and by the new hyperspectral missions that have been deployed. This chapter overviews the state-of-the-art of lossless and near-lossless compression of hyperspectral images, with a particular focus on approaches that comply with the requirements typical of real-world mission, in terms of low complexity and memory usage, error resilience and hardware friendliness. In particular, a very simple lossless compression algorithm is described, which is based on block-by-block prediction and adaptive Golomb coding, can exploit optimal band ordering, and can be extended to near-lossless compression. We also describe the results obtained with a hardware implementation of the algorithm. The compression performance of this algorithm is close to the state-of-the-art, and its low degree of complexity and memory usage, along with the possibility to compress data in parallel, make it a very good candidate for onboard hyperspectral image compression.

A. Abrardo (✉) • M. Barni
Dip. di Ingegneria dell'Informazione, Università di Siena, Siena, Italy
e-mail: abrardo@dii.unisi.it; barni@dii.unisi.it

A. Bertoli • R. Grimoldi
Carlo Gavazzi Space S.p.A., Milan, Italy
e-mail: abertoli@cgspace.it

E. Magli
Dip. di Elettronica, Politecnico di Torino, Torino, Italy
e-mail: enrico.magli@polito.it

R. Vitulli
European Space Agency – ESTEC TEC/EDP, Noordwijk, The Netherlands
e-mail: Raffaele.Vitulli@esa.int

B. Huang (ed.), *Satellite Data Compression*, DOI 10.1007/978-1-4614-1183-3_3,
© Springer Science+Business Media, LLC 2011

1 Introduction and Motivation

Compression of hyperspectral images has been an important and active research topic for a long time. All spectral imagers such as those of the multispectral, hyperspectral and ultraspectral type are generating increasing amounts of data, calling for image compression to reduce the data volume prior to transmission to the ground segment. Different spectral imagers differ in how much data are available in the spatial and spectral dimensions, potentially requiring different compression techniques. For example, multispectral images typically have very fine spatial resolution and coarse spectral resolution, and their compression mostly exploits the spatial correlation. The opposite is typically true of hyper- and ultraspectral imagers, in which the spectral correlation dominates. Compression is most useful for spaceborne sensors, as it is not possible to physically unmount or read the mass storage device that archives the acquired data, e.g. at the end of a measurement campaign. However, if the sensor resolution is very high or the mission is very long, compression may also be needed onboard an airborne platform.

Image compression techniques allow to transmit more data in the same amount of time, so much as more compression is applied. Several types of compression are possible. In lossless compression, the reconstructed image is identical to the original. In near-lossless compression, the maximum absolute difference between the reconstructed and original image does not exceed a user-defined value. In lossy compression, the reconstructed image is as similar as possible to the original "on average", i.e., typically in mean-squared error sense, given a target bit-rate. Lossless compression is highly desired to preserve all the information contained in the image; unfortunately, the best algorithms provide limited compression ratios, typically from 2:1 to 3:1 for 3D datasets. Near-lossless and lossy techniques yield larger size reductions, at the expense of some information loss. Although these kinds of compression have always received less attention because of the loss of quality and, even more, of the difficulty to assess the effect of the losses on the applications, they are becoming more and more important when the required compression ratio is large. E.g., in [1] it is shown that bit-rates of 0.5 and 0.1 bpp can be achieved with little or no impact on image classification performance.

This chapter overviews the requirements and the state-of-the-art of lossless image compression for hyperspectral images, with a particular focus on low-complexity approaches amenable for onboard implementation. It should be noted that most near-lossless algorithms can be seen as an extension of a predictive scheme, in which the prediction error is quantized in the DPCM loop. As a consequence, near-lossless compression can be obtained as a side-product of any predictive scheme, and for this reason we will not explicitly review existing near-lossless compression algorithms. However, in Sect. 3.4 we will show how the proposed algorithm can be extended to near-lossless compression, and what kind of performance can be expected.

This chapter is organized as follows. In Sect. 3.2 we review prior work in this area, including the existing lossless compression standards that are relevant to onboard compression. In Sect. 3.3 we outline the main requirements of onboard compression, and discuss the most relevant approaches. In Sect. 3.4 we describe a low-complexity approach that employs a block-based predictor. In Sect. 3.5 we show compression results and discuss the complexity of the proposed approach. Sect. 3.6 describes the hardware implementation, whereas in Sect. 3.7 we draw some conclusions and highlight important open problems.

2 Background

2.1 Compression Techniques

Lossless compression of hyperspectral images has mostly been based on the predictive coding paradigm, whereby each pixel is predicted from past data, and the prediction error is entropy coded [2, 3]. In [4] fuzzy prediction is introduced, switching the predictor among a predefined set using on a fuzzy logic rule. In [5] the prediction is improved using analysis of edges. In [6] classified prediction is introduced for near-lossless compression. Classified prediction is further developed in [7] for lossless and near-lossless compression. In [8] spectral prediction is performed using adaptive filtering. In [9] vector quantization is employed to yield lossless compression. In [10] the concept of clustered differential pulse code modulation (C-DPCM) is introduced. The spectra of the image are clustered, and an optimal predictor is computed for each cluster and used to decorrelate the spectra; the prediction error is coded using a range coder. In [11] it is proposed to employ a spectral predictor that is based on two previous bands. In [12] the spectral redundancy is exploited using a context matching method driven by the spectral correlation. In [13, 14] it is proposed to employ distributed source coding to achieve lossless compression with a very simple encoder. In [15] a simple algorithm is proposed, which encodes each image block independently. In [16] a low-complexity algorithm is introduced, based on linear prediction in the spectral domain. In [17] the performance of JPEG 2000 [18] is evaluated for lossless compression of AVIRIS scenes, in the framework of progressive lossy-to-lossless compression; lossy compression results are reported in [1]. Along the same lines, the CCSDS image data compression recommendation [19] uses wavelets for lossless compression, as does the algorithm in [20]. In [21] it is proposed to employ as predictor a pixel in the previous band, whose value is equal to the pixel co-located to the one to be coded; the algorithm is further refined in [22]. In [23] spectral correlation is exploited through context matching. Recent work has also borrowed ideas from distributed source coding to construct extremely simple and error-resilient algorithms [13, 14].

Band reordering has also been used in [24–26] to obtain improved performance; specifically, in band reordering the spectral channels of the image are reordered in such a way as to maximize the correlation of adjacent bands, optimizing the performance of the subsequent compression stage. In [27] it is shown that raw data have very different characteristics from calibrated data, and an algorithm able to exploit calibration-induced artifacts is proposed.

Finally, it should also be mentioned that lossy compression can be achieved via transform coding employing a reversible transform. This is the approach followed in JPEG 2000, allowing to achieve progressive lossy-to-lossless compression, although its performance is generally not as good as the best predictive approaches.

2.2 International Standards

Data and image compression are so important for space applications, that relevant international standards have been developed and approved. In the following we discuss the available lossless compression standards and their use for onboard image compression.

2.2.1 CCSDS 101: Lossless Data Compression

The first compression standard for space applications has been developed by the consultative committee for space data systems (CCSDS). The standard is titled "lossless data compression", meaning that it was developed for general purpose data compression, including but not specifically tailored to image compression. The standard defines a lossless compression algorithm that can be coupled with a user-defined or a default predictor, allowing the user to employ a predictor that has been developed for the target data, coupled with a standardized low-complexity entropy coding stage. The entropy coding stage consists of a mapper followed by an entropy coding stage. This latter operates on short blocks of mapped samples. For each block, all samples are coded using a Golomb power-of-two code [28] with the parameters that minimizes the bit-rate, provided that the other available options (zero-block, no compression) do not yield an even smaller bit-rate. A low-entropy option is also available (second extension), although it is covered by patent.

This entropy coder is somewhat suboptimal with respect to similar coders based on Golomb codes and developed specifically for images (e.g. [29]) that select a different coding parameter for each sample based on a window of previous samples. The performance loss is typically less than 0.2 bpp. The algorithm has very low complexity, and an ASIC implementation exists [30].

2.2.2 CCSDS 122: Image Data Compression

More recently, the CCSDS has developed a new standard for image compression [19]. The algorithm is somewhat similar in spirit to JPEG 2000, as it employs a 2D discrete wavelet transform followed by a simple entropy coder based on bit-planes. We mention this standard as, although it is natively a lossy compression standard, it also has a lossless mode based on a reversible wavelet transform. However, this standard is not suitable for the compression of 3D data, as it only captures the correlation in two dimensions. Repeated application of the algorithm to 2D slices of a 3D dataset is possible, but certainly suboptimal in terms of performance. Moreover, the algorithm complexity is significantly larger than that of the lossless data compression recommendation.

2.2.3 JPEG-LS

The JPEG-LS standard addresses both lossless and near-lossless compression of 2D images. It is based on a simple non-linear predictor, followed by a context-based entropy coder, and uses a quantizer in the prediction loop for near-lossless compression [29]. Although this standard has not been developed specifically for space applications, its low complexity and good performance make it suitable for spectral image compression. In several papers a "differential" JPEG-LS algorithm is used for compression of 3D images, which simply takes differences of adjacent bands and encodes them using JPEG-LS. To maximize coding efficiency, the same principle can be applied to spatial-spectral slices, with the difference being taken in the remaining spatial dimension.

3 Onboard Compression Requirements

Although compression performance is certainly a key aspect for onboard compression, there are other very important requirements that should also be taken into account. These requirements draw a scenario that is somewhat different from typical image compression, typically leading to the design of low-complexity and hardware-friendly compression algorithms.

Low encoder complexity. Since spectral imagers can generate very high data rates, it is of paramount importance that the encoder has low-complexity, in order to be able to operate in real time. It is worth mentioning that the computational power available for compression is generally limited, and nowhere near that yielded by a processor for workstation applications. A typical design involves the compression algorithm being implemented on an FPGA, although certain missions will afford the ad-hoc design of an application-specific integrated circuit (ASIC), while other missions may only use a digital signal processor. Typical clock frequencies are of the order of 100 MHz.

The low-complexity requirement generally rules out the application of transform coding methods because of their complexity. While these methods are suitable for 2D images, the 3D case would entail to apply 2D transform coding repeatedly after taking a transform in the third dimension. This is possible only with very significant computational resources.

The prediction plus entropy coding paradigm is amenable to low complexity. Although very sophisticated predictors exist, adequate performance can be obtained using relatively simple ones (e.g., as done in JPEG-LS). Traditionally, onboard compression algorithms avoid using arithmetic coding, as it is deemed to be a relatively complex coding scheme, especially for large alphabets. Instead, Golomb power-of-two codes are the preferred choice, as they achieve a good balance between performance and complexity. The use of simplified low-complexity arithmetic codes has not been evaluated extensively for space applications; such codes are used in JPEG 2000 and H.264/AVC compression standards, just to mention a few.

Error-resilience. Algorithms should be capable of contain the effect of bit-flippings or packet losses in the compressed file. These errors typically occur because of noise on the communication channel. Compressed data are known to be very sensitive to errors, to the extent that a single erroneous bit could prevent from decoding the remainder of the data. This problem can be alleviated in many ways, e. g. employing error-resilient entropy codes such as reversible variable-length codes [20], coset codes [14] or error-resilient arithmetic codes [31]. Another approach is to partition the data into units that are coded independently, in such a way that an error in one unit will not prevent from decoding other units. This latter solution is usually preferred, as it is indeed very simple, and guarantees that the scope of an error is limited to one partition.

As partitioning incurs a cost in terms of compression performance, these units should not be too small; the optimal size ultimately depends on how many errors or packet losses are expected on the communication channel. E.g., while typical Earth observation missions usually have very few losses, deep-space missions are subject to higher bit-error rates. The shape of the partition also plays some role. E.g. should the partition have a large spatial size and a low spectral size, or the opposite? This is also application-dependent, in that there are applications in which spatial information is critical and should be retained as much as possible while spectral information may be sacrificed, and vice versa.

Hardware friendliness. Since onboard compression algorithms are typically implemented on FPGA or ASIC, the algorithm has to be designed in such a way that its hardware implementation is simple, i.e., it must be able to operate using integer arithmetic, it must fit into a relatively small FPGA, and must use the available resources effectively, possibly not needing an external memory. Moreover, it is desirable that the algorithm can be parallelized in order to speed up the compression process for high data-rate sensors.

In this chapter we describe a compression solution that aims at fulfilling the criteria above. The algorithm performs lossless and near-lossless compression with very low complexity. The prediction stage is based on the design in [15]. This predictor has low complexity, as it computes a single optimal predictor for each 16×16 block of input samples. Working on 16×16 blocks, the predictor performs data partitioning, in that any 16×16 block can be decoded without reference to any other 16×16 block in different spatial locations in other bands. Thus, while there is a performance loss for working on a block-by-block basis, this is small with respect to pixel-based predictors as in [8]; it is possible to employ a block-based predictor along with coset coding to achieve increased error resilience [14], at the expense of an additional performance loss. The performance of the algorithm, for both lossless and near-lossless compression, is improved by using band reordering. Band reordering has never been given much attention for onboard compression, because it requires additional computations to obtain the optimal ordering. However, we show that, in a realistic application scenario, band reordering could be performed at the ground segment, requiring no additional computations onboard, providing most of the performance gain of optimal band reordering.

4 Block-Based Coding Approach

In the following we denote as $x_{m,n,i}$ the pixel of an hyperspectral image X in mth line, nth pixel, and ith band, with $m = 0, \ldots, M - 1$, $n = 0, \ldots, N - 1$ and $i = 0, \ldots, B - 1$. The algorithm compresses independent non-overlapping spatial blocks of size $N \times N$, with N typically equal to 16; each block is processed separately. This approach entails a small performance penalty, as all the prediction and entropy coding operations have to be reset at the end of each block. However, it has two important advantages.

- It allows parallelization, as each spatial block (with all bands) can be encoded as a completely independent data unit.
- It provides error resilience, as erroneous decoding of a block has no effect on the decoding of other blocks in different spatial positions.

Moreover, additional spectral error resilience and spectral parallelization can be obtained by periodically encoding a band *without* spectral prediction (i.e., intra-mode). The group of bands contained between two intra-coded bands, including the first and excluding the last, can be encoded and decoded independently and will generate an independent error containment segment.

4.1 Prediction

For the first band ($i = 0$), 2D compression is performed, as there is no "previous" band to be exploited. The choice of the 2D predictor is not critical, and several reasonably good predictors will work. The predictor we have used is defined as $x'_{m,n,0} = (x_{m-1,n,0} + x_{m,n-1,0}) \gg 1$, where $\gg$ denotes right-shift, i.e., it takes the average of the pixels on the top and left of the pixel to be predicted.

Except for the first sample of the block, all prediction error samples are mapped to nonnegative values using the following formula:

$$S = \begin{cases} 2|e| - 1 & \text{if } e > 0 \\ 2|e| & \text{if } e \leq 0 \end{cases}$$

where e is the prediction error value, and S is the corresponding mapped value. This mapping function is also used for all other bands. For all other bands, a $N \times N$ block in band i is processed as follows. The samples $x_{m,n,i}$ belonging to the block (with $m,n = 0, \ldots, N - 1$), are predicted from the samples $x_{m,n,1}$ of another already coded band (i.e., a spatially co-located block with index l different from i). In the following we assume that $l = i - 1$, i.e., the co-located block in the previous band $i - 1$ is used as reference to predict the block in the current band i. In Sect. 3.4.4 we will show how to set l to a value generally different from $i - 1$.

In order to make $x_{m,n,i-1}$ globally as "similar" as possible to $x_{m,n,i}$ in minimum mean-squared error sense, a least-squares estimator can be computed over the block as $\alpha = \alpha_N/\alpha_D$ [15], with $\alpha_N = \sum\limits_{m,n \in A} (x_{m,n,l} - m_l)(x_{m,n,i} - m_i)$ and $\alpha_D = \sum\limits_{m,n \in A} (x_{m,n,l} - m_l)(x_{m,n,l} - m_l)$, and m_l and m_i are average values of the block, as defined below.

Note that the original predictor [15] does not remove the mean value in the computation of α_N and α_D. However, we have found that removing the mean value provides improved performance, even though m_i has to be written in the compressed file. If α_D is equal to zero, then it is set to one to avoid dividing by zero.

In the equations above, the summations are computed over the set o indexes A. While in [15] A contains all indexes $m \in [0, \ldots, N - 1]$ and $n \in [0, \ldots, N - 1]$, we did not choose to do so. In fact, since α is going to be quantized, it is sufficient to use enough samples so that its precision is smaller than the quantization error. On the other hand, using fewer samples can greatly reduce the number of computations required to compute α. In particular, it has been empirically found that one fourth of the samples are enough to estimate α with sufficient accuracy. Therefore, we select A as a set of 64 positions uniformly picked at random in the block, as random positions yield slightly better performance than regularly spaced ones. The positions are the same for all spatial blocks and all bands, and need not be communicated to the decoder. m_i and m_l are the average values of the co-located blocks in bands i and l. They are also computed from the reduced set of samples A. Note that m_l does not need to be computed, as it had already been computed during the encoding of band l; only m_i has to be computed.

A quantized version of α is generated using a uniform scalar quantizer with 256 levels in the range [0,2]. In particular, the quantization yields $\alpha' = \text{floor}(128\alpha)$, and α' is then clipped between 0 and 255. The dequantized gain factor is obtained as $\alpha'' = \alpha'/128$.

A drawback of the original version of this predictor is that computing $\alpha = \alpha_N/\alpha_D$ requires a floating-point division, which is very undesirable in hardware. In practice, this division can be avoided. This is based on the fact that, after computing α, the obtained value will be quantized to α'' using 8 bits. That is, out of all the infinitely many possible values, only 256 values will actually be taken. Therefore, instead of computing the division, the encoder can simply select, out of all the 256 values, the one that is closest to α in squared-error sense. In particular, to speed up the search, the encoder can perform a dicotomic search over the 256 possible values of α'', seeking the value that minimizes $|\alpha''\alpha_D/\alpha_N|$. This is equivalent to minimizing $|\alpha'\alpha_D - 128\alpha_N|$.

Once α'' has been computed, the predicted values within the block are computed for all $m = 0, \ldots, N$ and $n = 0, \ldots, N$ as $x'_{m,n,i} = \text{round}[m_i + \alpha''(x_{m,n,l} - m_l)]$. The prediction error vector can be calculated in each block for all $m = 0, \ldots, N$ and $n = 0, \ldots, N$ as $e_{m,n,i} = x_{m,n,i} - x'_{m,n,i}$. Subsequently, the prediction error samples are mapped to nonnegative integers.

4.2 Near-Lossless Compression

Predictive schemes are suitable for near-lossless compression, since quantization in the DPCM feedback loop [32] produces the same maximum absolute error on each pixel, which is equal to half of the quantizer step size. Near-lossless compression can hence be achieved by means of a uniform scalar quantizer with step size $2\delta + 1$ and midpoint reconstruction, such that, letting $e_{m,n,i}$ be the prediction error, its reconstructed value $e'_{m,n,i}$ is given by

$$e'_{m,n,i} = \text{sign}(e_{m,n,i})(2\delta + 1)\text{floor}\left(\frac{|e_{m,n,i}| + \delta}{2\delta + 1}\right)$$

4.3 Entropy Coding

The proposed algorithm has two coding options, namely Golomb codes and Golomb power-of-two codes; the same context is employed for either code.

4.3.1 Golomb Coding

The $N \times N$ mapped prediction residuals of a block are encoded in raster order using a Golomb code [28], except for the first sample of the block that is encoded using an exp-Golomb code of order 0 [33].

The Golomb code parameter k_j for the jth sample of the block is computed from a running count Σ_j of the sum of the last 32 mapped values of the block; for samples with index less than 32, only the available mapped values are used. In particular, the following formula is used [34]: $k_j = \text{floor}[(0.693/J) \ \Sigma_j + 1]$, where $\Sigma_j = \sum_{k=j-32,k\geq 0}^{j-1} |S_k|$, 0.693 approximates $\log(2)$, $J \leq 32$ is the number of available samples in the running count Σ_j, S_j are the mapped values.

Note that computing k_j simply requires to update Σ_{j-1} as $\Sigma_j = \Sigma_{j-1} - |S_{j-33}| + |S_{j-1}|$, requiring at most two operations per sample.

4.3.2 Golomb Power-of-Two Coding

For Golomb power-of-two (GPO2) codes, the code parameter is computed in a similar way. The context is defined exactly as in the Golomb code case, except for the fact that the running count is taken over the magnitude of the unmapped prediction error residuals, rather than the mapped values. The parameter k_j is computed using the well-known C-language one-liner below [29], where D is the running count of sum of magnitudes of prediction residuals.

```
for (kj = 0; (j << kj) <= d;kj ++);
```

4.3.3 Coding of Predictor Parameters

Since the predictor is not causal, it is necessary to write in the compressed file two parameters, namely α' and m_i; it is not necessary to write m_l, as it can be computed by the decoder. For α', 8 bits are used to specify its binary representation. For m_i, a more sophisticated technique is employed. Letting the first band be denoted by index $i = 0$, for the second band ($i = 1$) m_i is written using 16 bits. For the other bands ($i > 1$), the difference $m_i - m_l$ is taken; that is, m_i is coded predictively in order to exploit its correlation across bands. The sign bit of the difference is written in the compressed file, followed by the exp-Golomb code of the magnitude.

4.3.4 File Format

For each block, the following information is written in the compressed file. For all bands other than the first one, the parameter α' and the coded value of m_i are written. Then, the coded values of all samples of the block are written in raster order.

4.4 Band Reordering

The principle of band reordering [24] is that the previous band is not necessarily the best metric for predicting the current band. This raises the problem of finding the best reference band l for the prediction of band i. The reference band has to be sought among the bands that have already been encoded. Band reordering is interesting for onboard compression in that, if the ordering is given, picking a reference band other than the previous band does not entail any additional operations or more hardware complexity, when all bands for a given pixel or line are buffered.

Solving the band reordering problem entails the definition of a similarity metric $r_{l,i}$ that describes how "similar" band l and band i are, and hence how good the prediction of band i using band l as reference band is expected to be. In [24] this metric is taken as the number of bits needed to encode band i from band l; however, this requires to perform the actual encoding, and is ad-hoc for a specific compression algorithm. In [25] $r_{l,i}$ is taken as the correlation coefficient between band l and band i. In [26] the correlation coefficient is also used, but band grouping is introduced to limit the complexity.

It is worth noting that band reordering is an additional option that the user may or may not use, without affecting the algorithm design and implementation. As will be seen in Sect. 3.5, not all datasets benefit from band reordering. In some cases the gain is negligible, while in other cases it is not. It is up to the user to decide whether the effort in calculating an optimal reordering is worth the performance gain.

For the algorithm described here, we have employed a band reordering scheme based on [25], improving it in a few aspects. The similarity metric is taken as the correlation coefficient. The metrics are used to construct a weighted undirected graph, in which every band is represented by a vertex, and $r_{l,i}$ is the weight of an arc linking bands l and i. To keep the number of arcs limited, each band i is "connected" with a maximum number M of previous and next bands (from $i - M$ to $i + M$), and considering that $r_{l,I} = r_{i,l}$. The optimal ordering is achieved solving a maximum weight tree (equivalent to a minimum spanning tree) problem over the graph. This can be done using Prim's algorithm as in [25].

For optimal band reordering, a specific ordering has to be computed and used for each image, possibly increasing complexity. Therefore, we also investigate a different approach. In this approach, a "good" band ordering is computed at the ground station based on sample data, and then it is uploaded to the satellite for subsequent use in the compression of all images. The motivation is that the optimal ordering arguably depends on both the sensor and the scene, with the former potentially dominating the ordering. If the contribution of the scene to the optimal ordering is small, then a per-sensor ordering would be almost as good as the optimal per-image ordering. In Sect. 3.5 we show that this is indeed the case. Thus, this approach achieves the twofold benefit of improving the performance via use of an almost-optimal ordering, and to do that without adding complexity to the algorithm, as all computations needed to solve the band reordering problem would be performed on the ground on sample data.

4.5   Complexity

We have analyzed the algorithm described above in order to calculate the average number of operations needed to encode one sample of input data. The prediction stage requires approximately nine operations per sample.

5   Compression Performance

5.1   Dataset Description

We present compression results for two different sensors.

The first is the well-known AVIRIS (Airborne Visible/Infrared Imaging Spectrometer), an airborne hyperspectral system that collects spectral radiance in 224 contiguous spectral bands with wavelengths from 400 to 2,500 nm. Most scientic papers in the field of hyperspectral image compression report results on five scenes from the 1997 missions, namely *Cuprite*, *Jasper Ridge*, *Moffett Field*, *Lunar Lake* and *Low Altitude*. However, these images are calibrated. It has been shown [27, 35] that the calibrated data have a peculiar histogram that can favor certain classes of algorithms, and which is not present on the raw data. Therefore, in this paper we employ the new corpus of raw AVIRIS images recently made available by NASA. These are five images acquired over Yellowstone park in 2006 (called *sc0*, *sc3*, *sc10*, *sc11*, *sc18*), available at http://compression.jpl.nasa. gov/hyperspectral/. Each image is a 512-line scene containing 224 spectral bands and 680 pixels per line.

The second is the AIRS (Atmospheric Infrared Sounder) instrument onboard the Aqua satellite. AIRS is used to create 3D maps of air and surface temperature, water vapor, and cloud properties. With 2,378 spectral channels, AIRS qualifies as an ultraspectral sensor. For the compression studies, ten granules have been simulated from the data obtained from NASA AIRS observations, removing 270 channels, converting radiances into brightness temperatures and scaling as unsigned 16-bit integers. The data are available via anonymous ftp (ftp://ftp.ssec.wisc.edu/pub/bormin/HES). For this study, we have considered only 1,501 bands, removing the unstable channels.

These AVIRIS and AIRS data are part of the dataset employed by the CCSDS (Consultative Committee for Space Data Systems) to assess the performance of multispectral and hyperspectral compression algorithms.

5.2   AVIRIS

Results for AVIRIS images are shown in Table 3.1. We compare the proposed algorithm using Golomb codes and GPO2 codes, but *without* band reordering, the LUT algorithm [10] and the FL algorithm [27], and the 3D-CALIC algorithm with BSQ ordering [36]. Results for the LUT algorithm are also taken from [27]. As can be seen the proposed algorithm is significantly better than LUT, and almost as good as FL, but with lower complexity. This is a very good result, as FL has a very competitive performance. LAIS-LUT [22] would score an average of 6.50 bpp, which is significantly larger than the proposed algorithm. The use of GPO2 codes does not significantly reduce performance.

Table 3.1 Compression performance (bit-rate in bpp) on AVIRIS yellowstone images

	Proposed (Golomb)	Proposed (GPO2)	LUT	FL	3D-CALIC (BSQ)
Sc0	6.44	6.45	7.14	6.23	6.41
Sc3	6.29	6.30	6.91	6.10	6.23
Sc10	5.61	5.62	6.26	5.65	5.62
Sc11	6.02	6.04	6.69	5.86	n.a.
Sc18	6.38	6.39	7.20	6.32	n.a.
Average	6.15	6.16	6.84	6.03	n.a.

Table 3.2 Compression performance (bit-rate in bpp) of band reordering on AVIRIS yellowstone images

	Proposed (Golomb)	BR (optimal)	BR (per-sensor)
Sc0	6.44	6.37	6.37
Sc3	6.29	6.21	6.22
Sc10	5.61	5.57	5.57
Sc11	6.02	6.00	5.97
Sc18	6.38	6.31	6.31
Average	6.15	6.09	6.09

Next, we analyze the performance of band reordering in the optimal and per-sensor mode. In this latter case, the optimal reordering has been computed on the sc0 image, and used for all five images. The band reordering computation has been performed using $M = 7$. Results are obtained using Golomb codes. The results are reported in Table 3.2. As can be seen, the improvement yielded by band reordering is very small (around 1%). Visual inspection of the optimal ordering shows that for these AVIRIS scenes it is almost identical to the sequential ordering. Therefore, for the AVIRIS dataset band reordering would not be useful. Interestingly, there is no performance loss in optimizing the ordering per-sensor instead of per-image.

5.3 AIRS

Compression results for AIRS granules are shown in Table 3.3. We compare the proposed algorithm using Golomb codes and GPO2 codes, but *without* band reordering, the LUT algorithm and the 3D-CALIC algorithm with BSQ ordering. Results for the LUT algorithm have been generated using the software kindly provided by the authors of [10].

Also in this case the GPO2 codes show a very limited performance loss with respect to Golomb codes. 3D-CALIC achieves better performance than the proposed algorithm, but exhibits a very high complexity. The LUT algorithm does not perform well on these data.

In Table 3.4 we analyze the performance of band reordering in the optimal and per-sensor mode. In this latter case, the optimal reordering has been computed on granule 120, and used for all the other granules. The band reordering computation has been performed using $M = 25$. Results are obtained using Golomb codes.

Table 3.3 Compression performance (bit-rate in bpp) on AIRS granules

	Proposed (Golomb)	Proposed (GPO2)	LUT	3D-CALIC (BSQ)
Gran9	4.58	4.59	5.32	4.34
Gran16	4.52	4.53	5.26	4.34
Gran60	4.80	4.81	5.65	4.52
Gran82	4.36	4.38	5.03	4.24
Gran120	4.65	4.66	5.40	4.40
Gran126	4.78	4.79	5.64	4.51
Gran129	4.44	4.45	5.17	4.24
Gran151	4.80	4.82	5.73	4.51
Gran182	4.89	.4.90	5.94	4.54
Gran193	4.83	4.84	5.62	4.53
Average	4.66	4.68	5.48	4.42

Table 3.4 Compression performance (bit-rate in bpp) of band reordering on AIRS granules

	Proposed (Golomb)	BR (optimal)	BR (per-sensor)
Gran9	4.58	4.25	4.27
Gran16	4.52	4.21	4.24
Gran60	4.80	4.38	4.46
Gran82	4.36	4.13	4.16
Gran120	4.65	4.32	4.32
Gran126	4.78	4.39	4.45
Gran129	4.44	4.15	4.19
Gran151	4.80	4.44	4.48
Gran182	4.89	4.48	4.53
Gran193	4.83	4.42	4.48
Average	4.66	4.32	4.36

As can be seen, the performance improvement obtained applying band reordering to AIRS data is significant (around 6%), and exceeds 0.3 bpp on average. Even more interestingly, the performance loss incurred by optimizing the reordering at the ground station is small (about 0.04 bpp), showing that band reordering provides a nice performance gain while not adding any complexity to the onboard compression algorithm. Therefore, for the AIRS dataset band reordering seems a viable option.

5.4 Near-Lossless Compression

For near-lossless compression, we show performance results on the AVIRIS sc0 scene. In particular, Fig. 3.1 reports rate-distortion curves for the proposed algorithm (without band reordering) and JPEG 2000 [37]. This latter is Part 2 of the standard, employing a spectral discrete wavelet transform in the spectral dimension, followed by a spatial wavelet transform, with full 3D post-compression rate-distortion optimization, and no line-based transform. As can be seen, the rate-distortion curve of the near-lossless algorithm shows the typical behavior of very good performance at high rates. The Golomb code (like a Huffman code) is unable to produce codewords

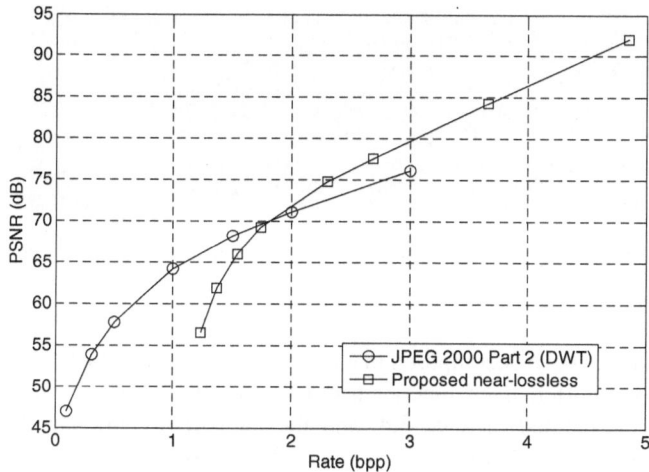

Fig. 3.1 Near-lossless compression results on the sc0 image and comparison with JPEG 2000

shorter than one bit, and this is the reason why the rate-distortion curve has a vertical asymptote at 1 bpp. To achieve low bit-rates it would be necessary to employ a different entropy code, e.g. an arithmetic code. Still, the algorithm is better than JPEG 2000 at bit-rates larger than 1.8 bpp, with significantly lower complexity, buffering and local memory requirements.

It should be noted that, unlike other block-based lossy compression algorithms such as JPEG, at low bit-rates the decoded images do not suffer from blockiness. In fact, JPEG blocking artifacts are due to quantization in the DCT domain, which acts as a vector quantizer over small blocks, which can generate discontinuities at the boundary of different blocks. Conversely, in the proposed scheme the prediction errors are quantized *independently* inside the block and between different blocks, yielding independent reconstruction errors. This is shown in Fig. 3.2 for band 63 of the sc0 image. On the left is a 64×64 crop of the original image. The center figure is the image decoded by the proposed algorithm at very low bit-rate, whereas the image on the right is a JPEG reconstruction. As can be seen, JPEG is very good at preserving structures, but it blurs the texture and introduces visible blocking artifacts. The proposed algorithm preserves some texture and does not exhibit any blockiness, even though the structures are partly compromised.

6 Hardware Implementation

A rapid prototyping implementation on hardware of the lossless compression algorithm has been performed. The design and modeling phase of the algorithm has been supported by Matlab/Simulink Xilinx system generator tool, a rapid prototyping environment for the design and implementation in FPGA.

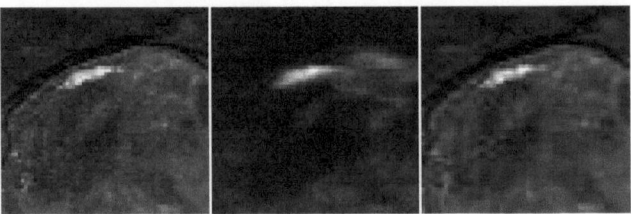

Fig. 3.2 Visual quality at very low bit-rate. *Left*: original; *center*: proposed algorithm; *right*: JPEG

The algorithm has been decomposed in elementary functional blocks communicating with ping-pong buffer. Each functional block is in charge of executing the macro computation as defined in Sect. 3.4, and the parallel execution in FPGA of the functional block is exploiting obtaining excellent results in term of samples rate processing of the algorithm.

System Generator is a DSP design tool from Xilinx that enables the use of The Mathworks model-based design environment Matlab-Simulink for FPGA design. The following key features have been exploited:

- Build and debug high-performance digital signal processing and generic algorithms in Simulink using the Xilinx blockset that contains functions for signal processing.
- Automatic code generation from Simulink for specific Xilinx IP cores from the Xilinx Blockset.
- Code generation option that allows to validate hardware and accelerate simulations in Simulink and MATLAB.
- Xilinx System Generator compatible board (i.e. ADM-XRC-II) available for hardware-in-the-loop for rapid prototyping and test.

In particular, the ADM-XRC-II is a high performance PCI Mezzanine Card format device designed for supporting development of applications using the Virtex-II series of FPGA's from Xilinx. Drivers and the software package including Xilinx System Generator blockset are available to interface the board with Matlab/Simulink/Sys generator host computer via PCI bus.

VHDL has been automatically generated starting from the rapid prototyping tool for two Xilinx FPGA components. The selected FPGA components also have an equivalent radiation-tolerant chip, and they are particularly interesting for space applications. Further, a second implementation step has been performed by using a high-level C- to VHDL converter tool applying the same approach used in the modelling phase. VHDL code has been generated and FPGA algorithm resources have been computed. Table 3.5 collects the data of the requested resources for the algorithm implementation in FPGA.

The algorithm can be implemented in a single FPGA requiring low total resources. Two instances of the algorithm can be included in the same FPGA, allowing parallel processing of two 16×16 blocks.

Table 3.5 Algorithm implementation on Xilinx FPGA

Device	xqr4vlx200	xq2v3000
Used LUT	10,306 (5%)	10,248 (35%)
Ram16s	21 of 336 (6%)	21 of 96 (22%)
Mult18 × 18s		9 of 96 (9%)
DSP48	9 of 96 (9%)	
Max freq. (MHz)	81	79
Throughput (Msamples/s)	70	69

In the space applications, an additional important constraint is the power consumption. An accurate estimation of the power consumption for Xilinx FPGA is given by the following formula:

$$P = 250\,mW + 20\,mW \times MSample/s$$

The power consumption has been also evaluated by using Xpower Xilinx tool for Xilinx xq2v3000 family. The tool yields an estimate of 1.1 W dynamic power consumption at throughput of 80 Msample/s on the xq2v3000 device. The static power consumption reported in the datasheet is less than 250 mW (typical) and less than 1.5 W (maximum). As can be seen, the proposed algorithm leads to a very efficient implementation with extremely low power consumption.

7 Conclusions

This chapter has overviewed current approaches to low-complexity lossless compression of hyperspectral images. We have discussed the requirements and possible solutions, and described a specific solution based on block-based compression and optimal band ordering. Experimental results on AVIRIS and AIRS data show that the algorithm has very competitive performance, with a minor performance loss with respect to the state-of-the-art, but significantly lower complexity. Moreover, it has been shown that band reordering can be effectively performed at the ground station using training data, with a small performance loss with respect to optimal reordering. This allows to avoid increasing the number of operations to be performed onboard, while still obtaining most of the advantages of reordering.

References

1. B. Penna, T. Tillo, E. Magli, and G. Olmo, "Transform coding techniques for lossy hyperspectral data compression," *IEEE Transactions on Geoscience and Remote Sensing*, vol. 45, no. 5, pp. 1408–1421, May 2007.
2. R.E. Roger and M.C. Cavenor, "Lossless compression of AVIRIS images," *IEEE Transactions on Image Processing*, vol. 5, no. 5, pp. 713–719, May 1996.
3. J. Mielikainen, A. Kaarna, and P. Toivanen, "Lossless hyperspectral image compression via linear prediction," Proc. SPIE, vol. 4725, 2002.
4. B. Aiazzi, P. Alba, L. Alparone, and S. Baronti, "Lossless compression of multi/hyperspectral imagery based on a 3-D fuzzy prediction," IEEE Transactions on Geoscience and Remote Sensing, vol. 37, no. 5, pp. 2287–2294, Sept. 1999.
5. S.K. Jain and D.A. Adjeroh, "Edge-based prediction for lossless compression of hyperspectral images," *Proc. IEEE Data Compression Conference*, 2007.
6. B. Aiazzi, L. Alparone, and S. Baronti, "Near-lossless compression of 3-D optical data," *IEEE Transactions on Geoscience and Remote Sensing*, vol. 39, no. 11, pp. 2547–2557, Nov. 2001.
7. B. Aiazzi, L. Alparone, S. Baronti, and C. Lastri, "Crisp and fuzzy adaptive spectral predictions for lossless and near-lossless compression of hyperspectral imagery," *IEEE Geoscience and Remote Sensing Letters*, vol. 4, no. 4, pp. 532–536, Oct. 2007.
8. M. Klimesh, "Low-complexity lossless compression of hyperspectral imagery via adaptive filtering," in *The Interplanetary Network Progress Report*, 2005.
9. M.J. Ryan and J.F. Arnold, "The lossless compression of AVIRIS images by vector quantization," *IEEE Transactions on Geoscience and Remote Sensing*, vol. 35, no. 3, pp. 546–550, May 1997.
10. J. Mielikainen and P. Toivanen, Clustered DPCM for the lossless compression of hyperspectral images, *IEEE Transactions on Geoscience and Remote Sensing*, vol. 41, no. 12, pp. 2943–2946, Dec. 2003.
11. E Magli, G Olmo, and E. Quacchio, "Optimized onboard lossless and near-lossless compression of hyperspectral data using CALIC," IEEE Geoscience and Remote Sensing Letters, vol. 1, no. 1, pp. 21–25, Jan. 2004
12. H. Wang, S.D. Babacan, and K. Sayood, "Lossless hyperspectral image compression using context-based conditional averages," *Proc. of IEEE Data Compression Conference*, 2005.
13. E. Magli, M. Barni, A. Abrardo, and M. Grangetto, "Distributed source coding techniques for lossless compression of hyperspectral images," *EURASIP Journal on Advances in Signal Processing*, vol. 2007, 2007.
14. A. Abrardo, M. Barni, E. Magli, and F. Nencini, "Error-resilient and low-complexity onboard lossless compression of hyperspectral images by means of distributed source coding," *IEEE Transactions on Geoscience and Remote Sensing*, Vol.48, No.4, pp.1892–1904, Apr. 2010.
15. M. Slyz and L. Zhang, "A block-based inter-band lossless hyperspectral image compressor," *Proc. of IEEE Data Compression Conference*, 2005, pp. 427–436.
16. F. Rizzo, B. Carpentieri, G. Motta, and J.A. Storer, "Low-complexity lossless compression of hyperspectral imagery via linear prediction," *IEEE Signal Processing Letters*, vol. 12, no. 2, pp. 138–141, Feb. 2005.
17. B. Penna, T. Tillo, E. Magli, and G. Olmo, "Progressive 3D coding of hyperspectral images based on JPEG 2000," *IEEE Geoscience and Remote Sensing Letters*, vol. 3, no. 1, pp. 125–129, Jan. 2006.
18. D.S. Taubman and M.W. Marcellin, *JPEG2000: Image Compression Fundamentals, Standards, and Practice*, Kluwer, 2001.
19. CCSDS-122.0-B-1 Blue Book, *Image Data Compression*, November 2005.
20. B. Huang, A. Ahuja, H.-L. Huang, and M.D. Goldberg, "Real-time DSP implementation of 3D wavelet reversible variable-length coding for hyperspectral sounder data compression," *Proc. of IEEE IGARSS*, 2006.

21. J. Mielikainen, "Lossless compression of hyperspectral images using lookup tables," *IEEE Signal Processing Letters*, vol. 13, no. 3, pp. 157–160, Mar. 2006.
22. B. Huang and Y. Sriraja, "Lossless compression of hyperspectral imagery via lookup tables with predictor selection," *Proc. SPIE, vol. 6365*, 2006.
23. H. Wang, S.D. Babacan, and K. Sayood, "Lossless hyperspectral-image compression using context-based conditional average," *IEEE Transactions on Geoscience and Remote Sensing*, vol. 45, no. 12, pp. 4187–4193, Dec. 2007.
24. S.R. Tate, "Band ordering in lossless compression of multispectral images," *IEEE Transactions on Computers*, vol. 46, no. 4, pp. 477–483, 1997.
25. P. Toivanen, O. Kubasova, and J. Mielikainen, "Correlation-based band-ordering heuristic for lossless compression of hyperspectral sounder data," *IEEE Geoscience and Remote Sensing Letters*, vol. 2, no. 1, pp. 50–54, Jan. 2005.
26. J. Zhang and G. Liu, "An efficient reordering prediction-based lossless compression algorithm for hyperspectral images," *IEEE Geoscience and Remote Sensing Letters*, vol. 4, no. 2, pp. 283–287, Apr. 2007.
27. A.B. Kiely and M.A. Klimesh, "Exploiting calibration-induced artifacts in lossless compression of hyperspectral imagery," *IEEE Transactions on Geoscience and Remote Sensing*, vol. 47, n. 8, pp. 2672–2678, Aug. 2009.
28. S.W. Golomb, "Run-length encodings," *IEEE Transactions on Information Theory*, vol. IT-12, no. 3, pp. 399–401, July 1966.
29. M.J. Weinberger, G. Seroussi, , and G. Sapiro, "The LOCO-I lossless image compression algorithm: Principles and standardization into JPEG-LS," *IEEE Transactions on Image Processing*, vol. 9, no. 8, pp. 1309–1324, Aug. 2000.
30. R. Vitulli, "PRDC: an ASIC device for lossless data compression implementing the Rice algorithm," *Proc. of IEEE IGARSS*,2006.
31. M. Grangetto, E. Magli, and G. Olmo, "Robust video transmission over error-prone channels via error correcting arithmetic codes," IEEE Communications Letters, vol. 7, no. 12, pp. 596–598, 2003.
32. N.S. Jayant and P. Noll, *Digital Coding of Waveforms*, Prentice-Hall, 1984.
33. J. Teuhola, "A compression method for clustered bit-vectors," *Information Processing Letters*, vol. 7, pp. 308–311, Oct. 1978.
34. M.J. Slyz and D.L. Neuhoff, "Some simple parametric lossless image compressors," *Proc. of IEEE ICIP*, 2000, pp. 124–127.
35. E. Magli, "Multiband lossless compression of hyperspectral images," *IEEE Transactions on Geoscience and Remote Sensing*, vol. 47, no. 4, pp. 1168–1178, Apr. 2009.
36. X. Wu and N. Memon, "Context-based lossless interband compression – extending CALIC," *IEEE Transactions on Image Processing*, vol. 9, no. 6, pp. 994–1001, June 2000.
37. Document ISO/IEC 15444–2, *JPEG 2000 Part 2 – Extensions*.

Chapter 4
FPGA Design of Listless SPIHT for Onboard Image Compression

Yunsong Li, Juan Song, Chengke Wu, Kai Liu, Jie Lei, and Keyan Wang

1 Introduction

Space missions are designed to leave Earth's atmosphere and operate in outer space. Satellite imaging payloads operate mostly with a store-and-forward mechanism, in which captured images are stored on board and transmitted to ground later on. With the increase of spatial resolution, space missions are faced with the necessity of handling an extensive amount of imaging data. The increased volume of image data exerts great pressure on limited bandwidth and onboard storage. Image compression techniques provide a solution to the "bandwidth vs. data volume" dilemma of modern spacecraft. Therefore, compression is becoming a very important feature in the payload image processing units of many satellites [1].

There are several types of redundancy in an image, including spatial redundancy, statistical redundancy, and human vision redundancy. Basically, compression is achieved by removing these types of redundancy. A typical image-compression system architecture consists of spatial decorrelation followed by quantization and entropy coding. According to the different methods used for decorrelation, compression systems can be classified into prediction, discrete cosine transform (DCT), and wavelet-based systems. Prediction-based compression methods include differential pulse-code modulation (DPCM) [2, 3], adaptive DPCM [4], and JPEG-LS [5]. DCT-based compression methods include JPEG-baseline [6] and specifically designed DCT compression methods.

Y. Li (✉) • J. Song • C. Wu • K. Liu • J. Lei • K. Wang
State Key Laboratory of Integrated Service Networks, Xidian University, Xi'an, China
e-mail: ysli@mail.xidian.edu.cn

B. Huang (ed.), *Satellite Data Compression*, DOI 10.1007/978-1-4614-1183-3_4,
© Springer Science+Business Media, LLC 2011

The introduction of wavelet transforms to image coding has brought about a great revolution in image compression because of the multiresolution analysis and high energy compactness offered by this approach. Since the 1990s, many image compression methods based on wavelet transforms have been developed, which make it possible to obtain satisfying reconstructed images at high compression ratios. Among these wavelet-based image compression methods, embedded image coding has become the mainstream approach because of its high compression efficiency and unique characteristics such as progressive transmission, random access to the bit stream, and region-of-interest (ROI) coding.

Typical embedded image-coding methods include the Embedded Zerotree Wavelet (EZW) [7] algorithm proposed by Shapiro, the Set Partition In Hierarchical Tree (SPIHT) [8] algorithm by Said and Pearlman, and the Embedded Block Coding with Optimized Truncation (EBCOT) [9] algorithm by Taubman. The EBCOT algorithm was finally adopted by the JPEG2000 [10] image-coding international standard (ISO/IECJTC1/SC29/WG1) published in December 2000. An image data compression group was established by the Consultative Committee for Space Data Systems (CCSDS) to research compression of remote sensing images. The CCSDS proposed a new image-data compression standard [11] and its implementation in November 2005 and October 2007 respectively. This standard is an improvement on SPIHT and has much lower complexity.

Because of the conflict between increased data volume and limited data transmission capability, many countries have made efforts in research and implementation of remote sensing image compression methods and onboard image compression systems. Utah State University proposed a "statistical lossless" image compression method that combined the advantages of vector quantization and lossless compression, and a corresponding CMOS VLSI chip was developed in 1989. It could process images in real time at 100 ns/pixel with a lossy compression ratio of 5:1 when compressing images from the NASA [12–14] "DE-1" satellite. The image compression system on the IKONOS2 [15] satellite launched in 1999 used the adaptive DPCM (ADPCM) method, which achieved a lossy compression ratio of approximately 4:1. SPOT2–SPOT4 [16], launched by France, used DPCM, and SPOT5 [17], launched in 2002, used a DCT-based method with a lossy compression ratio near 3:1. The Mars detectors "SPIRIT" and "OPPORTUNITY", called "MER-A" [18] and "MER-B" [19], which landed on Mars in 2004, used the efficient ICER [20] compression method developed by JPL, which is a simplification of EBCOT. This system achieved lossy compression ratios as high as 12:1 with a bit rate of 168 KB/s. It can be seen that the development of onboard compression systems has followed that of image-compression theory as it has evolved from DPCM, vector quantization, and JPEG to efficient wavelet-based image compression methods.

In this chapter, several typical embedded image-coding methods are first briefly introduced. Then the listless SPIHT [8] algorithm with lower memory requirement and the FPGA implementation proposed by the authors of this chapter are described in detail. The proposed methods have been successfully used in China's lunar exploration project. Finally, these related methods are compared, and conclusions are drawn.

2 SPIHT

Embedded coding yields two results after quantization: zero and nonzero values. The nonzero values, called significant values, determine the reconstructed image quality, and the zero values are used to reconstruct the original data structure, through which the position information of the significant values is determined. The term "significance map" is used to represent the significance of the quantized coefficients. The true cost of encoding the actual symbols is as follows:

Total Cost = Cost of Significance Map + Cost of Nonzero Values.

Given a fixed target bit rate, if a lower bit rate were used to create the significance map, more bits could be used to represent the significant values, and a higher reconstructed image quality could be achieved. Therefore, a simple and efficient model is needed to encode the significance map.

The embedded zero-tree wavelet (EZW) algorithm, proposed by Shapiro [7], included the definition of a new data structure called the *zero tree*. The zero tree is based on the hypothesis that if a wavelet coefficient at a coarse scale is insignificant with respect to a given threshold T, then all wavelet coefficients of the same orientation in the same spatial location at finer scales are likely to be insignificant. Zero-tree coding reduces the cost of encoding the significance map using this internal similarity.

Said and Pearlman [8] proposed a new, fast, and efficient image coding method called SPIHT. The SPIHT algorithm provided a new way of set partitioning which is more effective than previous implementations of EZW coding.

2.1 Progressive Image Transmission

Wavelet coefficients with larger magnitude should be transmitted first because they have greater information content than smaller coefficients. Figure 4.1 shows a schematic binary representation of a list of magnitude-ordered coefficients. Each

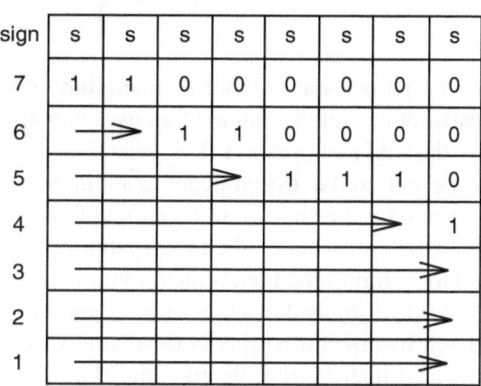

sign	s	s	s	s	s	s	s	s
7	1	1	0	0	0	0	0	0
6		→	1	1	0	0	0	0
5				→	1	1	1	0
4							→	1
3								→
2								→
1								→

Fig. 4.1 Binary representation of magnitude-ordered coefficients

column k in Fig. 4.1 contains the bits associated with a single coefficient. The bits in the top row indicate the sign of the coefficient. The rows are numbered from the bottom up, and the bits in the lowest row are the least significant.

In this way, a progressive transmission scheme can be presented as two concepts: ordering the coefficients by magnitude, and transmitting the most significant bits in each row, as indicated by the arrows in Fig. 4.1.

The progressive transmission method outlined above can be implemented as the following algorithm.

1. Initialization: output $n = \lfloor \log_2(\max_{(i,j)}\{|c_{i,j}|\}) \rfloor$ to the decoder.
2. Sorting pass: output the number m_n of significant coefficients that satisfy $2^n \leq |c_{i,j}| < 2^{n+1}$, followed by the pixel coordinates $h(k)$ and the sign of each of the m_n coefficients.
3. Refinement pass: output the nth most significant bit of all the coefficients with $c_{i,j} \geq 2^{n+1}$ (i.e., those that had their coordinates transmitted in previous sorting passes) in the same order used to send the coordinates.
4. Decrement n by one, and go to Step 2.

The algorithm stops at the desired rate or level of distortion.

2.2 Set Partitioning Sorting Algorithm

The sorting information represents the spatial position of the significant coefficients in the transformed image and is used to recover the original structure. Therefore, the performance of the sorting algorithm affects the efficiency of the whole coding algorithm.

The sorting algorithm does not need to sort all the coefficients. Actually, all that is needed is an algorithm that simply selects the coefficients such that $2^n \pounds |c_{i,j}| < 2^{n+1}$, with n decremented in each pass. Equation 4.1 is used to perform the magnitude test:

$$S_n(\tau) = \begin{cases} 1, \max_{(i,j) \in \tau} \{|c_{i,j}|\} \geq 2^n \\ 0, otherwise \end{cases} \tag{4.1}$$

To reduce the number of magnitude comparisons (sorting information), a set partitioning rule is defined that uses an expected ordering in the hierarchy defined by the subband pyramid. The objective is to create new partitions such that subsets expected to be insignificant contain a large number of elements and subsets expected to be significant contain only one element.

A tree structure, called the spatial orientation tree, naturally defines the spatial relationship in the hierarchical pyramid. Figure 4.2 shows how the spatial orientation tree is defined in a hierarchical pyramid. Each node of the tree corresponds to a pixel. Its direct offspring correspond to the pixels of the same spatial orientation at the next finer level of the pyramid. The tree is defined in such a way that each node has either no offspring, or four offspring which always form a group of 2×2

Fig. 4.2 Examples of parent-offspring dependencies in the spatial orientation tree

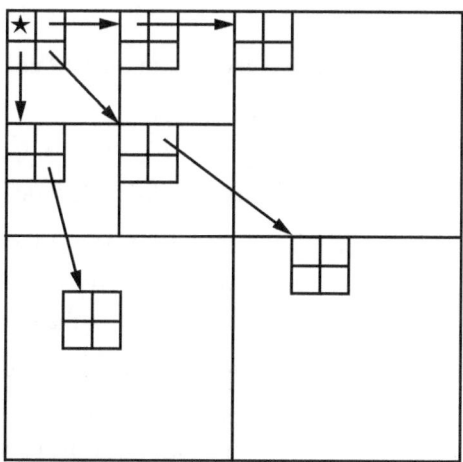

adjacent pixels. In Fig. 4.2, the arrows are oriented from the parent node to its four offspring. The pixels at the highest level of the pyramid are the tree roots and are also grouped in sets of 2×2 adjacent pixels. However, their offspring branching rule is different, and in each group, one of them (indicated by a *star* in Fig. 4.2) has no descendants.

In the SPIHT coding algorithm, if a set tests as insignificant, the whole set can be represented as one bit; thus, the expression of the set partition is simplified. Because of the internal similarity of the spatial orientation tree, this set partitioning method is more efficient than the EZW coding method.

2.3 Coding Algorithm

The following sets of coordinates and ordered lists are used to present the SPIHT coding method:

- $O(i,j)$: set of coordinates of all offspring of node (i,j).
- $D(i,j)$: set of coordinates of all descendants of node (i,j).
- H: set of coordinates of all spatial orientation tree roots (nodes at the highest level of the pyramid).
- $L(i,j) = D(i,j) - O(i,j)$.
- LIS: list of insignificant sets. An LIS entry is of type A if it represents $D(i,j)$ and of type B if it represents $L(i,j)$.
- LIP: list of insignificant pixels.
- LSP: list of significant pixels.

The SPIHT algorithm can be represented as follows:

1. *Initialization*: output $n = \lfloor \log_2(\max_{(i,j)}\{|c_{i,j}|\}) \rfloor$; set the LSP to be an empty list, add the coordinates $(i,j)\hat{I}H$ to the LIP, and add only those with descendants to the LIS as type A entries.

2. *Sorting Pass*:

    2.1. For each entry (i,j) in the LIP do

        2.1.1. Output $S_n(i,j)$
        2.1.2. If $S_n(i,j) = 1$ then move (i,j) to the LSP and output the sign of $c_{i,j}$

    2.2. For each entry (i,j) in the LIS do:

        2.2.1. If the entry is of type A then

- Output $S_n(D(i,j))$
- If $S_n(D(i,j)) = 1$ then

    - for each $(k,l)\hat{I}O(i,j)$do:

    - output $S_n(k,l)$;
    - if $S_n(k,l) = 1$ then add (k,l) to the LSP and output the sign of $c_{k,l}$;
    - if $S_n(k,l) = 0$ then add (k,l) to the end of the LIP;

    - if $L(i,j) \neq \phi$ then move (i,j) to the end of the LIS as an entry of type B, and go to Step 2.2.2; otherwise, remove entry (i,j) from the LIS;

        2.2.2. if the entry is of type B then

- output $S_n(L(i,j))$;
- if $S_n(L(i,j)) = 1$ then

    - add each $(k,l) \in O(i,j)$ to the end of the LIS as an entry of type A;
    - remove (i,j) from the LIS.

3. *Refinement Pass*: for each entry (i,j) in the LSP, except those included in the last sorting pass (i.e., with the same n), output the nth most significant bit of $|c_{i,j}|$.
4. *Quantization-Step Update*: decrement n by 1 and go to Step 2.

3 Listless SPIHT and FPGA Implementation

As illustrated above, SPIHT [22–24] is a very simple and efficient way to code a wavelet-transformed image. However, SPIHT needs to maintain three lists to store the image's zero-tree structure and significant information. These three lists represent a major drawback for hardware implementation because a large amount of memory is needed to maintain these lists. For example, for a 512×512 gray image, each entry in the list needs 18 bits of memory to store the row and column

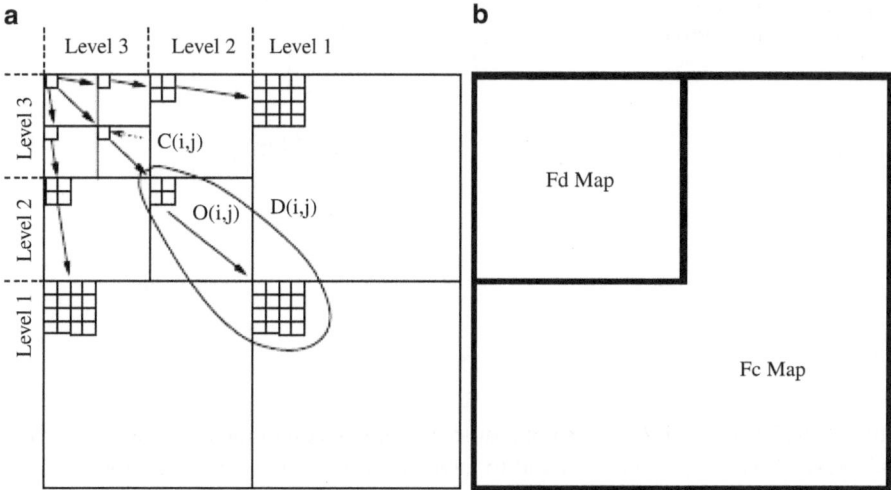

Fig. 4.3 (**a**) LZC tree structures; (**b**) Sizes of the F_D map and the F_C map

coordinates. Given that the total number of list entries for an image is approximately twice the total number of coefficients, the total memory required is 18 (bits) × 512(pixels) × 512(lines) × 2/8bits/1 K/1 K = 1.125 MB, and the memory requirement will increase if the bit rate increases. This large memory requirement means that SPIHT is not a cost-effective compression algorithm for VLSI implementation.

For this reason, researchers have tried to design listless SPIHT algorithms to reduce the memory requirement. Lin and Burgess proposed a new zero-tree coding algorithm called LZC [21] in which no lists are needed during encoding and decoding. Based on the LZC algorithm, the authors of this chapter have proposed an improved listless SPIHT algorithm and its FPGA implementation structure.

3.1   LZC

The zero-tree relations for the LZC algorithm [21] are shown in Fig. 4.3a. Unlike the SPIHT's zero-tree relations, a coefficient in the LL band has one child in each of the high-frequency bands: LH, HL, and HH. A coefficient in a high-frequency band has four children in the same spatial location at the next finer transform level. By using this tree relation in LZC, the image can be wavelet-transformed to several levels, leaving coefficients in the LL band while maintaining good coding results. Hence, LZC is a better algorithm than SPIHT for images that have relatively few dyadic wavelet transform levels. In addition, for this tree relation, LZC is also better

Fig. 4.4 Size of the F_L, F_D, and F_C maps in L-SPIHT

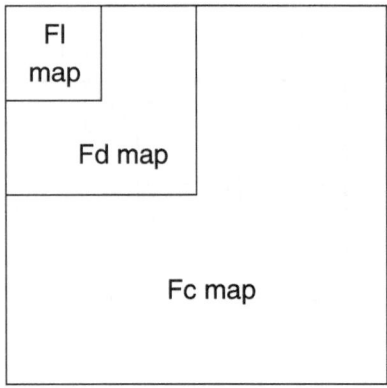

than SPIHT because LZC does not require the dimension of the LL band to be even, whereas SPIHT needs the LL band to have an even number of dimensions.

The maps used to indicate the significance of C and D are called the F_C map and the F_D map respectively, as shown in Fig. 4.3b. The size of the F_C map is the same as that of the image, whereas the F_D map is only one-quarter of the image size because coefficients at level 1 do not have any descendants. Therefore, for a 512×512 gray image, the total memory required to store the zero-tree structure is only $512 \times 512(F_C) + 256 \times 256(F_D)/8$ bit/1 K $= 40$ K for all bit rates. Compared with the 1.125-MB memory requirement for SPIHT, the memory requirement for LZC has been reduced significantly.

3.2 Proposed Listless SPIHT

To improve compression performance and further reduce the memory requirement, the authors have proposed a listless SPIHT (L-SPIHT) coding method [22]. In L-SPIHT, each independent spatial orientation tree, beginning with the root coefficients, is extracted from the whole wavelet-transformed image and encoded separately to reduce the memory requirement. Moreover, L-set partitioning is added to the LZC to express the significance map more efficiently and to exploit the correlation among the different scales more effectively. In this way, the L-SPIHT method can achieve performance comparable with that of SPIHT with a lower memory requirement.

As shown in Fig. 4.4, the size of the F_L map is only 1/16 that of a spatial orientation tree. Moreover, the size of a spatial orientation tree is determined by the level of the wavelet transform. Taking a four-level wavelet transform as an example, for a 512×512 gray image, the total memory required to store the zero-tree structure is only $16 \times 16(F_C) + 8 \times 8(F_D) + 4 \times 4(F_L)/8$ bit/1 K $= 42$ bytes.

The tree symbols used in the L-SPIHT zero tree are:

- $C(i,j)$: wavelet coefficient at coordinate (i,j);

- $p(i,j)$: parent node of $C(i,j)$;
- $O(i,j)$: set of child coefficients of $C(i,j)$;
- $D(i,j)$: set of descendant coefficients of $C(i,j)$, i.e., all offspring of $C(i,j)$;
- $L(i,j) = D(i,j) - O(i,j)$.
- $F_C(i,j)$: significant map of coefficient $C(i,j)$;
- $F_D(i,j)$: significant map of set $D(i,j)$;
- $F_L(i,j)$: significant map of set $L(i,j)$;
- $R(i,j)$: set of root coefficients in the LL band.

The detailed L-SPIHT algorithm can be described as follows:

1. Initialization: output $n = \lfloor \log_2(\max_{(i,j)}\{|c_{i,j}|\}) \rfloor$; the F_C, F_D, and F_L maps are set to be zero;
2. LL subband encoding:
   For each (i,j) $\in R(i,j)$ do:

   - if $F_C(i,j) = 0$, then do:

     - output $S_n(C(i,j))$;
     - if $S_n(C(i,j)) = 1$, then output the sign of C(i,j) and set $F_C(i,j)$ to 1;

   - if $F_C(i,j) = 1$, then output the nth most significant bit of C(i,j);
   - if $F_D(i,j) = 0$, then

     - output $S_n(D(i,j))$;
     - if $S_n(D(i,j)) = 1$;then set $F_D(i,j)$ to 1;

   - if $F_L(i,j) = 0$ and $F_D(i,j) = 1$, then:

     - output $S_n(L(i,j))$;
     - if $S_n(L(i,j)) = 1$,then set $F_L(i,j)$ to 1;

3. high-frequency-band coding:
   for each (i,j) in the high-frequency bands;

   - if $F_D(P(i,j)) = 1$ do:

     - if $F_C(i,j) = 0$ do:

       - output $S_n(C(i,j))$;
       - if $S_n(C(i,j)) = 1$, then output the sign of C(i,j) and set $F_C(i,j)$ to 1;

     - if $F_C(i,j) = 1$, then output the nth most significant bit of C(i,j);

   - if $F_L(P(i,j)) = 1$ do:

     - if $F_D(i,j) = 0$, then

       - output $S_n(D(i,j))$;
       - if $S_n(D(i,j)) = 1$;then set $F_D(i,j)$ to 1;

     - if $F_L(i,j) = 0$ and $F_D(i,j) = 1$,then

       - output $S_n(L(i,j))$;

- if $S_n(L(i,j)) = 1$, then set $F_L(i,j)$ to 1;

4. Quantization-Step Update: decrement n by 1 and go to Step 2.

3.3   Performance Analysis

The performance of the proposed scheme was tested on AVIRIS hyperspectral images and visible remote-sensing images. AVIRIS is a Jet Propulsion Laboratory instrument with 224 continuous bands ranging from the visible to the near-infrared regions (400–2500 nm) (the test images are available at http://aviris.jpl.nasa.gov). The spectral components were sampled with 12-bit precision; after radiometric correction, the data were stored as 16-bit signed integers. The resolution of one hyperspectral image scene is 512 rows by 614 columns by 224 bands. The hyperspectral images used for this test are shown in Figs. 4.5a, b, c, d. For convenience, each band was cropped to 512×512. The resolution of the visible remote sensing images shown in Figs. 4.5e, 5f is 1024 rows by 1024 columns. The image data are stored as 8-bit unsigned integers.

1. RD performance comparison
   Table 4.1 shows an RD performance comparison with SPIHT. The peak signal-to-noise ratio (PSNR) of L-SPIHT is approximately 0.5 dB lower than that of SPIHT at the lower bit rate and approximately 1 dB lower at the higher bit rate. In fact, the difference in MSE is much less, although that of the PSNR is much greater at the higher bit rate).
2. Lossless compression performance comparison
   The lossless compression performance of L-SPIHT has also been compared with that of SPIHT. The lossless bit rates achieved are shown in Table 4.2. It is clear that the lossless compression ratio of the proposed L-SPIHT is quite close to that of SPIHT.
3. Memory budget comparison
   Although the PSNR of L-SPIHT is a little lower than that of SPIHT, the required memory is much less than for SPIHT because three bit maps are used instead of three lists and because each spatial orientation tree beginning with each root coefficient is coded separately. It is acceptable for onboard compression to reduce the memory requirement at the cost of a slight loss in PSNR. The memory budget to store the significance map of one 512×512 image for each algorithm is shown in Table 4.3. It is evident that the required memory for L-SPIHT is much less than for SPIHT or LZC and that therefore the proposed L-SPIHT is highly preferable for onboard compression.

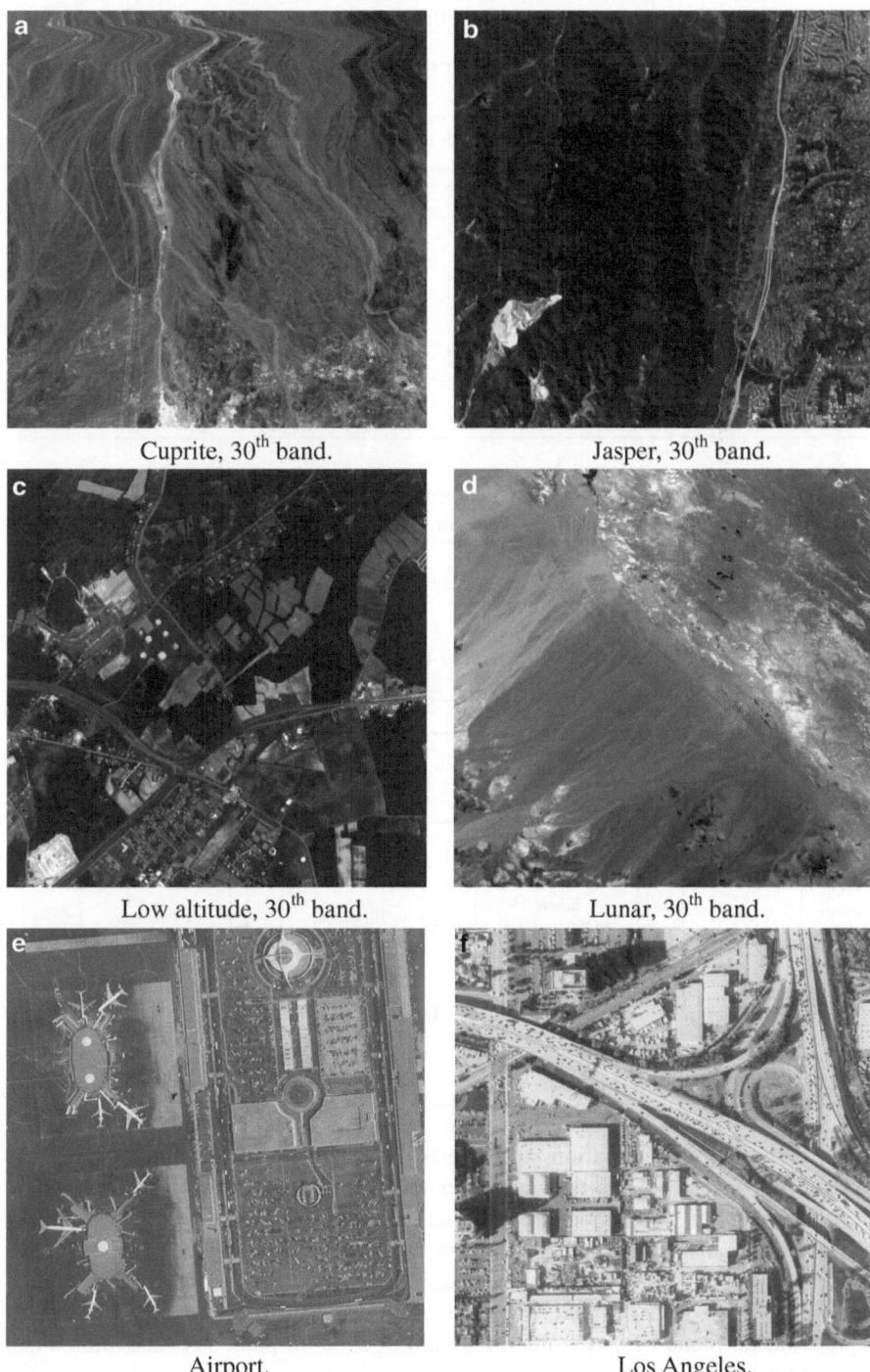

Cuprite, 30th band.

Jasper, 30th band.

Low altitude, 30th band.

Lunar, 30th band.

Airport.

Los Angeles.

Fig. 4.5 Test images

Table 4.1 RD performance comparison with SPIHT

Image	Compression ratio	16	8	6	4	2
Cuprite	L-SPIHT(dB)	71.06884	75.87616	79.11098	86.43692	122.3861
	SPIHT(dB)	71.4396	76.31545	79.70018	87.27964	123.7242
Jasper	L-SPIHT(dB)	66.74982	71.9254	75.40152	82.65513	128.0219
	SPIHT(dB)	66.20326	72.4554	75.91509	83.51281	130.2662
Low altitude	L-SPIHT(dB)	66.02848	71.20379	74.7096	81.86174	115.6257
	SPIHT(dB)	66.3446	71.66942	75.19085	82.60853	116.543
Lunar	L-SPIHT(dB)	71.57442	76.74754	79.96625	86.9421	123.8183
	SPIHT(dB)	71.85982	77.16701	80.52804	87.85067	125.7163
Airport	L-SPIHT(dB)	29.55	32.46	33.64	37.03	46.29
	SPIHT(dB)	29.83	32.57	34.01	37.09	47.59
Los Angeles	L-SPIHT(dB)	31.97	36.05	38.66	41.56	51.06
	SPIHT(dB)	32.42	36.57	38.7	42.3	52.14

Table 4.2 Lossless compression ratio comparison with SPIHT

Algorithm	Cuprite	Jasper	Low-altitude	Lunar	Airport	Los Angeles
SPIHT	2.224	2.049	2.007	2.257	1.399	1.635
L-SPIHT	2.222	2.047	2.006	2.255	1.397	1.634

Table 4.3 Memory budget to store the significance map of a single bit plane

Algorithm	Memory budget
SPIHT	1.125 MB
LZC	40 KB
L-SPIHT	42 bytes

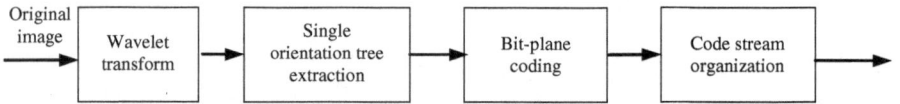

Fig. 4.6 Hardware implementation structure of L-SPIHT

3.4 FPGA Implementation

The hardware implementation structure of the L-SPIHT algorithm, as shown in Fig. 4.6, consists of four parts: wavelet transform, spatial orientation tree extraction, bit-plane encoding, and code-stream organization. First, the original input images are wavelet-transformed, and the wavelet coefficients obtained are reorganized according to the orientation-tree structure. Then the bit-plane encoding method described above is performed, proceeding from the most significant to the least significant bit planes. Finally, the bit stream of each bit plane is disassembled and reorganized.

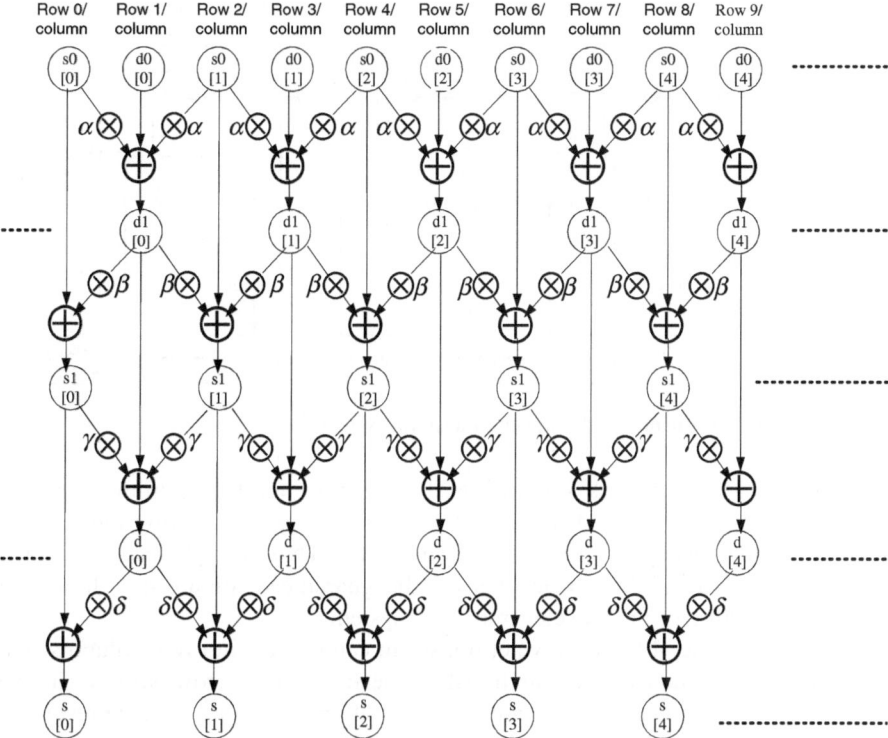

Fig. 4.7 Process of a row/column wavelet transform

3.4.1 Line-based Wavelet Transform Structure

The line-based wavelet transform [24] structure can not only process row and column data simultaneously, but can also process different levels at the same time. This approach can completely satisfy the requirement of real-time data processing. A 9/7-float wavelet is taken as an example to analyze the proposed high-speed parallel hardware implementation structure.

Figure 4.7 shows the 9/7-float wavelet transform of one row (or column) of data, in which row s0 (or d0) contains the one-dimensional input, rows d1and s1 represent the four listing steps, and rows d and s contain the final high- and low-frequency coefficients output from this one-dimensional wavelet transform.

It can be seen in Fig. 4.7 that in a row transform, the first lifting step can be performed once the third data point has been entered, and the final transform coefficient after the four lifting steps can be obtained when the fifth data point has been entered. This means that the first lifting step of the column transform can be performed once the data transform for the first three rows has been completed. The first level of the row and column transform for the first-row data can be performed when the transform of five rows of data has been completed. The coefficients of the first row can thus be obtained and sent to the second-level wavelet-transform module. In this way, the third and fourth levels of the wavelet

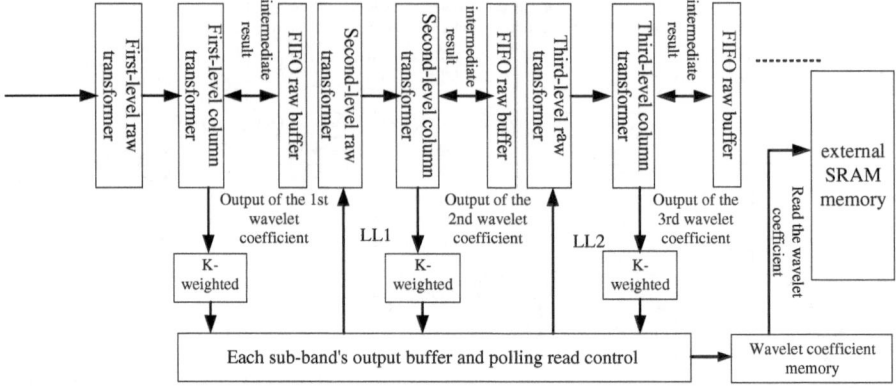

Fig. 4.8 Hardware structure of the wavelet-transform system

transform can start to work simultaneously, meaning that the multilevel wavelet transform will be completed in parallel with the process of entering the original images without occupying extra processing time. Thus, a high-speed real-time wavelet transform of the image data can be guaranteed. Figure 4.8 shows the hardware structure of the wavelet-transform system.

As shown in Fig. 4.8, the wavelet transform can process in rows, columns, and levels in parallel, and each column transform needs five or six row buffers with the capacity of the row width, which are used for buffering the intermediate results of the row transforms. With the expectation that the LL subband will be delivered to the next level of the wavelet transform, the other three subbands are sent to external memory.

3.4.2 Bit-Plane Parallel VLSI Architecture for the L-SPIHT

Before bit-plane coding, each independent spatial orientation tree, beginning with the root coefficients in the LL subband, is extracted from the whole wavelet-transformed image in depth-first search order [23].

Because only the significance status of the wavelet coefficients in each coded bit planes must be known before encoding the current bit plane, a bit-plane parallel processing algorithm should be implemented in the L-SPIHT code. A hardware implementation structure for this is shown in Fig. 4.9, in which an FPGA with two external SRAM modules is used.

A block diagram for each bit plane of the significance scanning procedure and the internal principle of the quantization coding is shown in Fig. 4.10:

The bit-plane encoder consists of a significance scanning module and a quantization coding module. The former is the core of the whole encoder and is responsible for obtaining the three sign bitmaps, F_C, F_D, and F_L, and the parent–child relations among them from the spatial orientation tree. The latter

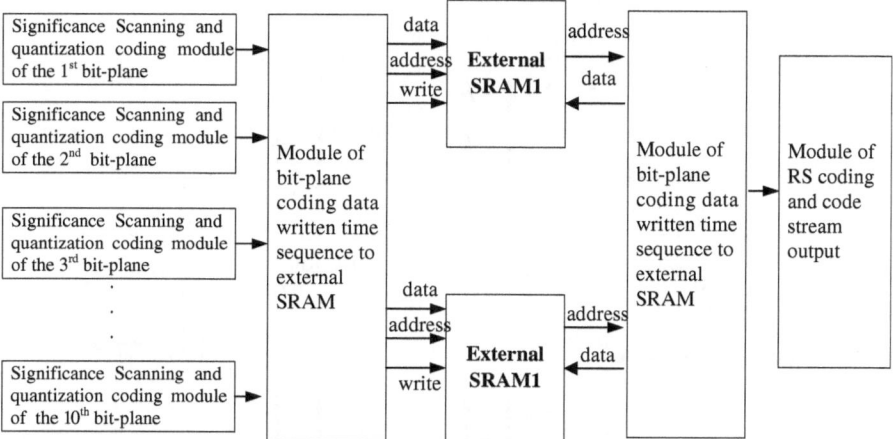

Fig. 4.9 Block diagram of hardware implementation of quantitative coding units

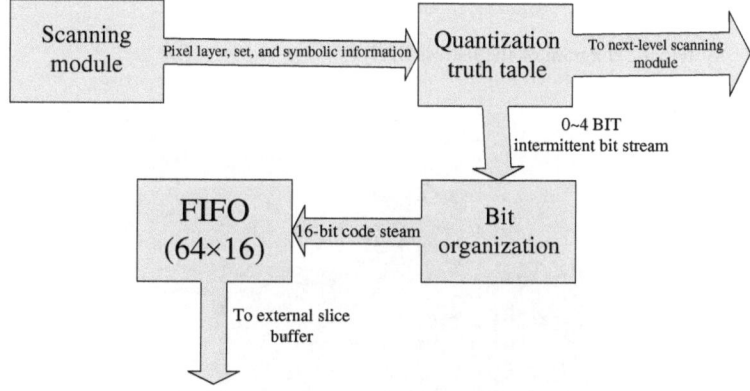

Fig. 4.10 Block diagram for each bit plane of the significance scanning procedure

forms the final code stream from the available significant information by means of a truth table search.

3.4.3 Hardware Implementation

A systematic diagram of the L-SPIHT is shown in Fig. 4.11. In the image compression system, the SPIHT core algorithm is executed on the FPGA, the external SRAM1 and SRAM2 are used to buffer the DWT coefficients by alternating processing, and the external SRAM3 and SRAM4 are used to buffer the coding streams by alternating processing. The coding streams are created from the bit-plane coding results. The input signals include the image data, the clock signal, the frame, and row synchronization signals, while the output signals include the clock

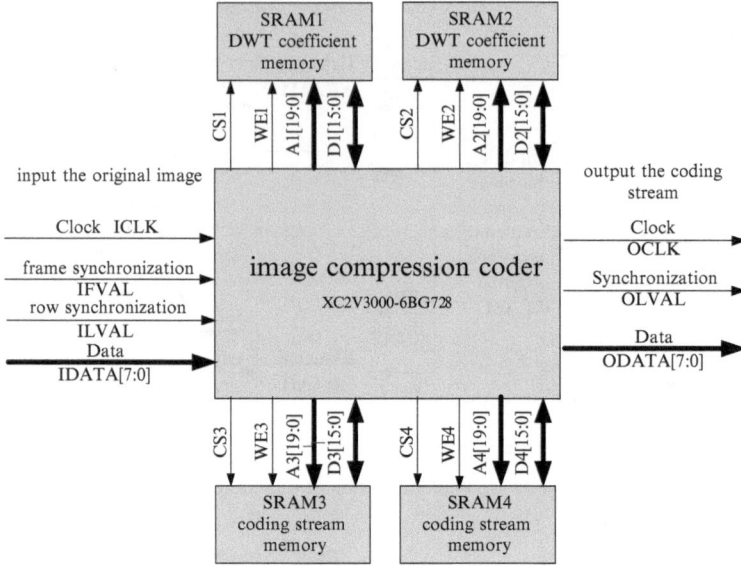

Fig. 4.11 Systematic diagram of the listless SPIHT

Fig. 4.12 FPGA verification system of the proposed L-SPIHT

signal, the data synchronization signal, and the code stream. The FPGA verification system of the L-SPIHT is shown in Fig. 4.12.

1. Simulation results of the wavelet transform implementation in hardware

   An FPGA device (XC2V3000-6BG728) is used to perform wavelet transform using VHDL. The block RAM (BRAM) inside the FPGA is used for storage and

Table 4.4 Simulation results of the hardware implementation of the wavelet transform

Target device	XILINX XC2V3000-6BG728
Programming language	VHDL
Synthesis tool	Synplify Pro
Logic unit (slice)	6,249
BRAM	70
System clock	75.236 MHz

Table 4.5 Simulation results for FPGA hardware implementation

Target device	XILINX XC2V3000-6BG728
Programming language	VHDL
Synthesis tool	Synplify Pro
Logic unit (slice)	3,326
BRAM	13
System clock	80.711 MHz

Table 4.6 Overall simulation results for hardware implementation of the compression system

Target device	XILINX XC2V3000-6BG728
Programming language	VHDL
Synthesis tool	Synplify Pro
Logic unit (slice)	10,181
BRAM	83
Image pixel precision supported	8 bits
Data throughput rate	560 Mbps
Power consumption	1.5 W

output organization of the wavelet coefficients. The input image resolution supported is 1024×1024, and the four-level float-97 wavelet transform is accomplished at high speed. Table 4.4 shows the simulation results of hardware implementation of the wavelet transform.

2. Simulation results of the hardware implementation for spatial orientation tree extraction and bit-plane coding

The hardware design of the L-SPIHT encoder using VHDL has been implemented on an FPGA device (xc2v3000-6BG728). The input image resolution supported is 1024×1024, and a four-level float 97 wavelet transform is used. The wavelet coefficients are then reconstructed as a hierarchical tree. A maximum of 13 bit planes can be coded in parallel in the proposed hardware design. Table 4.5 shows the simulation results for FPGA hardware implementation.

3. Simulation results for hardware implementation of the overall compression system

The image-compression encoding system using VHDL has been implemented on the XILINX (XC2V3000-6BG728) as a target device. The input image resolution supported is 8 bits, and the system data throughput rate is 560 Mbps. This image-compression encoding system uses 10181 logic units and 83 blocks of RAM with the Synplify Pro synthesis tool. A data throughput rate as high as 560 Mbps was achieved. Table 4.6 shows the overall specifications for hardware implementation of the compression system.

4   Conclusions

In this chapter, several typical embedded image-coding methods were first briefly introduced. Then the listless SPIHT algorithm and the FPGA implementation proposed by the authors have been described in detail.

In the proposed listless SPIHT algorithm, three bit maps, F_C, F_D, and F_L, are used instead of three lists to represent the significance of the set, which significantly reduces the memory requirement. Moreover, each independent spatial orientation tree beginning with the each root coefficient is extracted from the whole wavelet-transformed image and encoded separately to reduce the memory requirement further.

As for hardware implementation, a line-based real-time wavelet transform structure was proposed to perform horizontal and vertical filtering simultaneously. This structure makes it possible to complete the 2-D wavelet transform in parallel with image data input. A bit-plane-parallel VLSI structure for listless SPIHT was also proposed, in which the coding information for each bit plane can be obtained simultaneously. These structures make it possible to guarantee real-time implementation of the listless SPIHT image-compression system.

Experimental results show that the listless SPIHT algorithm needs much less memory and achieves an only slightly lower PSNR than the original SPIHT algorithm. The proposed listless SPIHT FPGA implementation can compress image data in real time and is highly suitable for onboard application. The proposed listless SPIHT and its FPGA implementation have been successfully used in China's lunar exploration project.

References

1. Yu, G., Vladimirova, T., Sweeting, M. (2009). Image compression systems on board satellites. *Acta Astronautica* 64: 988–1005.
2. Jayant, N.S., Noll, P. (1984). *Digital Coding of Waveforms: Principles and Applications to Speech and Video*. Prentice-Hall, Englewood Cliffs NJ.
3. Rabbani, M., Jones, P.W. (1991). *Digital Image Compression Techniques*. SPIE Press, Bellingham WA.
4. Brower, B.V., Couwenhoven, D., Gandhi, B., Smith, C. (1993). ADPCM for advanced LANDSAT downlink applications. *Conference Record of the 27th Asilomar Conference on Signals, System, and Computers*, vol. 2, November 1–3, 1993: 1338–1341.
5. Weinberger, M.J., Seroussi, G., Sapiro, G. (2000). The LOCO-I lossless image compression algorithm: principles and standardization into JPEG-LS. *IEEE Transactions on Image Processing* 9(8): 1309–1324.

6. Pennebaker, W.B., Mitchell, J.L. (1993). *JPEG Still Image Data Compression Standard.* Chapman & Hall, New York.
7. Shapiro, J.M. (1993). Embedded image coding using zero trees of wavelet coefficients. *IEEE Trans. on Signal Processing* 41(12): 3445–3462.
8. Said, A., Pearlman, W.A. (1996). A new, fast, and efficient image codec based on set partitioning in hierarchical trees. *IEEE Trans. on Circuits and Systems for Video Technology* 6(3): 243–250.
9. Taubman, D. (2000). High-performance scalable image compression with EBCOT. *IEEE Trans. on Image Processing* 9(7): 1158–1170.
10. *Information Technology—JPEG2000 Image Coding System—Part 1: Core Coding System.* ISO/IEC 15444–1, 2000.
11. CCSDS 120.1-G-1, *Image Data Compression.* CCSDS Recommendation for Space Data System Standards, June 2007.
12. Kremic, T., Anderson, D.J., Dankanich, J.W. (2008). NASA's in-space propulsion technology project overview and mission applicability. *IEEE Conference on Aerospace*, Big Sky, Montana, March 1–8: 1–10.
13. Shapiro, A.A. (2005). An ultra-reliability project for NASA. *IEEE Conference on Aerospace* Big Sky, Montana, March 5–12: 99–110 .
14. NASA's footprints in space for 50 years. *International Aviation* (12), 2008.
15. http://www.satimagingcrop.com/satellite-sensors/ikonos.html..
16. Baraldi, A., Durieux, L., Simonetti, D., Conchedda, G., Holecz, F., Blonda, P. (2010). Automatic spectral rule-based preliminary classification of radiometrically calibrated SPOT-4/-5/IRS, AVHRR/MSG, AATSR, IKONOS/QuickBird/OrbView/GeoEye, and DMC/SPOT-1/-2 imagery—Part II: classification accuracy assessment. *IEEE Transactions on Geoscience and Remote Sensing*,48 (3) :1326–1354 .
17. http://www.satimagingcrop.com/satellite-sensors/spot-5.html.
18. Steltzner, A., Kipp, D., Chen, A., Burkhart, D., Guernsey, C., Mendeck, G., Mitcheltree, R., Powell, R., Rivellini, T., San Martin, M., Way, D. (2006). Mars Science Laboratory entry, descent, and landing system. *IEEE Conference on Aerospace,* Big Sky, Montana, March 3–10:1–19.
19. Hurd, W.J., Estabrook, P., Racho, C.S., Satorius, E.H. (2002). Critical spacecraft-to-Earth communications for Mars Exploration Rover (MER) entry, descent, and landing. *IEEE Aerospace Conference Proceedings*, vol. 3: 1283–1292.
20. Kiely, A., Klimesh, M. (2003). *The ICER Progressive Wavelet Image Compressor.* IPN Progress Report 42–155, November 15, 2003.
21. Lin, W.K., Burgess, N. (1999). Low memory color image zero-tree coding. *Proceedings, Information, Decision, and Control (IDC 99) Conference*, Adelaide, SA , Australia: 91–95.
22. Chen, J., Li, Y., Wu, C. (2001). A listless minimum zero-tree coding algorithm for wavelet image compression. *Chinese Journal of Electronics* 10(2): 200–203.
23. Liu, K., Wu, C., Li, Y., et al. (2004). Bit-plane-parallel VLSI architecture for a modified SPIHT algorithm using depth-first search bit stream processing (in Chinese). *Journal of Xidian University* 31(5): 753–756.
24. Liu, K., Wang, K., Li, Y., Wu, C. (2007). A novel VLSI architecture for real-time line-based wavelet transform using lifting scheme. *Journal of Computer Science and Technology* 22(5): 661–672.

Chapter 5
Outlier-Resilient Entropy Coding

Jordi Portell, Alberto G. Villafranca, and Enrique García-Berro

Abstract Many data compression systems rely on a final stage based on an entropy coder, generating short codes for the most probable symbols. Images, multispectroscopy or hyperspectroscopy are just some examples, but the space mission concept covers many other fields. In some cases, especially when the on-board processing power available is very limited, a generic data compression system with a very simple pre-processing stage could suffice. The Consultative Committee for Space Data Systems made a recommendation on lossless data compression in the early 1990s, which has been successfully used in several missions so far owing to its low computational cost and acceptable compression ratios. Nevertheless, its simple entropy coder cannot perform optimally when large amounts of outliers appear in the data, which can be caused by noise, prompt particle events, or artifacts in the data or in the pre-processing stage. Here we discuss the effect of outliers on the compression ratio and we present efficient solutions to this problem. These solutions are not only alternatives to the CCSDS recommendation, but can also be used as the entropy coding stage of more complex systems such as image or spectroscopy compression.

J. Portell (✉)
Departament d'Astronomia i Meteorologia/ICCUB, Universitat de Barcelona, Barcelona, Spain

Institut d'Estudis Espacials de Catalunya, Barcelona, Spain
e-mail: jportell@am.ub.es

A.G. Villafranca • E. García-Berro
Institut d'Estudis Espacials de Catalunya, Barcelona, Spain

Departament de Física Aplicada, Universitat Politècnica de Catalunya, Castelldefels, Spain
e-mail: agonzalez@ieec.cat; garcia@fa.upc.edu

B. Huang (ed.), *Satellite Data Compression*, DOI 10.1007/978-1-4614-1183-3_5,
© Springer Science+Business Media, LLC 2011

1 Introduction and Motivation

Data compression systems for satellite payloads use to have tight restrictions on several aspects. First, the data block size should be rather small, in order to avoid losing large amounts of data when transmission errors occur [1, 2]. That is, data should be compressed in independent and small blocks of data, which contradicts with the fact that most of the adaptive data compression systems perform optimally only after a large amount of data is processed [3]. Second, the processing power for software implementations (or electrical power, in case of hardware implementations) is largely limited in space. Thus, the compression algorithm should be as simple and quick as possible. Finally, the required compression ratios are becoming larger as new missions are conceived and launched. When all these restrictions are combined with the need of a lossless operation, the design of such a data compression system becomes a true challenge.

The Consultative Committee for Space Data Systems (CCSDS) proposed a general-purpose compression solution [4, 5] based on a two-stage strategy – an otherwise typical approach. Firstly, a simple pre-processing stage changes the statistics of the data by applying a reversible function, often implemented as a data predictor followed by a differentiator. Secondly, a coding stage based on an entropy coder outputs a variable number of bits for each of the symbols calculated by the first stage. While no specific pre-processing method is included in the recommendation, the coding stage is mainly based on the Rice-Golomb codes [6, 7], which are simple to calculate and, hence, quick implementations are available – specially in hardware. This CCSDS recommendation (codenamed 121.0) operates with blocks of 8 or 16 samples, determining the best Rice code to be used for them. Summarizing, it is a quick algorithm that yields good compression ratios, and what is most important, it rapidly adapts to the statistical variations of the data to be compressed. Hence, this recommendation is widely used in scientific payloads [8], and it still remains the reference in terms of generic data compression for space missions, owing to its flexibility and speed. Although new techniques improving its performance have appeared, they mostly focus on a specific kind of data, such as image [9] or multi/hyperspectroscopy. However, it is important to realize that these new methods require more computational resources – and what is most important for the purpose of this chapter, they also require a final stage for entropy coding.

The CCSDS 121.0 recommendation is not exempt of problems. In particular, Rice codes offer excellent compression ratios for data following a geometric probability distribution, but any deviation from this leads to a decrease in the compression efficiency. One solution to this problem is to use a different kind of Golomb or prefix codes with a smoother growth for high values, but within the same framework and adaptive stage as CCSDS 121.0 [10]. Yet another solution is to develop a brand new set of codes, such as the Prediction Error Coder (PEC) [11, 12]. It is a semi-adaptive entropy coder that selects, from a given set of pre-configured codes, the best suitable one for each of the input values. This coder requires an also new adaptive stage, leading to what has been called the Fully Adaptive PEC (FAPEC) [12], which autonomously determines the nearly-optimal set of codes for a given block of data.

In this chapter we first discuss the limitations of the Rice coder and of the CCSDS 121.0 lossless compression standard. This is done in Sect. 5.2, while Sect. 5.3 presents the concept of outlier-resilient entropy coding, describing some examples inspired on the Rice codes as well as the new PEC coder. Section 5.4 describes some methods for analyzing these codes, including models for generating representative test data and also including illustrative results. The adaptive stages for these codes are described in Sect. 5.5, leading to nearly-optimal and autonomous data compressors. Finally, Sect. 5.6 shows some results that can be achieved using the synthetic tests previously mentioned, thus testing the entropy coders under a controlled environment. It also introduces a *data corpus* compiled from a variety of instruments useful to evaluate the compressors with real data, and it presents the results obtained with the several solutions previously described. The chapter finishes with a brief summary and some conclusions obtained from the developments and tests presented here.

2   Limitations of Rice and of CCSDS 121.0

Rice codes are optimal for data following discrete Laplacian (or two-sided geometric) probability distributions [13], which are expected after the CCSDS 121.0 pre-processing stage [4] – or, in general, after any adequate pre-processing stage. However, this assumes a correct operation of the predictor, which cannot be taken for granted as noisy samples and outliers can modify the expected distribution. This is especially true for the space environment, where prompt particle events (such as cosmic rays or solar protons) will affect the on-board instrumentation. Any deviation from the expected statistic can lead to a significant decrease in the resulting compression ratio.

As it is well known, the Rice-Golomb coder is based on a k parameter that must be chosen very carefully in order to obtain the expected compression ratios for a given set of data. Table 5.1 illustrates some Rice codes for small values and low k configurations. The lowest values of k lead to a rapid increase in the length of the output code – although such low k values are the ones leading to the shortest codes for small values. If we would use Rice codes statically (that is, manually calibrating the k value by means of simulations), an unacceptable risk would appear. That is, it could happen that we expect to code a dataset for which we always expect low values, and thus we select a low k such as 1. With this configuration, receiving a single high value such as 20,000 would lead to an output code of about 10,000 bits. Fortunately, the adaptive layer introduced by the CCSDS 121.0 recommendation automatically selects the best configuration for each given data block. It determines the total length of the coded block using $k = 1$ to $k = 13$, and then it selects the value of k leading to the shortest total length. Note that $k = 0$ is not considered, since it coincides with the Fundamental Sequence option already included in CCSDS 121.0. This automatic calibration significantly reduces the effect of outliers

Table 5.1 Some rice and subexponential codes, and bit length differences between them

	$k = 0$			$k = 1$		
n	Rice	Subexp	Diff.	Rice	Subexp	Diff.
0	0|	0|	0	0|0	0|0	0
1	10|	10|	0	0|1	0|1	0
2	110|	110|0	+1	10|0	10|0	0
3	1110|	110|1	0	10|1	10|1	0
4	11110|	1110|00	+1	110|0	110|00	+1
5	111110|	1110|01	0	110|1	110|01	+1
6	1111110|	1110|10	−1	1110|0	110|10	0
7	11111110|	1110|11	−2	1110|1	110|11	0
8	111111110|	11110|000	−1	11110|0	1110|000	+1
9	1111111110|	11110|001	−2	11110|1	1110|001	+1
10	11111111110|	11110|010	−3	111110|0	1110|010	0
...	...			...		
31	1...(×31)...10|	111110|1111	−22	1...(×15)...10|1	11110|1111	−8

present in the data, leading to acceptable ratios even in case of rapidly changing statistics. Nevertheless, this is done at the expense of increasing k when such outliers are found. For example, in a data block where all the values are small (or even zero), a single high value makes CCSDS 121.0 to select a high value of k, thus leading to a small compression ratio. Our goal is to reduce the effect of such outliers even within a data block, making possible to select smaller values of k and, thus, increasing the achievable compression ratios.

3 Outlier-Resilient Codes

Data compression through entropy coding basically assigns very short codes to the most frequent symbols, while letting less frequent symbols generate longer codes. If a code is not carefully designed, less frequent symbols can lead to prohibitively long codes – affecting the overall compression ratio. Here we define *outlier-resilient entropy coding* as this careful design of an entropy coder in the sense that less frequent symbols lead to relatively short codes, typically of the order of just twice the original symbol size or even less.

The goal of the codes presented here is to be *resilient* in front of outliers in the data. It is very important to emphasize at this point that we do not refer to error-resiliency or data integrity here. We refer to *outlier-resiliency* as the ability of achieving high compression efficiencies despite of having large amounts of outliers in a data block – that is, despite of compressing data which does not strictly follow an expected probability distribution. It should also be emphasized that the developments shown here require an adequate pre-processing stage. This is crucial to data compression, because a well-designed pre-processing algorithm can boost the final ratios, while the second stage is limited by the entropy level achieved at the output of the first stage. Besides, if the second stage is largely affected by outliers in the pre-processed data, the ratios are further limited. This effect is what we intend to minimize in this chapter.

The coders presented here can be used as the coding stage of elaborated data compression systems, such as those used in imaging or hyperspectroscopy. It is also worth mentioning that these coders are especially applicable when high compression ratios are needed but the processing resources available are low – which is the usual case in satellite data compression. Other entropy coders such as adaptive Huffman [14] or arithmetic coding [15] could provide better results, but at the expense of more processing resources. Adaptive Huffman could even be not applicable at all, since it requires large block lengths in order to get optimal results. Finally, a lossless operation is assumed throughout the whole chapter, which is the most frequent premise of an entropy coder. Lossy compression is typically implemented at the pre-processing stage, although an entropy coder could be modified in order to operate in a near-lossless mode – such as ignoring some least-significant bits from the input data.

3.1 Subexponential Codes

The main reason for the CCSDS 121.0 performance to drop abruptly when noise or outliers are introduced is that Rice codes are not intended to be used with noisy data. This limitation is due to the fact that the length of Rice codes grows too fast for large input values, especially when low values are assigned to the k parameter. On the other hand, there are some Golomb codes the length of which grows slowly in case of outliers. One example is the Exponential-Golomb coding [16], which provides shorter lengths than Rice codes for large values. However, smooth code growth for small data values provided by the Rice codes is lost. Simulations with synthetic data reveal that the compression gain obtained when coding large values does not compensate the loss when compressing smaller values, introducing undesirable effects in the resulting compression ratio.

Subexponential codes are prefix codes used in the Progressive FELICS coder [3, 17]. Similarly to the Golomb codes, the subexponential coder depends on a configuration parameter k, with $k \geq 0$. Actually, subexponential codes are related to both Rice and exponential Golomb codes. The design of this coder is supposed to provide a smoother growth of the code lengths, as well as a smoother transition from the inherent CCSDS 121.0 strategies (Zero Block, Second Extension or Fundamental Sequence) to the prefix coding strategy. In particular, for small dispersions, moving from these strategies to subexponential coding does not imply a significant increase in the output code length and, thus, we avoid the poor performance that the exponential Golomb coder has in this region.

Essentially, subexponential codes are a combination of Rice and exponential-Golomb codes. These two coding strategies are used depending on the value being coded and the value of k. When $n < 2^{k+1}$, the length of the code increases linearly with n, while for $n \geq 2^{k+1}$ the length increases logarithmically. The first linear part resembles a Rice coding strategy and maintains a slow code growth for small values, while the second part resembles the exponential-Golomb code. Table 5.1 shows some subexponential codes for several values of n and k. These two different coding strategies combined provide an advantage with respect to both Rice and exponential-Golomb codes. In particular, this strategy allows obtaining similar code lengths to Rice for small input values, and additionally, in the case of outliers or large values, the code length is shorter than that of Rice due to the exponential steps in the second stage. This is also shown in Table 5.1, which includes the difference in code length (in bits) between Rice and subexponential. While this second exponential behavior is also present in the exponential-Golomb coder, the average code length is estimated to be shorter, since smaller values have higher probabilities. Specifically, in those scenarios where there are few or no outliers, the coder is expected to deliver higher compression ratios than the exponential-Golomb coder while at the same time providing robustness against outliers.

3.2 Limited-Length Signed Rice Codes

In some specific cases, an adaptive data compressor such as the CCSDS lossless compression standard may not even be applicable due to very tight limitations in the processing resources. Let us give a good example of this. The payload of Gaia [18], the astrometric mission of the European Space Agency, will generate a complex set of data which will contain, among other fields, a large number of tiny images of the stars observed by the satellite. The data compression solutions initially proposed for the mission [1], although in the good direction, were not applicable when considering the flight-realistic limitations of the mission. More specifically, the available processing power on-board was not enough for compressing a data stream of about 5 Mbps or more (which is the requirement for that mission). Even a quick algorithm like the CCSDS 121.0 lossless compression technique required a slightly excessive percentage of the on-board processor capacity, and moreover a software implementation was mandatory in this case – thus making useless the efficient hardware implementations available for CCSDS 121.0. A possible solution would be to remove the adaptive routines, which are probably the most time-consuming ones. That is, to use only the Rice coder – the core of the CCSDS recommendation – with some fixed (pre-calibrated) value of k for each set of samples. However, this solution is not reliable at all. As it is well known, the value of k must be chosen very carefully in order to obtain the expected compression ratios for a given set of data. Furthermore, and most important, a single outlier can lead to large expansion ratios, as previously explained. This is a very important limitation of the Rice coder,

Table 5.2 An example of limited-length signed Rice codes, for $k = 0$ and 16-bit values

n	$k = 0$
0	S|0|
1	S|10|
2	S|110|
3	S|1110|
...	...
17	S|11 1111 1111 1111 1110|
≥ 18	1|0|S|XXXX XXXX XXXX XXXX

and is obviously one of the reasons why the CCSDS introduced the adaptive layer – which significantly compensates any outlier found in a data block.

This limitation of the isolated Rice coder can be solved if we get rid of the mapping stage as well (called PEM in the CCSDS recommendation). That is, separately storing the sign bits coming from the pre-processing stage. Then, the Rice codes are calculated only for the moduli of the samples, and the sign bits are simply inserted at the beginning of the codes. By removing the PEM stage we save further processing resources, while at the same time we introduce an interesting feature. This is, the possibility of generating a -0 ("minus zero") code. This special Rice coding was devised within the frame of the GOCA Technology Research Programme (Gaia Optimum Compression Algorithm), proposing to generate such special code when detecting a value that would lead to a large Rice code. This code is then followed by the value in its original coding, thus using the -0 code as an "escape sequence". The JPEG-LS coding stage uses a similar feature [3]. This very simple method can be called *Limited-Length Signed Rice Coding* (or *signed Rice coding with escape sequence*), and it makes the Rice codes much more resilient in front of noise and outliers, and hence it is highly recommended to apply it when only the Rice core is to be used.

Table 5.2 illustrates an example of Limited-Length Signed Rice code for $k = 0$ (the most "dangerous" case) and 16-bit values. The sign bits coming from the pre-processing stage are indicated with an S here. As we can see, when the code reaches a too large size, the *minus zero* special code is generated (assuming $S = 1$ as the negative sign), followed by the signed value in its 16-bit (plus sign) original format. It limits the total code length to just 19 bits in this case – or, in general, to $19 + k$ bits. Alternatively, the original value (prior to pre-processing) without the sign bit could be output after the -0 code, thus saving one additional bit for outliers – assuming that such original value is available to the coder.

3.3 Prediction Error Coder (PEC)

Rice codes are the adequate solution when the data to be compressed follow a geometric statistical distribution, which often arise after an adequate pre-processing stage [13]. However, any deviation from this statistic can lead to a significant

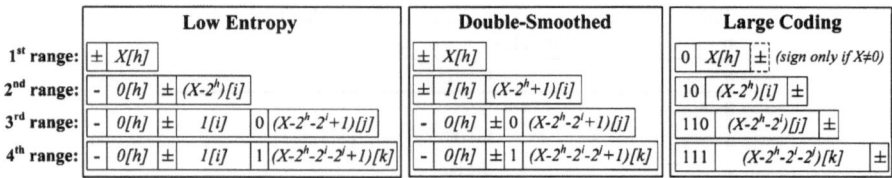

Fig. 5.1 Illustration of the three PEC coding strategies, for data with low, medium and high entropy levels

decrease in the final compression ratio. Despite of the adaptive stage, the CCSDS recommendation also suffers from this limitation. Additionally, and most importantly, even with such adaptive stage we have found important decreases in the compression efficiency when realistic fractions of outliers are found in the data, as it will be demonstrated in Sect. 5.6.1 below. Improvements based on the codes described in Sects. 5.3.1 and 5.3.2 can mitigate it, but a more flexible and robust solution is preferable.

The Prediction Error Coder [11, 12] has been developed to solve this weakness of the Rice-like codes. As its name indicates, PEC is focused on the compression of prediction errors, and hence a pre-processing stage outputting signed values is required – such as a data predictor plus a differentiator. Similarly as the Limited-Length Signed Rice coding, PEC was devised and developed within the frame of the GOCA Technology Research Programme of ESA. Efforts were put on the development of a very fast and robust compression algorithm, and PEC was the outcome – a partially adaptive entropy coder based on a segmentation strategy.

PEC is composed of three coding options, namely, Low Entropy (LE), Double-Smoothed (DS) and Large Coding (LC). All of them are segmented variable-length codes (VLC). LC also makes use of unary prefix codes [3], while LE and DS rely on the "minus zero" feature of signed prediction errors, similarly as the Limited-Length Signed Rice codes previously described. In Fig. 5.1 a schematic view of PEC is shown and the coding strategy of each of the options and segments is unveiled. The number of bits per segment (h, i, j and k) are independent parameters, which fix the compression ratio achievable for each range of input values. In the first range (or segment) PEC outputs $X[h]$, that is, the h least-significant bits of the value (X). It obviously means that the first range is only applicable to values smaller than 2^h (or smaller than 2^{h-1}, in the DS option). For other ranges (that is, for higher values), an adequate number of zeroes or ones are output if using the LE or DS option, then subtracting an adequate value to X and finally outputting the i, j or k least-significant bits of the result. The sign bit of the value is indicated with '$\pm$' in the figure, while a fixed negative bit is noted with '$-$'.

As can be seen, the coding scheme is completely different to that of the Rice coder. The three coding options share the same principles: the dynamic range of the data to be coded is split into four smaller ranges (or segments). The size of each segment is determined by its corresponding coding parameter (h, i, j or k), which indicates the number of bits dedicated to code the values of that segment. This set of parameters is called *coding table*. Then, for each value to be coded, the appropriate segment and number of bits to code the value are chosen, following the procedure indicated in Fig. 5.1. PEC follows the assumption that most values to be coded are close to zero – although this premise is not really mandatory for a successful operation of PEC. When this is true, the coding parameters must be fixed in a way that the first segments are significantly smaller than the original symbol size, while the last segments are slightly larger. This obviously leads to a compressed output, while the ratio will be determined by the probability density function (PDF) of the data combined with the selected coding table. Additionally, one of the main advantages of PEC is that it is flexible enough to adapt to data distributions with probability peaks far from zero. With an adequate choice of parameters, good compression ratios can still be reached with such distributions.

It is worth emphasizing that a pre-processing stage is required when using PEC, and that separate sign bits shall be used for its outcome – that is, avoiding any mapping stage. Although it adds some redundancy because of the existence of $+0$ and -0 codes, they are used as a key feature in both the LE and DS options. The last value of a segment (all bits set to one) is also used as an escape value in these two options. Thus, we use these sequences to implicitly indicate the coding segment to be used. On the other hand, the LC option simply uses the unary coding to indicate the segment used, and it avoids the output of the sign bit when coding a zero. If required, PEC could also be used for coding unsigned values, by using only the LC option without outputting any sign bit.

An adequate coding table and coding option must be selected for the operation of PEC. In order to easily determine the best configuration for each case an automated PEC calibrator can be used, which just requires a representative histogram of the values to be coded. It exhaustively analyzes such histogram and determines the optimum configuration of PEC. This is done by testing each of the possible PEC configurations on the histogram of values and selecting the one offering the highest compression ratio – that is, a trial-and-error process. Although this calibration process is much quicker than it may seem (less than 1 s on a low-end computer for a 16-bit histogram), it is too expensive in computational terms for being included in an on-board coder. In the case of space missions the calibrator must be run on-ground with simulated data before launch. PEC is robust enough for offering reasonable compression ratios despite of variations in the statistics of the data. Nevertheless, the calibration process should be repeated periodically during the mission, re-configuring PEC with the data being received in order to guarantee the best results.

PEC can be considered a *partially adaptive* algorithm. That is, the adequate segment (and hence the code size) is selected for each one of the values. This is obviously an advantage with respect to the Rice coder, which uses a fixed parameter for all the values – at least within a given coding block, in the case of the CCSDS recommendation. Another advantage with respect to Rice is that, by construction, PEC limits the maximum code length to twice the symbol size in the worst of the cases (depending to the coding table). Nevertheless, despite of these features, it is true that PEC must be trained for each case in order to get the best compression ratios. Therefore, if the statistics of the real data significantly differ from those of the training data, the compression ratio will decrease.

4   Analysis of the Codes

In this section we introduce the necessary formulation and metrics to evaluate the performance of an entropy coder, both under controlled conditions (with synthetic data) and on real data. We also show here some performances obtained with the coders previously described.

4.1   Modeling the Input Data

It is widely known that the output of a pre-compression (or pre-processing) stage can usually be well approximated by a geometric distribution. Exponential distributions, the continuous equivalent to the geometric distribution, are very common in some processes, as they are related with homogeneous Poisson processes. In the case of signed predictions (as required by PEC, for example), the corresponding distribution would be a discrete Laplacian. This distribution naturally arises when subtracting two realizations of random samples following geometric distributions. Hence, this result is expected when dealing with data following the exponential distribution, sampled and then passed through a prediction error filter.

The continuous Laplacian distribution can be seen as a two-sided exponential – that is, one exponential coupled to an inverse exponential. When the mean of the distribution is 0, it follows the equation:

$$f(x|b) = \frac{1}{2b} \exp\left(-\frac{|x|}{b}\right) \tag{5.1}$$

This distribution can be compared with the exponential one:

$$f(x|\lambda) = \left\{ \begin{array}{c} \lambda e^{-\lambda x}, \ x \geq 0 \\ 0, \ x < 0 \end{array} \right\} \tag{5.2}$$

Rather obviously, it turns out that there is a correspondence between λ and b, $\lambda = 1/b$. Additionally, there is a 1/2 factor, which is the scale factor applied due to the two sides of the Laplacian and the single side of the exponential. Thus, the frequency of the positive values in an exponential distribution is twice that of a Laplacian distribution. However, it is important to realize that in the case of our coders, which deal with discrete distributions, there is one exception. This exception is the frequency of the value 0, which is the same for both distributions, due to the fact that a -0 value does not exist. It is illustrated in the right panel of Fig. 5.2. The left panel of the figure shows a simulated Laplacian compared to the actual Laplacian realization following 5.3 below. The right panel of the figure, on the other hand, shows the modulus of the same simulated Laplacian compared to an actual exponential realization given by 5.2 for which we adopted $\lambda = 1/b = 0.1$. We can clearly see the divergence in the probability of value 0.

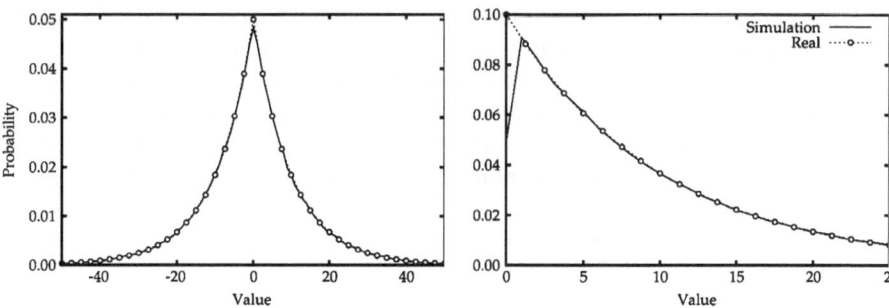

Fig. 5.2 Simulated Laplacian distribution with $b = 10$ (*left panel*) and its modulus (*right panel*)

Tests with real data assess that entropy coders usually receive this kind of distributions in pre-processed data. Thus, it seems reasonable to use them to evaluate the compression efficiency under controlled conditions. In some cases – especially when using too simple pre-processing stages, or when dealing with certain data sources – the data may be better modeled by a Gaussian distribution, so it should be included in the testbed for entropy coders. Finally, we should also include a model for the outliers. This is more difficult to model, since they could either appear as Gaussian-like samples (with a very wide distribution) added to the original data, or even as saturated samples, or also as values spread rather uniformly over the whole dynamic range. As a first approach, we will model here the outliers as a simple uniform distribution superimposed to the original function. The following equations summarize the models used in our initial tests, for Laplacian and Gaussian distributions respectively:

$$l[i] = (1 - P)\left(\frac{1}{2b} \exp\left(-\frac{|i|}{b}\right)\right) + P\left(\frac{1}{A}\right) \tag{5.3}$$

$$g[i] = (1 - P)\left(\frac{1}{\sqrt{2\pi\sigma^2}} \exp\left(\frac{-i^2}{2\sigma^2}\right)\right) + P\left(\frac{1}{A}\right) \tag{5.4}$$

where P is the fraction of noise and A the normalization factor. Experience tells us that the typical conditions found in space leads to noise fractions between 0.1% and 10%. For example, in the Gaia mission it is estimated to find about 2% of pixels affected by outliers, but in other missions values as high as 10% can be reached [19]. We emphasize that a uniformly distributed noise is just a first approach to the different sources of noise that can be found in many space instruments. For example, Prompt Particle Events (PPEs), which can be caused by cosmic rays and solar protons, increase the accumulated charge of the pixels in CCDs, leading to high or even saturated readout values. Nevertheless, the resulting effect highly depends on the actual response of the instrument. Tests with saturating outliers have been carried out, revealing results very close to those obtained using flat noise.

Finally, regarding the width of the data distribution (that is, b or σ), the whole significant parameter range shall be explored, covering from very small to very large entropies. Additionally, converting the b or σ parameter to an entropy level gives a better interpretation of the results. As an illustrative example, in the results shown hereafter we test our coders on 16-bit samples for entropy levels of 2–12 bits – which translate into maximum theoretical ratios of 8.0–1.3 respectively.

4.2 Performance Metrics

The most common and well-understood evaluation of the performance of a data compressor is the compression ratio that it achieves on some given test data. Nevertheless, as previously mentioned, it highly depends on the kind of data and on the pre-processing stage used. We should objectively evaluate the performance of an entropy coder, and hence an adequate metric shall be used. If we fix both the input data and the pre-processing stage we obtain a given entropy value for the values to be coded. In other words, we can obtain the Shannon theoretical compression limit – which is the original symbol size (S_S) divided by the entropy (H). We define the *compression efficiency* as the compression ratio achieved by the entropy coder divided by the Shannon limit of the pre-processed data block. The goal is, obviously, a compression efficiency of 100% – that is, reaching the compression limit fixed by the Information Theory [20]. This result must be evaluated under different conditions, such as different entropy levels and different outliers (or noise) levels. Alternatively, we can also use the *relative coding redundancy* (h), which gives the difference between the Shannon limit and the actual compression ratio obtained using a given coder, obviously targeting a result of 0%:

$$
h_\lambda^C = \frac{\overline{L_\lambda^C} - H_\lambda}{H_\lambda} = \frac{\overline{L_\lambda^C}}{-\sum_{i=-2^{S_s}+1}^{2^{S_s}-1} P[i] \cdot \log_2(P[i])} - 1 \tag{5.5}
$$

where $\overline{L_\lambda^C}$ is the average coding length for a given coder (C) and distribution (λ), H_λ is the entropy of the pre-processed data, and S_S is the symbol size in bits. Both the coding redundancy and the compression efficiency can be evaluated for either real or synthetic data (e.g., generated with the models described in the previous section). If using synthetic data, considering its stochastic nature, it is highly recommended to run several realizations (for each entropy and noise level) and average the result, in order to obtain a more reliable metric. This process has been followed in the results shown hereafter. Something similar happens when we wish to evaluate the *computing performance* (or processing requirements) of an entropy coder: several repetitions shall be done, but in this case we should only consider the shortest execution time. The reason is that typical domestic and research computers usually run non-realtime operative systems together with many other programs and routines

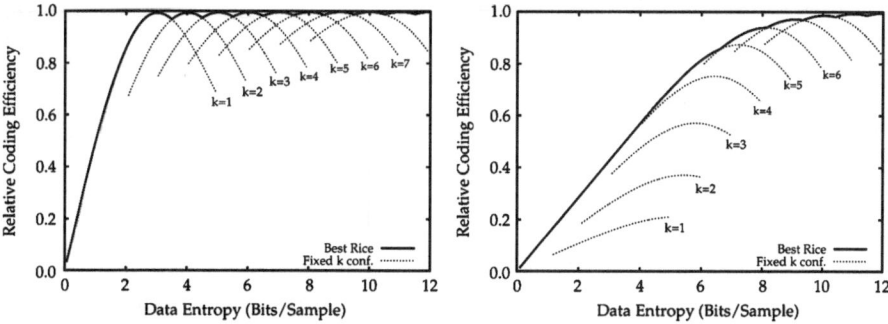

Fig. 5.3 Expected compression efficiency of the Rice coder on Laplacian data, clean (*left panel*) and with 0.1% of outliers (*right panel*)

(such as the graphical environment). Considering only the shortest time is a good approach to what we can get in an optimized implementation onboard a satellite payload.

4.3   Rice-Golomb Codes

Figure 5.3 illustrates the compression efficiency that can be obtained with Rice-Golomb codes on synthetic data following a Laplacian (or two-sided geometric) distribution, using the models shown in Sect. 5.4.1 on 16-bit samples. Thick solid lines indicate the best result that can be obtained, while thin dashed lines show the efficiencies for each given configuration. The right panel features an additional 0.1% of uniformly distributed noise, modeling outliers. As we can see in the left panel, this coder performs almost optimally when the data perfectly follows a Laplacian distribution without any outlier. The efficiency decreases below 90% only for highly compressible data (below 2 bits of entropy), for which case the CCSDS 121.0 recommendation chooses more "aggressive" compression options such as the Fundamental Sequence, the Second Extension or even the Zero Block [4]. On the other hand, when adding just 0.1% of outliers (right panel of the figure), the Rice coder cannot reach a desirable 90% of efficiency except for high entropies (such as 7 bits and above), and the low k configurations lead to very poor efficiencies (typically below 50%). This is caused by the high sensitivity to outliers previously indicated in Sect. 5.2.

4.4   Subexponential Codes

Subexponential codes are equivalent to Rice codes when the data to be compressed perfectly follows a Laplacian distribution. It is demonstrated in the left panel of Fig. 5.4, which illustrates the best compression efficiency achievable for both

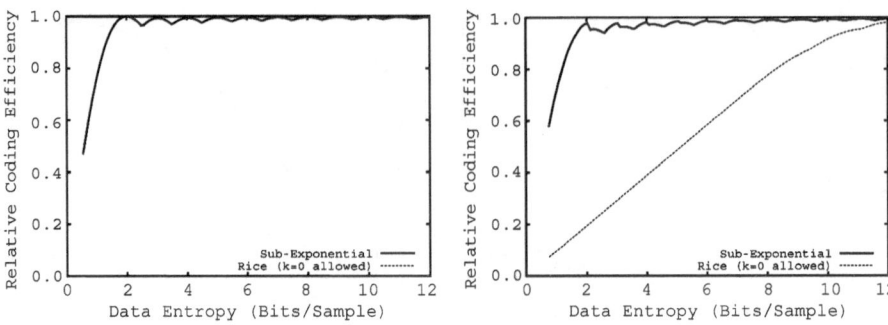

Fig. 5.4 Expected compression efficiency of subexponential and Rice codes on Laplacian data, clean (*left panel*) and including 1% of outliers (*right panel*)

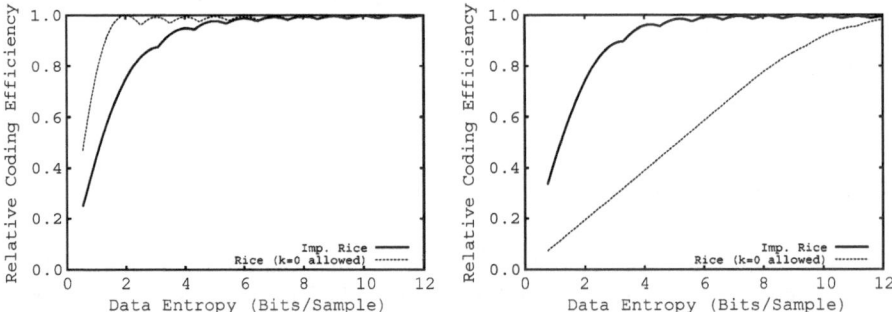

Fig. 5.5 Expected compression efficiency of signed Rice codes with escape sequence, on clean Laplacian data (*left panel*) and on data with 1% of outliers (*right panel*)

subexponential and Rice codes. Here we allow $k = 0$ for Rice, which is equivalent to the Fundamental Sequence of the CCSDS 121.0 recommendation. On the other hand, the right panel of Fig. 5.4 reveals the resiliency of subexponential codes when dealing with outliers. Here we have increased the level of outliers to 1% in order to better appreciate the resiliency. Rice codes offer even a worse efficiency than in the right panel of Fig. 5.3, being unable to reach 90% for almost any entropy level. On the other hand, subexponential codes keep performing better than 90% even for entropies as low as 1.5 bits.

4.5 Limited-Length Signed Rice Codes

As can be seen in the left panel of Fig. 5.5, the inclusion of the sign bit in the Limited-Length Signed Rice codes leads to worse compression efficiencies than standard Rice when applied to clean Laplacian data. This decrease can only be seen for entropies lower than 6 bits, but it can reach -25% for entropies as low as 1 bit. But when including 1% of outliers (right panel of Fig. 5.5), this improved Rice coder offers an almost identical efficiency throughout the whole range of data

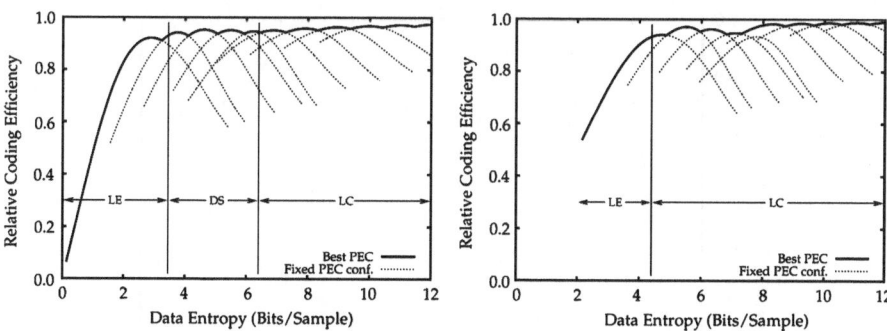

Fig. 5.6 Expected compression efficiency of PEC on Laplacian data, clean (*left panel*) and with 10% of outliers (*right panel*)

entropy. Actually, in some cases the efficiency is slightly better than in the clean case. This is caused by the excellent adaptability of this coder to large outliers in the data. Nevertheless, the overall efficiency for this entropy coder is still smaller than that of the subexponential coder, although tests with their adaptive versions will reveal the ultimate efficiency that we can expect from them.

4.6 PEC Codes

For PEC we have essayed again the same test as in the previous subsections. That is, evaluating the compression efficiency for different entropy levels on clean and noisy data. The difference, in this case, is that we have directly tried with 10% of outliers instead of the modest 0.1% used in the Rice tests or the 1% used in the improved Rice-based coders. Figure 5.6 shows the results, combining the best results into the thick solid lines while thin dashed lines show the efficiencies for each given configuration, as in Sect. 5.4.3. In the left panel we can see that PEC is not as efficient as the Rice family coders when dealing with ideal data distributions, although the results are very good. Efficiencies of 90–95% are obtained for the most common ranges – that is, for entropy levels higher than 2 bits. The figure also highlights the optimality ranges of each PEC option. On the other hand, when analyzing the results of PEC with 10% of outliers (right panel), we can clearly see that PEC is an outlier-resilient entropy coder. While Rice was almost unable to reach high efficiencies even with tiny fractions of outliers, PEC still can perform above 80% for the most typical dispersions even with large amounts of contamination. The reason lies in the segmentation strategy used by PEC. Figure 5.7 better illustrates this advantage of PEC with respect to other prefix codes. The results with Rice have been included in the figure for a better comparison. In this test, both Rice and PEC have been calibrated (or configured) to only three entropy levels (3, 5 and 11 bits), and the results for each configuration on the whole entropy range are shown. The figure reveals the relatively narrow range for which a given Rice configuration is optimal. As can be seen, the Rice efficiency quickly drops even below 30% when the

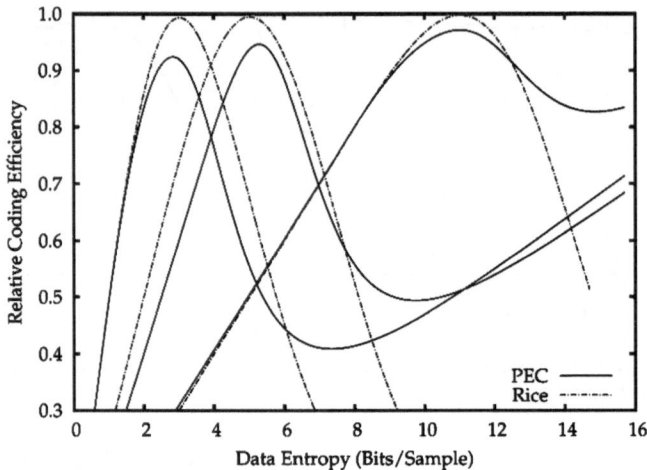

Fig. 5.7 Coding efficiency of Rice and PEC when using only three fixed calibrations

entropy level unexpectedly increases by just 4 bits. On the other hand, PEC always offers efficiencies better than 40% – except for very low entropy levels. It is worth mentioning that the apparent "recovery" seen for very high entropies does not mean that PEC is able to compress better there. Instead, it just indicates that the expansion ratio is limited by construction (typically below 2), while Rice can reach huge expansion ratios as previously shown. Summarizing, although PEC cannot reach the excellent compression efficiencies of Rice-like coders on clean data, it significantly mitigates the effects of outliers on the final compression ratio even under very unfavorable conditions.

5   Adaptive Coders

Section 5.4 has introduced the several *base codes* for entropy coding, each with different outlier resiliency levels. Even in the most robust of the solutions (that is, PEC), an adaptive layer automatically selecting the best configuration for each data block is highly recommended. Here we briefly describe the most suitable implementations for each case.

5.1   Adaptive Subexponential

The case of subexponential codes is the easiest one: they can be directly integrated within the CCSDS 121.0 framework, substituting the Rice coder. The rest of the recommendation can be kept unchanged, also including the Prediction Error Mapper (PEM). Only one change is required, namely, the $k = 0$ option must be allowed, in order to lead to a smooth transition between the Fundamental Sequence option and the subexponential coding. On the other hand, the $k = 13$ option is not required

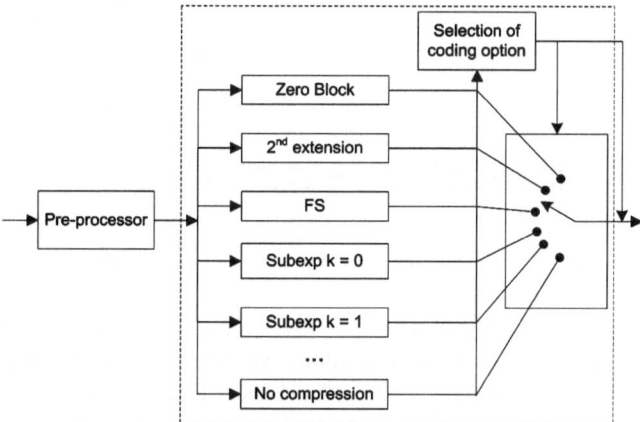

Fig. 5.8 Implementation of the adaptive subexponential coder within the CCSDS 121.0 framework

anymore (at least for 16-bit samples), so the configuration header output at the beginning of each data block can be of the same size. Figure 5.8 illustrates this integration of the subexponential coder within the CCSDS 121.0 framework.

Another change is recommended in order to improve further the compression efficiency. CCSDS 121.0 originally works with very small blocks of data (8 or 16 samples). Owing to the better resiliency of subexponential in front of outliers, 32-sample blocks can be used, slightly decreasing the overhead introduced by the configuration header.

5.2   Adaptive Limited-Length Signed Rice Coder

For the signed Rice coder with escape sequence, the PEM stage of the 121.0 recommendation must be removed in order to generate signed prediction errors. The remainder of the recommendation can be kept identical, including the automatic selection of the adequate k configuration for each data block – or the selection of a low-entropy option. In case an additional improvement is desired, the original data should be made available to the coder so that it can be directly output without the sign bit when an outlier is found, as described in Sect. 5.3.2. Finally, the block size can also be slightly increased in order to reduce the header overhead.

5.3   Fully Adaptive PEC (FAPEC)

An adaptive algorithm for PEC has been designed and implemented which solves the previously commented limitations of the coder. The solution, protected by a

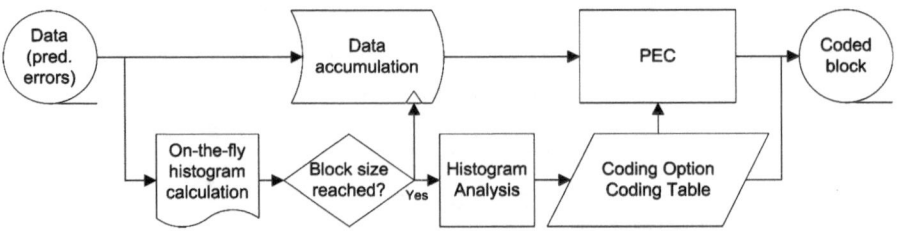

Fig. 5.9 Operation of the fully adaptive prediction error coder (FAPEC)

patent, has been called *Fully Adaptive Prediction Error Coder* (FAPEC) [12]. Similarly to the CCSDS recommendation, where an adaptive stage selects the most appropriate value of k for a given data block, FAPEC adds an adaptive layer to PEC in order to configure its coding table and coding option according to the statistics of each data block. In this way, nearly-optimal compression results can be achieved without the need of any preliminary configuration of PEC, and without requiring any knowledge of the statistics of the data to be compressed. The block length is configurable and not restricted to a power of two, with typical (recommended) values of 250–500 samples. One of the main premises in the design of FAPEC was the quickest possible operation, even if at the expense of a slight decrease in the optimality of the PEC configuration – and hence a slight decrease in the compression ratio. The intrinsic robustness of PEC guarantees that such decrease will be negligible.

FAPEC accumulates the values to be coded and, at the same time, a histogram of their moduli is calculated on-the-fly. This is a logarithmic-like histogram, in the sense that higher sample values (which are less frequent) are grouped and mapped to fewer bins, and values close to zero are mapped to independent bins. This reduces the memory required for the histogram and, most important, the time required to analyze it. This logarithmic-like resolution in the statistics is enough for our case. Once the required amount of values has been loaded, an algorithm analyzes the histogram and determines the best coding option (LE, DS or LC) and coding table, based on the accumulated probability for each value and a set of probability thresholds. A default threshold configuration has been fixed in the algorithm, determined heuristically using simulations of two-sided geometric distribution with outliers, looking for the highest possible ratios. Despite of this specific training set, this default configuration offers excellent ratios for almost any dataset with a decreasing trend in its PDF as we move towards higher values, such as the abovementioned two-sided geometric, bigamma [21] or Gaussian. The results shown later confirm this assertion. Nevertheless, such thresholds could be modified if necessary. In this way, FAPEC could be fine-tuned to better adapt to other statistical distributions if really needed. This is an interesting feature that the CCSDS recommendation does not have. Finally, once the coding option and the corresponding parameters have been determined, they are output as a small header followed by all the PEC codes for the values of that block. Explicitly indicating the PEC configuration makes possible to change the FAPEC decision algorithms without requiring any modification in the receiver. Figure 5.9 illustrates the overall operation of FAPEC.

6 Testing the Coders

In this section we illustrate the compression results obtained with software implementations of the coders previously described. Both the compressors and decompressors have been implemented, so that we have also been able to assess their lossless operation.

6.1 Synthetic Tests

The first recommended step to test an entropy coder is through synthetic data, that is, random data following a given probability density function (PDF). The models described in Sect. 5.4.1 make possible to test a coder under the most typical scenarios that we can expect for space-borne instrumentation. It also makes possible to objectively compare different entropy coders, such as those presented in the previous sections. Here we illustrate this testing procedure using Laplacian (or two-sided geometric) and Gaussian distributions, with outlier levels of 0.1%, 1% and 10%. Figure 5.10 shows the results obtained for CCSDS 121.0, Adaptive Limited-Length Signed Rice (titled "C-ImRi" in the figure) and FAPEC. Adaptive Subexponential is not included here for the sake of readability, but its results are very similar to those of the adaptive improved Rice, except for better efficiencies at high entropies. For the sake of completeness we also include the "Best PEC" results, that is, the best possible ratios achievable by PEC (when adequately calibrated to each entropy level). While we have previously seen figures showing the compression efficiency, here we illustrate the results as coding redundancy.

The figure demonstrates that the CCSDS 121.0 recommendation is too sensitive to outliers in the data, despite of the adaptive stage added to Rice. Not only this,

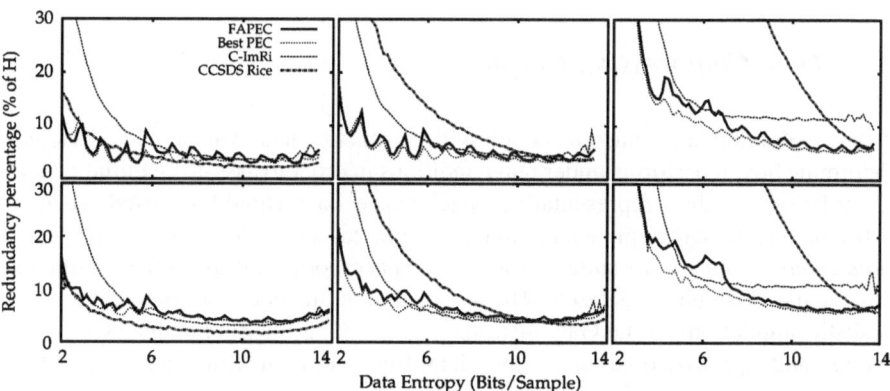

Fig. 5.10 Coding redundancy of some entropy coders on Gaussian (*top panels*) and Laplacian (*bottom panels*) distributions. From left to right, 0.1%, 1% and 10% of outliers

but also the sensitivity to non-Laplacian distributions has been assessed. Even with just 0.1% of outliers in the data, CCSDS 121.0 can be beaten by PEC and FAPEC at low entropies. Thus, the current standard is only optimal for Laplacian data with fractions of outliers as low as 0.1%. When just 1% of the data is uniformly distributed throughout the whole dynamic range, CCSDS 121.0 typically keeps at least 10% of the original data redundancy – that is, it performs less than 90% of the Shannon limit. Finally, when 10% of the data is outliers, the current lossless standard almost cannot remove any redundancy from the data – except for very high entropies. It translates into CCSDS 121.0 being unable to reach ratios higher than 2 for data with high amounts of outliers.

Regarding the outlier-resilient coders introduced here, both Rice-based improvements offer typical efficiencies above 90% of the Shannon limit – that is, typically less than 10% of redundancy is left. For high amounts of outliers the adaptive subexponential coder performs better than the adaptive Limited-Length Signed Rice coder, which cannot remove the last 10% of redundancy.

Finally, both PEC and FAPEC appear as excellent outlier-resilient coders. Their efficiency is almost identical when moving from 0.1% to 1% of outliers, always performing at about 90–95% of the Shannon limit – even for Gaussian distributions. Only with high amounts of outliers they decrease a bit their efficiency (specially at low entropies), but the modest 15% redundancy kept at an entropy of 3 bits is much better than the results obtained with the other coders – specially when compared to CCSDS 121.0. It is also worth mentioning the correct operation of the FAPEC adaptive algorithm, since the FAPEC performance is always very close to the best results achievable with PEC. Actually, in some cases, FAPEC can outperform an optimally configured PEC. It can happen on real data with varying statistics, where FAPEC adapts to each data block while PEC is configured for the whole data set. Also, in very low entropies, FAPEC features some low-entropy improvements similar to those existing in CCSDS 121.0, such as the handling of sequences with many zeroes.

6.2   Data Compression Corpus

After evaluating any entropy coder with synthetic data, we should obviously determine how it performs under real conditions in order to assess the initial results. In order to do this, representative space-borne data should be used, such as astronomical imaging, planetary imaging, spectroscopy, etc. Here we present a *data compression corpus* suitable for testing space-borne entropy coders, compiled from real and simulated datasets. The testing dataset includes astronomical images, realistic simulations of the Gaia instruments [22], GPS measurements data, photometric and spectroscopic data, and data from some instruments of the Lisa PathFinder mission (LPF). Seismogram data have also been included in order to obtain a more complete range of instrument types. The dataset has been compiled by the authors and it is available upon request.

It is worth mentioning that a pre-processing stage was obviously required in order to perform these tests, as previously mentioned. Very simplistic stages have been used here, typically using just a differentiator – that is, predicting each sample as equal to the previous one. In other words, the entropy coders have just received the differences between the consecutive values to be compressed. In some cases we have essayed up to third-order filtering (within the prediction loop), thus smoothing the quick variations or noise that the data may have. The pre-processing stage leading to the best ratios has been used for each of the data files, but it is obviously far from the elaborated stages that we can find for imaging, spectroscopy or GPS, just to mention some examples. Thus, the ratios shown here can be seen as the "worst result" that we can get for a given file.

In some cases, large symbol sizes are used in the data files, such as 32 or even 64 bits per sample. The implementations used in the tests of the entropy coders described here can handle up to 28 bits per sample, and thus a *sample splitting* process has been applied. That is, 32-bit samples have been split into two, alternatively compressing the most and least significant 16-bit words of the prediction errors. Similarly, in case of 64-bit samples, four 16-bit words have been used. The effect on the coders is that the compressor receives values with very different statistics. The most significant portions of the samples appear as very low-entropy values, while the least significant portions resemble uniformly distributed noise and thus appear as outliers. In other words, huge amounts of outliers appear in the data. This operation has interesting effects on the results, as we will see hereafter, since PEC and FAPEC are able to deal with this situation without significantly affecting the overall performance.

Starting with imaging data, the corpus includes FOCAS data [23] which is a standardized dataset of astronomical data, commonly used for testing calibration methods that must be able to deal with very different astronomical images. Images simulated by GIBIS [22] are also included, which is a very detailed simulator of the images that will be seen by Gaia [18]. The FITS images generated by the simulator are used here. Finally, a miscellaneous group of data files is included, covering extended sources such as galaxies, stellar fields, accretion disks, nebulae, or microphotographs of ground samples. Table 5.3 summarizes the most relevant results for each coder, also indicating the Shannon Limit (S_L) and the word length or sample size in bits (W_L). The best result for each file is highlighted in bold face, and its corresponding compression efficiency is shown in the last column. Further tests on the remainder of imaging data of the corpus can be found in [24].

As we can see, the optimality of each coder depends on the kind of data, but generally speaking the CCSDS 121.0 recommendation is beaten by the outlier-resilient entropy coders introduced in this chapter. The most elaborated solutions, that is, PEC and FAPEC, use to offer the best results. We remind that PEC has been optimally calibrated for each of these data files. If we wish to focus only on the adaptive systems, then FAPEC is probably the best option. In the worst of the cases it offers ratios 5% smaller than CCSDS 121.0, while in the best of the cases in can double the ratios obtained with such system. Such extremely good results are due to the sample splitting procedure previously mentioned, applied when the sample size

Table 5.3 Compression ratios achieved by several entropy coders on real imaging data

File	CCSDS 121.0	Adaptive Subexp	ALLSR	PEC	FAPEC	S_L	W_L (bits)	Best Eff. (%)
FOCAS								
tuc0004	**3.18**	3.01	3.08	3.00	3.07	3.22	16	99
com0001	**2.00**	1.95	**2.00**	1.98	**2.00**	2.07	16	97
ngc0001	1.79	**1.84**	1.81	1.81	1.83	1.92	16	96
for0001	**3.13**	2.99	3.08	3.04	**3.13**	3.22	16	97
gal0003	3.59	3.74	3.51	**4.25**	4.09	4.98	32	85
sgp0002	4.16	4.27	4.14	**4.54**	4.41	5.44	32	83
GIBIS								
7135_SM1_6	3.48	3.39	3.40	**3.62**	3.59	3.77	16	96
5291_AF1_5	2.12	2.22	2.20	2.51	**2.62**	2.83	32	93
5291_RP1_2	2.29	2.37	2.38	2.63	**2.89**	3.04	32	95
5291_RVS1_5	1.94	2.02	2.05	2.39	**2.61**	2.69	32	97
Miscellaneous								
stellar_field	1.12	2.14	2.49	**2.87**	2.63	5.14	64	56
noisy_source	1.01	1.03	1.00	1.12	**1.15**	1.19	32	97
Galaxy	1.14	2.17	2.39	**2.85**	2.58	4.84	64	59
nebula_stellar	1.02	1.19	1.59	**1.75**	1.64	3.78	64	46
ground_2	**1.92**	1.86	1.81	1.70	1.83	1.81	8	106

is too large for the current implementations, leading to a situation equivalent to having even more than 50% of outliers. In such cases the best efficiency is around 50%, an otherwise expected result for such high amounts of outliers. The worst efficiency is 46% of the Shannon limit (obtained with PEC), but in that case the CCSDS 121.0 is only able to offer an efficiency of 27%.

We can also see that FAPEC indeed achieves nearly-optimal calibrations for its PEC core, even offering better results – due to the reasons previously explained. In the worst of the cases FAPEC achieves a performance around 5% worse than PEC, but we remind that this is achieved with a completely adaptive operation. Regarding the modifications to CCSDS 121.0 (Adaptive Subexponential and Adaptive Limited-Length Signed Rice), they also use to outperform the current standard, although the improvement uses to be smaller than with PEC or FAPEC. Finally, it is worth mentioning the results with the 8-bit file, where we can see that all the adaptive compressors are able to surpass the Shannon theoretical limit. It is achieved by using the low-entropy extensions that both the CCSDS 121.0 framework and FAPEC include, specially the zero block extension.

The data corpus continues with GIBIS telemetry data [22], which correspond to highly realistic simulations of Gaia [18], roughly following the same data format that the on-board Video Processing Units will deliver during the mission. Some results are shown in Table 5.4, where we can see that FAPEC outperforms CCSDS 121.0 again – although with a slightly smaller margin this time. The best improvement is for low-resolution images (Sky Mapper, SM) with a 7%. We must mention that the Rice-based coders with outlier resiliency improvements also offer very good results, specially for the SM data mentioned and for spectroscopy (Radial Velocity

Table 5.4 Compression ratios achieved by several entropy coders on simulated telemetry data from CCD images

File	CCSDS 121.0	Adaptive Subexp	ALLSR	PEC	FAPEC	S_L	W_L (bits)	Best Eff. (%)
SM_L90b40	2.21	**2.39**	2.31	2.34	2.38	2.53	16	94
AF_L10b70	1.68	**1.69**	1.68	1.61	**1.69**	1.73	16	98
BP_L170b60	3.56	3.52	3.47	3.38	**3.58**	3.68	16	97
RP_L10b70	3.29	3.24	3.23	3.10	**3.30**	3.30	16	100
RVS_L1b1	2.15	2.19	**2.24**	2.20	2.21	2.35	16	95

Table 5.5 Compression ratios achieved by several entropy coders on real GPS data

File	CCSDS 121.0	Adaptive Subexp	ALLSR	PEC	FAPEC	S_L	W_L (bits)	Best Eff. (%)
Raw GPS data								
global_S1	**2.30**	2.25	2.29	2.22	2.29	2.35	16	98
global_L1	1.57	1.64	1.65	**1.68**	1.67	1.76	24	95
global_C1	1.74	1.82	1.84	**1.85**	1.84	1.93	24	96
Treated GPS data								
is07_lat	3.61	3.56	3.47	**3.79**	3.68	4.40	16	86
nun2_height	2.99	2.92	2.96	3.04	**3.06**	3.22	16	95
nun2_lon	4.45	4.45	4.48	4.61	**4.64**	4.93	24	94

Table 5.6 Compression ratios achieved by several entropy coders on real and simulated LPF data

File	CCSDS 121.0	Adaptive Subexp	ALLSR	PEC	FAPEC	S_L	W_L (bits)	Best Eff. (%)
kp30_row2	3.86	3.85	3.91	3.88	**4.00**	4.12	24	97
kp30_row5	**1.76**	1.74	**1.76**	1.74	**1.76**	1.80	24	98
kp30_row10	12.83	**15.34**	11.07	11.85	13.76	20.35	24	75
acc_intrf	**1.02**	1.00	**1.02**	1.01	**1.02**	1.21	24	84

Spectrometer, RVS), although the difference with respect to FAPEC is a modest 1%. Finally, in this case we have reached again the Shannon limit (on Red Photometer data), owing to the low entropy coding improvements.

We have also included GPS data in the corpus because the signal model and the kind of data generated are adequate for this study. A more sophisticated data compression system for GPS data for geophysics applications – specifically a better pre-processing stage – can be found in [25]. There are two sub-groups of GPS data in the corpus, namely, raw data of a GPS observation file and data already processed from the GPS system. Table 5.5 shows some results, where we can see again that PEC and FAPEC outperform the other systems (especially CCSDS 121.0) in most cases.

The Lisa PathFinder space mission (LPF) is a technology demonstrator specifically designed to test and assess the key technologies that will be used in the LISA mission. Table 5.6 shows some results obtained on data kindly provided by the IEEC LPF team, generated from instrumentation for accurate temperature measurements and from a simulation of the nominal acceleration of one test mass

Table 5.7 Compression ratios achieved by several entropy coders on real spectroscopic data

File	CCSDS 121.0	Adaptive subexp	ALLSR	PEC	FAPEC	S_L	W_L (bits)	Best Eff. (%)
observ_irrad	**2.79**	2.75	2.77	2.50	2.72	2.68	16	104
er_spec	**1.62**	1.61	**1.62**	1.56	1.61	1.63	24	99
all_relative_stars	**2.09**	2.06	2.08	2.00	2.05	2.12	16	98
bkg-1o0235_freq_lin	1.30	**1.31**	**1.31**	1.26	1.29	2.63	24	50
ganimedes_freq_log	1.05	**1.06**	**1.06**	**1.06**	1.05	2.41	24	44

with respect to the spacecraft. Here the differences between the several coders are typically smaller, except for very low entropies – where the low-entropy extensions of the architectures play a major role. FAPEC reveals to deal nicely with such low dispersions, although the CCSDS 121.0 framework combined with subexponential codes offer the best results. We can also see that the last file cannot be compressed at all, either caused by the too high noise levels or by the too-simple pre-processing stage. This is a clear demonstration of the importance of an adequate pre-processing algorithm adapted to each kind of data.

The table with the results on seismogram data is not included for the sake of brevity, since all the coders offer very similar results, obtaining ratios of 1.85–2.87 depending on the seismic event or conditions. Finally, Table 5.7 shows some results obtained on spectroscopic data, including the solar spectral atlas and stellar libraries, both with high and low spectral resolutions. Here FAPEC is not the best solution anymore, although it is always at less than 3% below CCSDS 121.0. Actually, all the coders offer very similar results – as in the case of seismic data – with differences of only 2.5% at most. This is most probably caused by the naïve pre-processing stages used here. It is worth mentioning that some tests have revealed that more elaborated algorithms always benefit more to FAPEC than to the CCSDS recommendation. Again, this is caused by the ability of FAPEC to reduce the effect of outliers. A more elaborated pre-processing stage may lead to a steeper statistic but will probably keep similar portions of outliers in the data. As we have seen in the synthetic tests, FAPEC easily takes advantage of steep statistics (that is, low entropies) without being significantly affected by outliers.

7   Summary and Conclusions

In this chapter we have discussed the importance of an adequate entropy coder in order to build an efficient data compression solution, either for simple and generic compressors, or for specific and elaborated systems such as imaging, hyperspectroscopy or various instrumental data. The adequate models and metrics for evaluating an entropy coder have been presented and applied to several systems, including the current lossless compression standard for space (CCSDS 121.0) and also a brand new solution with promising results (FAPEC). Table 5.8 summarizes the main features of the entropy coders discussed here.

Table 5.8 Summary of the main features of the entropy coders discussed in this chapter

Entropy coding solution	Typical efficiency	Pros	Cons
Rice-Golomb	~0% to ~100%	Very quick Optimal for Laplacian distributions and "clean" data	Too dangerous to be used in isolation (adaptive stage mandatory) It can generate huge codes
CCSDS 121.0	~50% to ~99%	Quick Adaptive Good enough in many cases Efficient hardware implementations available	Too sensitive to outliers in the data (efficiency can decrease below 50%) and to non-Laplacian data
Subexponential codes	~70% to ~98%	Very quick High efficiencies achievable Reduced expansion ratio for outliers Good integration within CCSDS 121.0 framework	Adaptive stage highly recommended Sensitive to non-Laplacian distributions
Limited-length signed rice codes	~80% to ~95%	Very quick Expansion ratio <2 Could run in isolation if needed Good integration within CCSDS 121.0 framework	Efficiency limited by the additional sign bit Adaptive stage highly recommended
PEC	~85% to ~95%	Very quick Semi-adaptive Robust enough for being used in isolation Adequate for most typical distributions	Efficiency at low entropies limited by the sign bit Adaptive stage recommended Four configuration parameters
FAPEC	~85% to ~97%	Quick Adaptive Good results for any case Excellent for large sample sizes Low-entropy extensions Fine-tuning possible. Hardware prototype available	Efficiency slightly limited by the sign bit Completely different to CCSDS 121.0

The availability of a data compression corpus has made possible to evaluate the several coders under real conditions. The compression ratios achieved with these tests reveal that the new FAPEC algorithm is a reliable alternative to the CCSDS 121.0 recommendation. Its software implementation has been evaluated, indicating very similar processing requirements than those for CCSDS 121.0. Additionally, a hardware prototype implemented on an FPGA device is available, which assesses the applicability of FAPEC in space missions.

Alternatively, the CCSDS 121.0 recommendation could be improved by substituting the Rice-Golomb codes by outlier-resilient codes, such as subexponential or a Limited-Length Signed Rice coder. Although the results with such systems are often slightly worse than those of FAPEC, they are resilient enough in front of outliers in the data, and thus appear as another reliable alternative.

In general, it is highly recommended to use outlier-resilient entropy coders when data from space-borne instrumentation is to be compressed. While guaranteeing almost the same compression ratios as the current standards under any situation, they can take much more advantage of the data redundancies when large amounts of outliers appear – such as those caused by Prompt Particle Events or artifacts in the data or in the instruments.

References

1. J. Portell, E. García-Berro, X. Luri, and A. G. Villafranca, "Tailored data compression using stream partitioning and prediction: application to Gaia," *Experimental Astronomy* 21, 125–149 (2006).
2. CCSDS-101.0-B-5 Blue Book, *Telemetry channel coding*, 2001.
3. D. Solomon, *Data Compression. The complete reference*, Springer, 2004.
4. CCSDS-121.0-B-1 Blue Book, *Lossless data compression*, 1993.
5. CCSDS-120.0-G-2 Informational Report, *Lossless data compression*, 2006.
6. R. F. Rice, "Some practical universal noiseless coding techniques," *JPL Technical Report* 79–22 (1979).
7. S. W. Golomb, "Run-lengths encodings," *IEEE Transactions on Information Theory* 12, 399–401 (1966).
8. P.-S. Yeh, "Implementation of CCSDS lossless data compression for space and data archive applications," *Proc. CCSDS Space Operations Conf.*, 60–69, 2002.
9. P.-S. Yeh, P. Armbruster, A. Kiely, B. Masschelein, G. Moury, C. Schaefer, and C. Thiebaut, "The new CCSDS image compression recommendation," *IEEE Aerospace Conf.*, 4138–4145, 2005.
10. M. Clotet, J. Portell, A. G. Villafranca, and E. García-Berro, "Simple resiliency improvement of the CCSDS standard for lossless data compression," *Proc. SPIE 7810*, 2010.
11. J. Portell, A. G. Villafranca, and E. García–Berro, "Designing optimum solutions for lossless data compression in space," *Proc. ESA On-Board Payload Data Compression Workshop*, 35–44, 2008.
12. J. Portell, A. G. Villafranca, and E. García-Berro, "A resilient and quick data compression method of prediction errors for space missions," *Proc. SPIE 7455*, 2009.
13. P.-S. Yeh, R. Rice, and W. Miller, "On the optimality of code options for a universal noiseless coder," *JPL Technical Report* 91–2 (1991).
14. D. Huffman, "A method for the construction of minimum redundancy codes," *Proc. IRE* 40, 1098–1101 (1952).
15. I. H. Witten, R. M. Neal, and J. G. Cleary, "Arithmetic coding for data compression," *Communicat. ACM* 30, 520–540 (1987).
16. Teuhola, J., "A compression method for clustered bit-vectors," *Information Processing Letters* 7(6), 308–311 (1978).
17. Howard, P. and Vitter, J., "Fast progressive lossless image compression," in *Image and Video Compression Conference*, SPIE, 98–109 (1994).
18. M. A. C. Perryman, K. S. de Boer, G. Gilmore, E. Hoeg, M. G. Lattanzi, L. Lindegren, X. Luri, F. Mignard, O. Pace, and P. T. Zeeuw, "Gaia: Composition, formation and evolution of the Galaxy," *Astronomy & Astrophysics* 369, 339–363 (2001).

19. Nieto-Santisteban, M. A., Fixsen, D. J., Offenberg, J. D., Hanisch, R. J. & Stockman, H. S., "Data Compression for NGST", in *Astronomical Data Analysis Software and Systems VIII*, vol. 172 of Astronomical Society of the Pacific Conference Series, 137–140 (1999).
20. C. E. Shannon, "A mathematical theory of communication," *Bell system technical journal*, vol. 27, 1948.
21. A. Kiely and M. Klimesh, "Generalized Golomb codes and adaptive coding of wavelettransformed image subbands," *JPL Technical Report*, IPN 42–154 (2003).
22. C. Babusiaux, "The Gaia Instrument and Basic Image Simulator," in *The Three-Dimensional Universe with Gaia*, ESA SP-576, 125–149 (2005).
23. F. Murtagh and R. H.Warmels, "Test image descriptions," in *Proc. 1st ESO/ST-ECF Data Analysis Workshop*, 17(6), 8–19 (1989).
24. Portell, J., Villafranca, A. G., and García-Berro, E., "Quick outlier-resilient entropy coder for space missions," *Journal of Applied Remote Sensing* 4 (2010).
25. A. G. Villafranca, I. Mora, P. Ruiz-Rodríguez, J. Portell, and E. García-Berro, "Optimizing GPS data transmission using entropy coding compression", *Proc. SPIE 7810*, 2010.

Chapter 6
Quality Issues for Compression of Hyperspectral Imagery Through Spectrally Adaptive DPCM

Bruno Aiazzi, Luciano Alparone, and Stefano Baronti

Abstract To meet quality issues of hyperspectral imaging, differential pulse code modulation (DPCM) is usually employed for either lossless or near-lossless data compression, i.e., the decompressed data have a user-defined maximum absolute error, being zero in the lossless case. Lossless compression thoroughly preserves the information of the data but allows a moderate decrement in transmission bit rate. Lossless compression ratios attained even by the most advanced schemes are not very high and usually lower than four. If strictly lossless techniques are not employed, a certain amount of information of the data will be lost. However, such an information may be partly due to random fluctuations of the instrumental noise. The rationale that compression-induced distortion is more tolerable, i.e., less harmful, in those bands, in which the noise is higher, and vice-versa, constitutes the *virtually lossless* paradigm.

1 Introduction

Technological advances in imaging spectrometry have lead to acquisition of data that exhibit extremely high spatial, spectral, and radiometric resolution. To meet the quality issues of hyperspectral imaging, differential pulse code modulation

B. Aiazzi (✉)
IFAC-CNR, Via Madonna del Piano 10, 50019 Sesto F.no, Italy
e-mail: b.aiazzi@ifac.cnr.it

L. Alparone
Department of Electronics & Telecommunications, University of Florence,
Via Santa Marta 3, 50139 Florence, Italy
e-mail: alparone@lci.det.unifi.it

S. Baronti
IFAC-CNR, Via Madonna del Piano 10, 50019 Sesto F.no, Italy
e-mail: s.baronti@ifac.cnr.it

B. Huang (ed.), *Satellite Data Compression*, DOI 10.1007/978-1-4614-1183-3_6,
© Springer Science+Business Media, LLC 2011

(DPCM) is usually employed for either lossless or near-lossless data compression, i.e., the decompressed data have a user-defined maximum absolute error, being zero in the lossless case. Several variants exist in prediction schemes, the most performing being adaptive [2, 6, 42].

When the hyperspectral imaging instrument is onboard a satellite platform, data compression is crucial. Lossless compression thoroughly preserves the information of the data but allows a moderate decrement in transmission bit rate to be achieved. Compression ratios attained even by the most advanced schemes are lower than four, with respect to plain PCM coding of raw data [10, 11, 27]. Thus, the bottleneck of downlink to ground stations may severely hamper the wide coverage capabilities of modern satellite instruments. If strictly lossless techniques are not employed, a certain amount of information of the data will be lost. However, such a statistical information may be partly due to random fluctuations of the instrumental noise. The rationale that compression-induced distortion is more tolerable, i.e., less harmful, in those bands, in which the noise is higher, and vice-versa, constitutes the *virtually lossless* paradigm [29].

In the literature, there exist several distortion measurements, some of which are suitable for quality assessment of decompressed hyperspectral data. Mean square error (MSE), maximum absolute deviation (MAD), i.e., peak error, average and maximum spectral angle, are usually adopted to measure the distortion of lossy compressed hyperspectral data. The problem is that they measure the distortion introduced in the data, but cannot measure the consequences of such a distortion, i.e., how the information loss would affect the outcome of an analysis performed on the data.

In this perspective, discrimination of materials is one of the most challenging task, in which hyperspectral data reveal their full potentiality. In fact, if remote sensing imagery is analyzed with the goal of recognizing broad classes of land cover, like vegetation, bare soil, urban, ice, etc., also data acquired by multispectral instruments are effective. Instead, if more specific tasks are concerned, such as minerals identification or geological inspections, especially on coastal waters, in order to identify the presence of chlorophyll, phytoplankton or dissolved organic materials, the high spectral resolution captured by hyperspectral instruments is beneficial.

2   Lossless/Near-Lossless Image Compression Algorithms

Considerable research efforts have been recently spent in the development of lossless image compression techniques. The first specific standard has been the lossless version of JPEG [36, 39], which may use either Huffman or arithmetic coding. A more interesting standard, which provides also near-lossless compression is JPEG-LS [49]. It is based on an adaptive nonlinear prediction and exploits context modeling followed by Golomb-Rice entropy coding. A similar context-based algorithm named CALIC has also been successively proposed [52].

The simple adaptive predictors used by JPEG-LS and CALIC, however, the *median adaptive predictor* (MAP) and the *gradient adjusted predictor* (GAP), respectively, are both empirical. Thorough comparisons with more advanced methods [8] have revealed that their performance is actually limited and still far from the entropy bounds. It is noteworthy that, unlike a locally MMSE linear prediction, a nonlinear prediction, like GAP of CALIC and MAP of JPEG-LS, that may occur to minimize the *mean absolute error* (MAE), does not ensure local entropy minimization [31]. Therefore only linear prediction, yet adaptive, will be concerned for a 3D extension suitable for multi/hyperspectral data.

A number of integer-to-integer transforms, e.g., [1, 5, 40, 44], are capable of ensuring a perfect reconstruction with integer arithmetics. Their extension to multiband data is straightforward, if a spectral decorrelation is preliminarily performed [17]. However, the drawback of all critically-subsampled multiresolution transforms, is that they are suitable for L_2-constrained compression only. Thanks to Parceval's theorem, if the transformation is orthogonal, the MSE, or its square root (RMSE), namely the L_2 distortion between original and decoded data, is controlled by the user, up to possibly yield lossless compression, by resorting to the aforementioned integer-to-integer transforms. The problem is that L_∞-constrained, i.e., near-lossless, compression is not trivial and, whenever feasible [14] is not rewarding in terms of L_∞-bitrate plots with respect to DPCM. Indeed, DPCM schemes, either *causal* (prediction-based) or *noncausal*, i.e., interpolation-based or *hierarchical* [12], are suitable for L_∞-constrained compression, that is either lossless or near-lossless. The latter is recommended for lower quality compression (i.e., higher CR), the former for higher-quality, which is the primary concern in remote sensing applications.

Eventually, it is worth mentioning that Part I of the JPEG2000 image coding standard [46] incorporates a lossless mode, based on reversible integer wavelets, and is capable of providing a scalable bit stream, that can be decoded from the lossy (not near-lossless) up to the lossless level. However, image coding standards are not suitable for the compression of 3D data sets: in spite of their complexity, they are not capable of exploiting the 3D signal redundancy featured, e.g., by multi/hyperspectral imagery.

2.1   Prediction-Based DPCM

Prediction-based DPCM basically consists of a decorrelation followed by entropy coding of the outcome prediction errors according to the scheme of Fig. 6.1 that outlines the flowcharts of the encoder, featuring context modeling for entropy coding, and of the decoder. The quantization noise feedback loop at the encoder allows the L_∞ error to be constrained, by letting prediction at the encoder be carried out from the same distorted samples that will be available at the decoder.

The simplest way to design a predictor, once a *causal* neighborhood is set, is to take a linear combination of pixel values within such a neighborhood, in particular

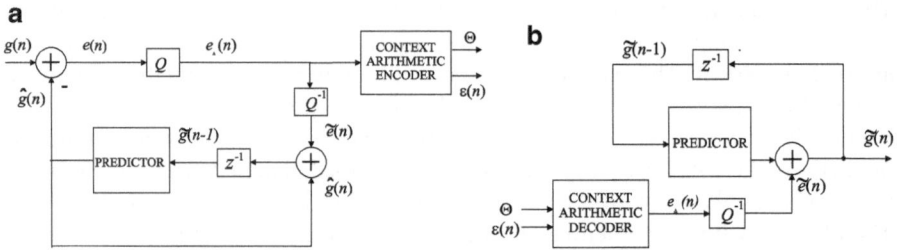

Fig. 6.1 Flowchart of DPCM with quantization noise feedback loop at the encoder, suitable for near-lossless compression: (**a**) encoder; (**b**) decoder

with coefficients optimized in order to yield *minimum mean square error* (MMSE) over the whole image. Such a prediction, however, is optimum only for stationary signals. To overcome this drawback, *adaptive* DPCM (ADPCM) [39] has been proposed, in which the coefficients of predictors are continuously recalculated from the incoming new data at each pixel location. A more significant alternative is *classified* DPCM [22], in which a number of statistical classes are preliminary recognized and an optimized predictor (in the MMSE sense) is calculated for each class and utilized to encode the pixels of that class [9]. Alternatively, predictors may be adaptively combined [21], also based on a fuzzy-logic concept [8], to attain an MMSE space-varying prediction. These two strategies of classified prediction will be referred to as *adaptive selection/combination of adaptive predictors* (ASAP/ACAP). In the ACAP case, the linearity of prediction makes it possible to formulate the problem as an approximation of the optimum space-varying linear predictor at each pixel through its projection onto a set of nonorthogonal prototype predictors capable of embodying the statistical properties of the data. Two algorithms featuring the ASAP/ACAP strategies, respectively, have been developed in [9, 8] and named *Relaxation Labeled Prediction* (RLP) and *Fuzzy Matching Pursuit* (FMP), respectively.

To enhance the entropy coding performance, both RLP and FMP schemes may use context modeling (see Sect. 2.2) of prediction errors followed by arithmetic coding. It is noteworthy that the original 2D FMP [8] achieves lossless compression ratios 5% better than CALIC and 10% than JPEG-LS, on an average. Although 2D RLP encoder [9] is slightly less performing than 2D FMP, its feature of real-time decoding is highly valuable in application contexts, since an image is usually encoded only once, but decoded many times.

2.2   Context Modeling for Entropy Coding

A notable feature of all advanced image compression methods [19] is the statistical context modeling for entropy coding. The underlying rationale is that prediction errors should be similar to stationary white noise as much as possible. As a matter of fact, they are still spatially correlated to a small extent and especially are non-stationary, which means that they exhibit space-varying variance. The better the prediction, however, the more noise-like prediction errors will be.

Following a trend established in the literature, first in the medical field [37], then for lossless coding in general [44, 49, 52], and recently for *near-lossless* coding [7, 51], prediction errors are entropy coded by means of a classified implementation of an entropy coder, generally arithmetic [50] or Golomb-Rice [41]. For this purpose, they are arranged into a predefined number of statistically homogeneous classes based on their spatial *context*. If such classes are statistically discriminated, then the entropy of a *context-conditioned* model of prediction errors will be lower than that derived from a stationary memoryless model of the decorrelated source [48].

3   Hyperspectral Data Compression Through DPCM

Whenever multiband images are to be compressed, advantage may be taken from the spectral correlation of the data for designing a prediction that can be both *spatial* and *spectral*, from a causal neighborhood of pixels [43, 47]. Causal means that only previously scanned pixels on the current and previously encoded bands must be used for predicting the current pixel value. This strategy is more effective when the data are more spectrally correlated, as in the case of hyperspectral data. If the *interband* correlation of the data is weak, as it usually occurs for data with few and sparse spectral bands, a 3D prediction may lead to negligible coding benefits, unless the available bands are reordered in such a way that the average correlation between two consecutive bands is maximized [45]. In this case, however, advantage may be taken from a *bidirectional* spectral prediction [38], in which once the $(k - 1)$st band is available, the kth band is skipped and the $(k + 1)$st band is predicted from the $(k - 1)$st one; then, both these two bands are used to predict the kth band in a spatially causal but spectrally noncausal fashion. In practice, the bidirectional prediction is achieved by applying a causal prediction to a permutation of the sequence of bands. This strategy, however, is not rewarding when hyperspectral data have to be compressed [6].

When hyperspectral data are concerned, the non-stationarity characteristics of the data in both spatial and spectral domains, together with computational constraints, make the jointly spatial and spectral prediction to take negligible extra advantage from a number of previous bands greater than two. In this case, a different and more effective approach is to consider a purely spectral prediction. A first attempt has been introduced by [33, 34], and provides compression ratios among the very best in the literature. This method classifies the original hyperspectral pixel vectors into spatially homogeneous classes, whose map must be transmitted as side information. Then a purely spectral prediction is carried out on pixel spectra belonging to each class, by means of a large set of linear spectral predictors of length up to 20, i.e., spanning up to 20 previous bands. However, such drawbacks exist as the computational effort for pre-classifying the data, as well as a crucial adjustment of such parameters as number of classes and length of predictors (one for each wavelength of each class), which determine a large coding overhead. Finally, since the cost of overhead (classification map and set of spectral predictors) is independent of the target compression ratio, the method seems to be not recommendable for near-lossless compression, even if it might be achieved in principle.

A different situation involves the FMP and RLP algorithms, which can be easily adapted to spatial/spectral or purely spectral modalities, by simply changing the prediction causal neighborhood. In this way, their features of near-lossless and virtually-lossless working modalities are perfectly preserved. In the spatial/spectral case the FMP encoder has been extended by the authors to 3D data [13], same as the RLP encoder [6], by simply changing the 2D neighborhood into a 3D one spanning up to three previous bands. The obtained algorithms will be referred as 3D-FMP and 3D-RLP, respectively. In a similar way, a purely spectral prediction can be achieved by considering a 1D spectral causal neighborhood spanning up to 20 previous bands, as in the case of [33, 34]. The two algorithms will be referred as S-FMP and S-RLP [11]. The hyperspectral versions of FMP and RLP will be reviewed in the following Sect. 4. Eventually, we wish to remind that a forerunner of the ACAP paradigm has been proposed in the fuzzy 3D DPCM method [2], in which the prototype MMSE spatial/spectral linear predictors constituting the dictionary were calculated on *clustered* data, analogously to [33].

Another interesting approach specific to hyperspectral images is the extension of 3D CALIC [53], originally conceived for color images, having few spectral bands, to image data having a larger number of highly correlated bands. The novel algorithm, referred to as M-CALIC [30] significantly outperforms 3D CALIC, to which it largely sticks, with a moderately increased computational complexity and absence of setup parameters crucial for performances.

Eventually, it is important to remind that in satellite on-board compression computing power is limited and coding benefits must be traded off with computational complexity [30, 42]. In this framework, fast and simple DPCM methods have been recently proposed by researchers involved in sensor development [27, 28].

4   Crisp/Fuzzy Classified 3D/Spectral Prediction

Figure 6.2 depicts the classified DPCM encoders (ACAP/ASAP paradigms) in the case of hyperspectral compression. The initialization phase is the same for the crisp and fuzzy-based methods, while the differentiation between the two strategies is apparent in the refinement and prediction phases. The switch from the spatial/spectral methods (3D-FMP and 3D-RLP) to the purely spectral ones (S-FMP and S-RLP) is simply obtained by considering the spectral neighborhood instead of the 3D neighborhood for calculating the predicted values. Being S-FMP and S-FMP the most performing methods, they are described in this section, while a complete description of 3D-RLP and 3D-FMP can be found in [6, 13], respectively.

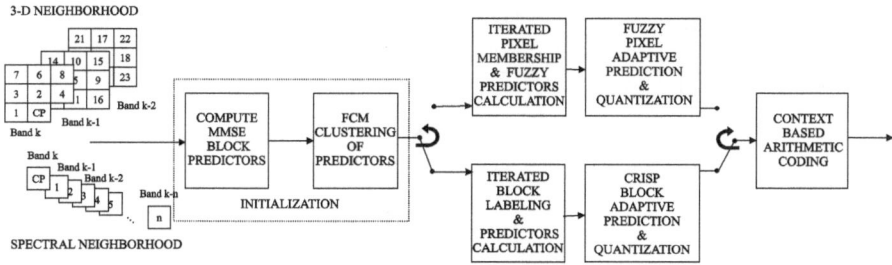

Fig. 6.2 Flowchart of the classified DPCM encoders. Prediction is accomplished either on a 3D neighborhood (jointly spatial and spectral) or on an 1D neighborhood (purely spectral), whose pixels are labeled by increasing Euclidean distance from the current pixel (*CP*). The encoder is switchable between fuzzy prediction (upper branch) and crisp prediction (lower branch)

4.1 S-FMP and S-RLP Basic Definitions

Let $g(x, y, z)$ $(0 \leq g(x, y, z) < g_{fs})$ denote the intensity level of the sequence at row x $(1 \leq x \leq N_x)$, column y $(1 \leq y \leq N_y)$, and band z $(1 \leq z \leq N_z)$. For a fixed wavelength z, the discrete grey scale image $\{g(x, y)\}$ may be scanned left to right and top to bottom by successive lines, so as to yield a 1D set of coordinates $\{n\}$. For a fixed pixel in the n position, a purely spectral causal neighborhood can be defined by taking the coordinates of the pixels in the same position n from a fixed number of spectral bands previously transmitted, namely W, so as to yield a vector of length W. Let us denote this vector as $\mathcal{N}^W(n)$.

Prediction is obtained as a linear combination of pixels whose coordinates lay in $\mathcal{N}^W(n)$. Let us define as prediction support of size S of the current pixel n and denote it as $\mathcal{P}_s(n)$ the subset of $\mathcal{N}^W(n)$ that includes these coordinates. Let $\Psi(n)$ denote the vector containing the grey levels of the S samples laying within $\mathcal{P}_S(n)$ sorted by increasing spectral distance from the pixel n. Let also $\phi = \{\phi_k \in \mathbb{R}, \ k = 1, \cdots, S\}^T$ denote the vector comprising the S coefficients of a linear predictor operating on $\mathcal{P}_S$. Thus, a linear prediction for $g(n)$ is defined as $\hat{g}(n) = \sum_{k=1}^{S} \phi_k \cdot \psi_k(n)$, in which $< \cdot, \cdot >$ indicates scalar (inner) product.

4.2 Initialization

The determination of the initial dictionary of predictors is the key to the success of the coding process. It starts from observing that patterns of pixel values, occurring within $\mathcal{P}_S(n)$, $n = 1, \cdots, N - 1$ $(\mathcal{P}_S(0) = \oslash)$, reflect local spectral properties and that such patterns are spatially correlated due to the scene features, e.g., homogeneous areas, edges and textures. A proper prediction should be capable of reflecting such features as much as possible. An effective algorithm consists in preliminarily

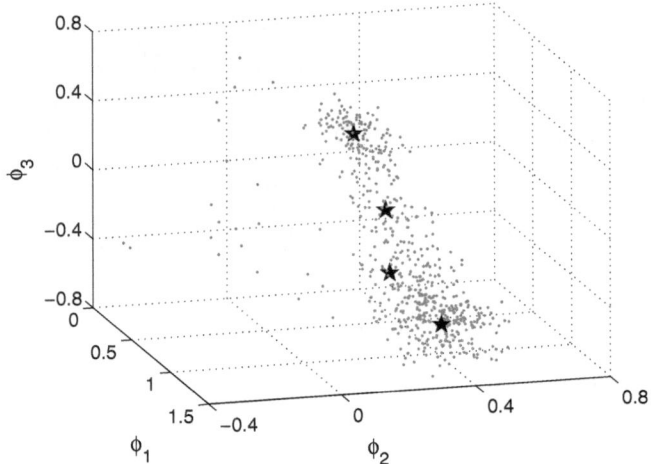

Fig. 6.3 Example of block predictors of length three (*dots*) and four clusters (*stars*) obtained through the fuzzy-C-means algorithm

partitioning the input image into square blocks, e.g., 16×16, and calculating the S coefficients of an MMSE linear predictor for each block by means of a least squares (LS) algorithm. Specifically, if B denotes one block of the partition, the LS algorithm is fed by the pairs $\{(\Psi(n), g(n)) \mid n \in B\}$ to yield the related predictor ϕ_B.

The above process produces a large number of predictors, each optimized for a single block. The S coefficients of each predictor are arranged into an S-dimensional space. It can be noticed that statistically similar blocks exhibit similar predictors. Thus, the predictors found previously tend to cluster when represented on a hyperplane, instead of being uniformly spread.

Figure 6.3 shows how predictors calculated from a band of a true hyperspectral image tend to cluster in the hyperspace, in the case of $S = 3$. A reduced set of M representative predictors, four in this case and plotted as stars, with M defined by the user or empirically adjusted, is thus obtained by applying a clustering algorithm to the optimal block predictors. Such "dominant" predictors are calculated as *centroids* of as many clusters in the predictors' space, according to a vector Euclidean metrics. Although a variety of fuzzy clustering algorithms exists [16], the widespread Bezdek's *Fuzzy C Means* (FCM) algorithm [18] was used because it yields centroids that speed-up convergence of the successive refinement and training procedure, as experimentally noticed. Thus, an $S \times M$ matrix $\Phi^{(0)} = \{\phi_m^{(0)}, m = 1, \cdots, M\}$ containing the coefficients of the M predictors is produced. The superscript $^{(0)}$ highlights that such predictors are start-up values of the iterative refinement procedure which will yield the final "dictionary" of predictors.

4.3 Iterative Refinement of Predictors

The initial predictors obtained by the FCM procedure are used as the input for the iterative refinement procedure. This procedure will be crisp for the S-RLP algorithm, or fuzzy for the S-FMP method. In the former case, each image block is assigned to one of the M classes by considering how each predictor performs on the pixels of the block. In the latter case, instead, the prediction is fuzzy, by considering a cumulative predictor given to the sum of all the predictors, each of them weighted by its membership to the pixel to be predicted.

4.3.1 S-FMP Membership Function and Training of Predictors

For the S-FMP method the M predictors found out through fuzzy clustering are used to initialize a training procedure in which firstly pixels are given degrees of membership to predictors, then each predictor is recalculated based only on pixels whose membership to it exceeds a threshold μ pre-defined by the user.

The choice of the fuzzy-membership function is crucial for optimizing performances. It must be calculated from a *causal* subset of pixels, not necessarily identical to the prediction support. A suitable fuzzy membership function of the nth pixel to the mth predictor was devised as the reciprocal of the weighted squared prediction error, produced by the mth predictor on a causal neighborhood of pixel n, raised to an empirical exponent γ, and incremented by one to avoid divisions by zero.

The causal neighborhood adopted is a 2D spatial square and, thus, uniquely defined by its side R as $\mathcal{N}_\infty^R(n)$, being R user defined. $\mathcal{M}_R \triangleq \mathcal{N}_\infty^R$ will be referred to as *membership support* in the following.

The *weighted* squared prediction error produced by the mth predictor on the nth pixel is defined as

$$\bar{d}_m^2(n) = \frac{\sum\limits_{k \in \mathcal{M}_R(n)} \delta_k^{-1} \cdot [g(k) - <\phi_m, \psi(k)>]^2}{\sum\limits_{k \in \mathcal{M}_R} \delta_k^{-1}}. \tag{6.1}$$

The weight of each squared prediction error is taken to be inversely proportional to the distance δ_k from the current pixel n. Thus, closer neighbors will contribute more than farther ones. The weighted squared error (6.1) is normalized to the sum of its weights. Thus, its magnitude is roughly independent of the neighborhood size.

The membership of the nth pixel to the mth predictor will be

$$U_m(n) = \frac{1}{1 + [\bar{d}_m^2(n)]^\gamma}. \tag{6.2}$$

As a matter of fact, (6.2) measures the capability of ϕ_m to predict the grey levels of the closest *causal neighbors* of the current pixel n. By a fuzzy inference, it also reflects the ability of ϕ_m to predict the value $g(n)$ itself. If the outputs of the mth predictor exactly fit the grey levels within the membership support of pixel n, then $\bar{d}_m^2(n)$ will be zero and, hence, $U_m(n) = 1$. The membership exponent γ rules the degree of fuzziness of the membership function; it was adjusted empirically.

Since the fuzzy membership will be used to measure a *projection pursuit*, same as a scalar product, the *absolute* membership given by (6.2) is normalized to yield a *relative* membership

$$\tilde{U}_m(n) = \frac{U_m(n)}{\sum_{m=1}^{M} U_m(n)} \tag{6.3}$$

suitable for a *probabilistic* clustering.

With reference to the flowchart of Fig. 6.2, the iterative procedure is outlined in the following steps:

- **Step 0**: for each pixel n, $n = 1, \cdots, N - 1$, calculate the initial membership array, $\tilde{U}_m^{(0)}(n)$, $m = 1, \cdots, M$, from the initial set of predictors $\Phi^{(0)} = \{\phi_m^{(0)}, m = 1, \cdots, M\}$ by using (6.1)–(6.3); set the iteration step $h = 0$ and a membership threshold μ.
- **Step 1**: recalculate predictors $\{\phi_m^{(h+1)}, m = 1, \cdots, M\}$ from those pixels whose membership $\tilde{U}_m^{(h)}(n)$ exceeds μ; weight by $\tilde{U}_m^{(h)}(n)$ the contribution of the pair $(\Psi(n), g(n))$ to $\phi_m^{(h+1)}$ in the LS algorithm.
- **Step 2**: recalculate memberships to the new set of predictors, $\tilde{U}_m^{(h+1)}(n)$, $m = 1, \cdots, M$, $n = 1, \cdots, N - 1$.
- **Step 3**: check convergence; if realized, stop; otherwise, increment h by one and go to **Step 1**.

Convergence can be checked by thresholding the decrement in cumulative *mean square prediction error* (MSPE) associated to the current iteration. Another iteration is executed if such an amount exceeds a preset threshold. Such an *open loop* check is ruled by thresholds that can be calculated once through a *closed loop* procedure, in which the coder of Fig. 6.2 is enabled to produce code bits at each iteration.

Notice that the standard LS algorithm has been modified to account for the memberships of pixels to predictors at the previous iteration. Pixels having larger degrees of memberships to one predictor will contribute to the determination of that predictor more than pixels having smaller degrees. Furthermore, depending on the threshold μ, a pixel may contribute, though with different extents, to more predictors, in the fuzzy-logic spirit. The membership threshold μ is non-crucial for coding performances.

Eventually, an $S \times M$ matrix $\Phi = \{\phi_m, m = 1, \cdots, M\}$, containing the coefficients of the M predictors after the last iteration stage is produced and stored in the file header.

4.3.2 S-RLP Relaxation Block Labeling and Predictors Refinement

The initial guess of classified predictors is delivered to an iterative labelling procedure which classifies image blocks, simultaneously refining the associated predictors. All the predictors are transmitted along with the label of each block.

The image blocks are assigned to M classes, and an optimized predictor is obtained for each class, according to the following procedure.

Step 0: classify blocks based on their MSPE. The label of the predictor minimizing MSPE for a block is assigned to the block itself. This operation has the effect of partitioning the set of blocks into M classes that are best matched by the predictors previously found out.

Step 1: recalculate each of the M predictors from the data belonging to the blocks of each class. The new set of predictors is thus designed so as to minimize MSPE for the current block partition into M classes.

Step 2: reclassify blocks: the label of the new predictor minimizing MSPE for a block is assigned to the block itself. This operation has the effect of moving some blocks from one class to another, thus repartitioning the set of blocks into M new classes that are best matched by the current predictors.

Step 3: check convergence; if found, stop; otherwise, go to **Step 1**.

Convergence is checked in the same way as for S-FMP.

Conversely from S-FMP, the prediction is now crisp by considering only the predictor representative to the class to which the pixel belongs. The final sets of refined predictors, one per wavelength, are transmitted as side information together with the set of block labels.

4.3.3 Prediction

Although the concept of *fuzzy* prediction, as opposed to a classified or *crisp* prediction is not novel, the use of *linear* predictions makes it possible, besides an LS adjustment of predictors, to formulate the fuzzy prediction as a problem of *matching pursuit* (MP).

In fact, by the linearity of prediction, a *weighted sum* of the outputs of predictors is equal to the output of a linear combination of the same predictors with the same weights, that is to calculate an adaptive predictor at every pixel:

$$\phi(n) \overset{\Delta}{=} \sum_{m=1}^{M} \tilde{U}_m(n) \cdot \phi_m \qquad (6.4)$$

in which the weights are still provided by $\tilde{U}_m(n)$, i.e., (6.3), with (6.1) calculated from the predictors $\{\phi_m, \ m = 1, \cdots, M\}$ after the last iteration stage. The predictor (6.4) will yield the adaptive linear prediction as $\hat{g}(n) = < \phi(n), \ \Psi(n) >$.

Equivalently, each pixel value $g(n)$ can be predicted as a *fuzzy switching*, i.e., a *blending*, of the outputs of all the predictors, which are defined as

$$\hat{g}_m(n) = <\phi_m, \Psi(n)> \tag{6.5}$$

with the *fuzzy* prediction, $\hat{g}(n)$, given by

$$\hat{g}(n) = \text{round}\left[\sum_{m=1}^{M} \tilde{U}_m(n) \cdot \hat{g}_m(n)\right] \tag{6.6}$$

The right term of (6.6) is rounded to integer to yield integer valued prediction errors, i.e., $e(n) = g(n) - \hat{g}(n)$, that are sent to the entropy coding section.

Concerning S-RLP, once blocks have been classified and labeled, together with the attached optimized predictor, each band is raster scanned and predictors are *activated* based on the classes of crossed blocks. Thus, each pixel belonging to one block of the original partition, $g(x, y)$, is predicted as a $\hat{g}(x, y)$ by using the one out of the M predictors that was found to better fit the statistics of that class of data block in the MMSE sense. The integer valued prediction errors, viz., $e(x, y) = g(x, y) - \text{round}[\hat{g}(x, y)]$, are delivered to the *context-coding* section, identical to that of FMP.

4.4 Context Based Arithmetic Coding

As evidenced in Fig. 6.2, a more efficient coding of the prediction errors can be obtained if a context classification of the residuals is put ahead the arithmetic coding block. A context function may be defined and measured on prediction errors laying within a causal neighborhood, possibly larger than the prediction support, as the RMS value of prediction errors (RMSPE). The context function should capture the non-stationary of prediction errors, regardless of their spatial correlation. Again, causality of neighborhood is necessary in order to make the same information available both at the encoder and at the decoder. At the former, the probability density function (PDF) of RMSPE is calculated and partitioned into a number of intervals chosen as equally populated; thus, contexts are equiprobable as well. This choice is motivated by the use of adaptive arithmetic coding for encoding the errors belonging to each class. Adaptive entropy coding, in general, does not require previous knowledge of the statistics of the source, but benefits from a number of data large enough for training, which happens simultaneously with coding. The source given by each class is further split into sign bit and magnitude. The former is strictly random and is coded as it stands, the latter exhibits a reduced variance in each class; thus, it may be coded with fewer bits than the original residue. It is noteworthy that such a context-coding procedure is independent of the particular method used to decorrelate the data. Unlike other schemes, e.g., CALIC [52], in which context-modeling is embedded in the decorrelation procedure, the method [7] can be applied to any DPCM scheme, either lossless or near-lossless, without adjustments in the near-lossless case, as it happens to other methods [51].

5 Lossless Hyperspectral Image Compression Based on LUTs

In a recent published paper [32], Mielikainen introduced a very simple prediction method, referred as LUT-NN, which predicts the current pixel by taking the value of its *nearest neighbor* (NN), defined as the pixel previously transmitted in the current band, having the following two properties: (1) the pixel in the previous band, at the same position of the NN pixel, has the same value of the pixel in the previous band at the same position of the current pixel, and (2) among all the pixels fulfilling the previous property, the NN pixel is the spatially closest to the current pixel along the scan path. Such an algorithm can be effectively implemented by means of a dynamically updated lookup table (LUT). The prediction value taken from the LUT is that of the cell indexed by the value of the pixel at the current pixel position in the previous band and corresponds to the value of the NN pixel, which has been inserted in this cell in a previous updating pass [32]. Surprisingly, such a simple method, notwithstanding only one previous band is considered, performs comparably to the most advanced DPCM methods on the 1997 AVIRIS data set (16-bit radiance format). However, this improvement occurs only if some particular calibrated sequences are considered. Actually, in the case of the 1997 AVIRIS data set, the calibration procedures produces for each band radiance histograms that are very unbalanced, i.e., some values are much more frequent than others. So, the LUT-NN method forces its predictions to yield the most frequent values in the band to be compressed, unlike conventional prediction strategies usually do.

In any case, this artificial efficiency has suggested the investigation of more sophisticated versions of LUT-NN. A first attempt was the method proposed by Huang et al., which was named LAIS-LUT [23]. This algorithm utilizes two LUTs, respectively containing the values of the two closest NNs to the current pixel. The prediction of the current pixel is obtained by choosing the value that is more similar to a reference prediction, which takes into account the interband gain between the current and the previous bands, as it is indicated by its acronym *Locally Average Interband Scaling* (LAIS). The LAIS-LUT algorithm yields significant coding benefits over LUT-NN at a moderate extra cost, thanks to its better accounting of the spatial correlation, even if the exploitation of the spectral correlation is still limited to the previous band only.

The idea underlying LUT-NN and LAIS-LUT can be further generalized by considering a more advanced reference prediction and by exploiting also the spectral correlation of the sequences. The adoption of the S-RLP and S-FMP methods as reference predictors has brought to the two generalizations proposed in [10] and denoted as S-RLP-LUT and S-FMP-LUT.

These two algorithms feature a complete exploitation of spatial and spectral correlations, because the prediction value is obtained by considering more than two LUTs for bands, say M, where M is usually chosen equal to 4, and an arbitrary number of previous bands, say N, where N may reach 20. The decision among the $M \cdot N$ possible prediction values is based on the closeness of the candidate predictions to the S-FMP or S-RLP reference predictions, which can span the same N previous bands. The advantage of considering a large number of LUTs is strictly connected to the utilization of S-FMP and S-RLP as reference predictors. In fact, by adopting the simpler LAIS predictor as reference, the advantages of more than one band for LUT-based prediction and of more than two LUTs for each band quickly vanish as M and N increase.

Figure 6.4 shows how S-FMP-LUT and S-RLP-LUT work, i.e., how an algorithm based on multiple LUTs can be connected with the S-RLP and S-FMP advanced DPCM predictors. By using the same notation of Sect. 4.3.1, let $g(x, y, z)$ ($0 \leq g$ $(x, y, z) < g_{fs}$) denote the intensity level of the sequence at row x ($1 \leq x \leq N_x$), column y ($1 \leq y \leq N_y$), and band z ($1 \leq z \leq N_z$). Let also $L_{m,n,z}[\cdot]$, $m = 1, \cdots M$, $n = 1, \cdots N$ indicate the set of $N \cdot M$ LUTs used for prediction of band z. All LUTs are of length g_{fs}. Eventually, let $\hat{g}_r(x, y, z)$ be the reference prediction for pixel (x, y) in band z, obtained by means of S-RLP or S-FMP. The multiband LUT prediction is given by

$$\hat{g}(x, y, z) = \hat{g}_{\hat{m}, \hat{n}}(x, y, z) \tag{6.7}$$

in which

$$\{\hat{m}, \hat{n}\} = \arg \min_{\substack{m=1,\cdots M \\ n=1,\cdots N}} \left\{ \left| \hat{g}_{m,n}(x, y, z) - \hat{g}_r(x, y, z) \right| \right\} \tag{6.8}$$

and

$$\hat{g}_{m,n}(x, y, z) = L_{m,n,z}[g(x, y, z - n)],$$
$$m = 1, \cdots M, n = 1, \cdots N \tag{6.9}$$

are the $N \cdot M$ possible prediction values among which the final prediction $\hat{g}(x, y, z)$ is chosen. The values of $\hat{g}_{m,n}(x, y, z)$ belong to the current band and have been stored in the set of multiband LUTs during the previous updating steps.

After the final prediction value has been produced according to (6.7), the set of multiple LUTs is updated, analogously to [23, 32] in the following way:

$$L_{m,n,z}[g(x, y, z - n)] = L_{m-1,n,z}[g(x, y, z - n)],$$
$$m = M, M - 1, \cdots 2; \quad n = 1, \cdots N$$
$$L_{1,n,z}[g(x, y, z - n)] = g(x, y, z),$$
$$n = 1, \cdots N. \tag{6.10}$$

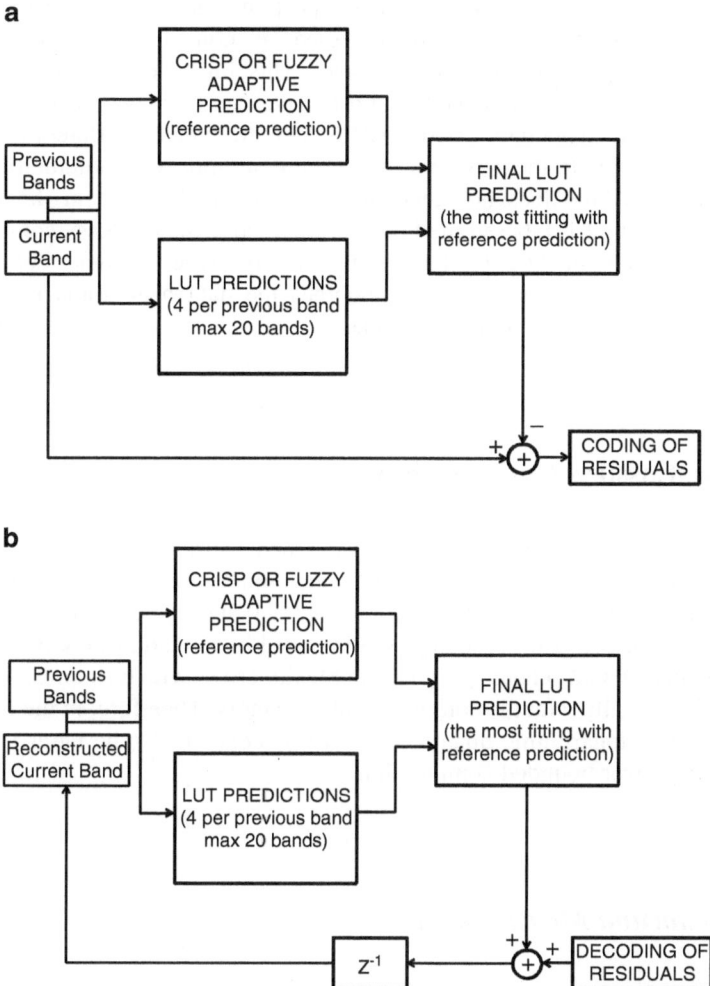

Fig. 6.4 General scheme for the adaptation of the basic LUT method to S-RLP and S-FMP algorithms (**a**) encoder; (**b**) decoder. The crisp or fuzzy adaptive DPCM algorithms work as advanced reference predictors for an optimal selection among the multiple LUT-based prediction values

All LUTs are initialized with the g_{fs} value that is outside the range of the data. Whenever such a value is returned by (6.7), i.e., no pixel exists fulfilling the two NN properties, the reference prediction $\hat{g}_r(x, y, z)$ is used instead of $\hat{g}_{\hat{m},\hat{n}}(x, y, z)$. At the decoder, the set of multiband LUTs and the reference prediction are calculated from the previously decoded lossless data, by following the same procedure as at the encoder.

The performances of S-FMP-LUT and S-RLP-LUT on the 1997 AVIRIS data set are impressive and shows an average improvement with respect to LAIS-LUT of more than 20%. However, in the same paper [10], the authors note that LUT-based methods lose their effectiveness on unprocessed, i.e., non-calibrated, images, which do not contain calibration-induced artifacts. This limitation of the LUT-based algorithms has been strengthened in [27], where it is shown that LUT-based methods are ineffective on the 2006 AVIRIS data set. In fact, a different calibration procedure is applied to the new data, that exhibit an improved radiometric resolution. Such procedure does not generate recurrent pixel values in the radiance histogram. Finally, LUT-based methods are ineffective also in lossy compression, and in particular in near-lossless and virtually-lossless cases, because the quantization of the predictor errors smoothes the data histograms from which these algorithms take the prediction values.

6   Near-Lossless Compression

So far, quantization in the FMP and RLP schemes was not addressed; that is, lossless compression was described. Quantization is necessarily introduced to allow a reduction in the code rate to be achieved [24] at the price of some distortions in the decompressed data. Although trellis coded quantization may be optimally coupled with a DPCM scheme [25], its complexity grows with the number of output levels and especially with the complexity of predictors. Therefore, in the following only linear and logarithmic quantization will be concerned. The latter is used to yield relative error bounded compression.

6.1   Distortion Measurements

Before discussing quantization in DPCM schemes, let us review the most widely used distortion measurements suitable for single-band image data (2D) and multiband image data (3D), either multispectral or hyperspectral.

6.1.1   Radiometric Distortion

Let $0 \leq g(x, y) \leq g_{fs}$ denote an N-pixel digital image and let $\tilde{g}(x, y)$ be its possibly distorted version achieved by compressing $g(x, y)$ and decompressing the outcome bit stream. Widely used distortion measurements are reported in the following.

Mean absolute error (MAE), or L_1 norm,

$$\text{MAE} = \frac{1}{N} \sum_x \sum_y |g(x,y) - \tilde{g}(x,y)|; \tag{6.11}$$

Mean Squared Error (MSE), or $L_2{}^2$,

$$\text{MSE} = \frac{1}{N} \sum_x \sum_y [g(x,y) - \tilde{g}(x,y)]^2;$$ (6.12)

Root MSE (RMSE), or L_2,

$$\text{RMSE} = \sqrt{\text{MSE}};$$ (6.13)

Signal to Noise Ratio (SNR)

$$\text{SNR}_{(\text{dB})} = 10 \cdot \log_{10} \frac{\overline{g^2}}{\text{MSE} + \frac{1}{12}};$$ (6.14)

Peak SNR (PSNR)

$$\text{PSNR}_{(\text{dB})} = 10 \cdot \log_{10} \frac{g_{fs}^2}{\text{MSE} + \frac{1}{12}};$$ (6.15)

Maximum absolute distortion (MAD), or *peak error*, or L_∞,

$$\text{MAD} = \max_{x,y}\{|g(x,y) - \tilde{g}(x,y)|\};$$ (6.16)

Percentage maximum absolute distortion (PMAD)

$$\text{PMAD} = \max_{x,y}\left\{\frac{|g(x,y) - \tilde{g}(x,y)|}{g(x,y)}\right\} \times 100.$$ (6.17)

In both (6.14) and (6.15) the MSE is incremented by the variance of the integer roundoff error, to handle the limit lossless case, when MSE=0. Thus, SNR and PSNR will be upper bounded by $10 \cdot \log_{10}(12 \cdot \overline{g^2})$ and $10 \cdot \log_{10}(12 \cdot g_{fs}^2)$, respectively.

When multiband data are concerned, let $v_z \triangleq g_z(x, y)$, $z = 1, \cdots, N_z$, denote the zth component of the original multispectral pixel vector **v** and $\tilde{v}_z \triangleq \tilde{g}_z$ (x, y), $z = 1, \cdots, N_z$ its distorted version. Some of the radiometric distortion measurements (6.11)–(6.17) may be extended to vector data.

Average vector RMSE (VRMSE), or $L_1(L_2)$ (the innermost norm L_2 refers to vector space (z), the outer one to pixel space (x, y))

$$\text{VRMSE}_{\text{avg}} = \frac{1}{N} \sum_{x,y} \sqrt{\sum_z [g_z(x,y) - \tilde{g}_z(x,y)]^2};$$ (6.18)

Peak VRMSE, or $L_\infty(L_2)$,

$$\text{VRMSE}_{\max} = \max_{x,y} \sqrt{\sum_z [g_z(x,y) - \tilde{g}_z(x,y)]^2}; \qquad (6.19)$$

SNR

$$\text{SNR} = 10 \cdot \log_{10} \frac{\sum_{x,y,z} g_z^2(x,y)}{\sum_{x,y,z} [g_z(x,y) - \tilde{g}_z(x,y)]^2}; \qquad (6.20)$$

PSNR

$$\text{PSNR} = 10 \cdot \log_{10} \frac{N \cdot N_z \cdot g_{fs}^2}{\sum_{x,y,z} [g_z(x,y) - \tilde{g}_z(x,y)]^2}; \qquad (6.21)$$

MAD, or $L_\infty(L_\infty)$,

$$\text{MAD} = \max_{x,y,z} \{|g_z(x,y) - \tilde{g}_z(x,y)|\}; \qquad (6.22)$$

PMAD

$$\text{PMAD} = \max_{x,y,z} \left\{ \frac{|g_z(x,y) - \tilde{g}_z(x,y)|}{g_z(x,y)} \right\} \times 100. \qquad (6.23)$$

In practice, (6.18) and (6.19) are respectively the average and maximum of the Euclidean norm of the distortion vector. SNR (6.20) is the extension of (6.14) to the 3D data cube. PSNR is the maximum SNR, given the full-scales of each vector component. MAD (6.22) is the maximum over the pixel set of the maximum absolute component of the distortion vector. PMAD (6.23) is the maximum percentage error over each vector component of the data cube.

6.1.2 Spectral Distortion

Given two spectral vectors $\mathbf{v}$ and $\tilde{\mathbf{v}}$ both having L components, let $\mathbf{v} = \{v_1, v_2, \cdots, v_L\}$ be the original spectral pixel vector $v_z = g_z(x, y)$ and $\tilde{\mathbf{v}} = \{\tilde{v}_1, \tilde{v}_2, \cdots, \tilde{v}_L\}$ its distorted version obtained after lossy compression and decompression, i.e., $\tilde{v}_z = \tilde{g}_z(x,y)$. Analogously to the *radiometric* distortion measurements, *spectral* distortion measurements may be defined.

The spectral angle mapper (SAM) denotes the absolute value of the spectral angle between the couple of vectors:

$$\text{SAM}(\mathbf{v}, \tilde{\mathbf{v}}) \overset{\Delta}{=} \arccos\left(\frac{<\mathbf{v}, \tilde{\mathbf{v}}>}{||\mathbf{v}||_2 \cdot ||\tilde{\mathbf{v}}||_2}\right) \qquad (6.24)$$

in which $< \cdot, \cdot >$ stands for scalar product. SAM can be measured in either degrees or radians. Another measurement especially suitable for hyperspectral data (i.e., for data with large number of components) is the spectral information divergence (SID) [20] derived from information-theoretic concepts:

$$SID(\mathbf{v}, \tilde{\mathbf{v}}) = D(\mathbf{v}||\tilde{\mathbf{v}}) + D(\tilde{\mathbf{v}}||\mathbf{v}) \tag{6.25}$$

with $D(\mathbf{v}||\tilde{\mathbf{v}})$ being the Kullback-Leibler distance (KLD), or entropy divergence, or *discrimination* [24], defined as

$$D(\mathbf{v}||\tilde{\mathbf{v}}) \overset{\Delta}{=} \sum_{z=1}^{L} p_z \log\left(\frac{p_z}{q_z}\right) \tag{6.26}$$

in which

$$p_z \overset{\Delta}{=} \frac{v_z}{||\mathbf{v}||_1} \quad \text{and} \quad q_z \overset{\Delta}{=} \frac{\tilde{v}_z}{||\tilde{\mathbf{v}}||_1} \tag{6.27}$$

In practice SID is equal to the symmetric KLD and can be compactly written as

$$SID(\mathbf{v}, \tilde{\mathbf{v}}) = \sum_{z=1}^{L} (p_z - q_z) \log\left(\frac{p_z}{q_z}\right) \tag{6.28}$$

which turns out to be symmetric, as one can easily verify. It can be proven as well that SID is always nonnegative, being zero iff. $p_z \equiv q_z$, $\forall z$, i.e., if $\mathbf{v}$ is parallel to $\tilde{\mathbf{v}}$. The measure unit of SID depends on the base of logarithm: *nat/vector* with natural logarithms and *bit/vector* with logarithms in base two.

Both SAM (6.24) and SID (6.28) may be either averaged on pixel vectors, or the maximum may be taken instead, as more representative of spectral quality. Note that radiometric distortion does not necessarily imply spectral distortion. Conversely, spectral distortion is always accompanied by a radiometric distortion, that is minimal when the couple of vectors have either the same Euclidean length (L_2) for SAM, or the same city-block length (L_1), for SID.

6.2 Linear and Logarithmic Quantization

In this subsection, linear and logarithmic quantization are described. Quantization has the objective of reducing the transmission rate. Linear quantization is capable of controlling near-lossless coding while logarithmic quantization is used to yield relative error bounded compression.

6.2.1 Linear Quantization

In order to achieve reduction in bit rate within the constraint of a near-lossless compression, prediction errors are quantized, with a quantization noise feedback loop embedded into the encoder, so that the current pixel prediction is formulated from the same "noisy" data that will be available at the decoder, as shown in Fig. 6.1a,b. Prediction errors, $e(n) \triangleq g(n) - \hat{g}(n)$, may be linearly quantized with a step size Δ as $e_\Delta(n) = \text{round}[e(n)/\Delta]$ and delivered to the *context-coding* section, as shown in Fig. 6.1a. The operation of inverse quantization, $\tilde{e}(x, y) = e_\Delta(x, y) \cdot \Delta$ introduces an error, whose variance and maximum modulus are $\Delta^2 / 12$ and $\lfloor \Delta / 2 \rfloor$, respectively. Since the MSE distortion is a quadratic function of the Δ, an odd-valued step size yields a lower L_∞ distortion for a given L_2 than an even size does; thus, odd step sizes are preferred for near-lossless compression. The relationship between the target peak error, i.e., $\varepsilon \in \mathbb{Z}^+$, and the step size to be used is $\Delta = 2\varepsilon + 1$.

6.2.2 Logarithmic Quantization

For the case of a relative-error bounded compression a rational version of prediction error must be envisaged. Let us define the *relative prediction error* (RPE) as ratio of original to predicted pixel value:

$$r(n) \triangleq \frac{g(n)}{\hat{g}(n)} \tag{6.29}$$

The *rational* nature of RPE, however, makes linear quantization unable to guarantee a strictly user-defined relative-error bounded performance.

Given a step size $\Delta \in \mathbb{R}$ ($\Delta > 0$, $\Delta \neq 1$), define direct and inverse *logarithmic quantization* (Log-Q) of $t \in \mathbb{R}$, $t > 0$, as

$$\mathcal{Q}_\Delta(t) \triangleq \text{round}[\log_\Delta(t)] = \text{round}[\log(t)/\log(\Delta)]$$

$$\mathcal{Q}_\Delta^{-1}(l) = \Delta^l \tag{6.30}$$

Applying (6.30) to (6.29) yields

$$\mathcal{Q}_\Delta[r(n)] = \text{round}\left[\frac{\log(g(n)) - \log(\hat{g}(n))}{\log \Delta}\right] \tag{6.31}$$

Hence, Log-Q of RPE is identical to Lin-Q of $\log(g(n)) - \log(\hat{g}(n))$ with a step size $\log\Delta$. If a Log-Q with a step size Δ is used to encode pixel RPE's (6.29), it can be proven that the ratio of original to decoded pixel value is strictly bounded around one

$$\min\left\{\sqrt{\Delta}, \frac{1}{\sqrt{\Delta}}\right\} \leq \frac{g(n)}{\tilde{g}(n)} \leq \max\left\{\sqrt{\Delta}, \frac{1}{\sqrt{\Delta}}\right\} \tag{6.32}$$

From (6.32) and (6.29) it stems that the percentage pixel distortion is upper bounded

$$\text{PMAD} = \max\left\{\sqrt{\Delta} - 1, 1 - \frac{1}{\sqrt{\Delta}}\right\} \times 100 \tag{6.33}$$

depending on whether $\Delta > 1$, or $0 < \Delta < 1$. Hence, the relationship between the target percentage peak error, ρ, and the step size will be, e.g., for $\Delta > 0$, $\Delta = (1 + \rho/100)^2$.

7 Virtually Lossless Compression

The term *virtually lossless* indicates that the distortion introduced by compression appears as an additional amount of noise, besides the intrinsic observation noise, being statistically independent of the latter, as well as of the underlying signal. Its first order distribution should be such that the overall probability density function (PDF) of the noise corrupting the decompressed data, i.e., intrinsic noise plus compression-induced noise, closely matches the noise PDF of the original data. This requirement is trivially fulfilled if compression is lossless, but may also hold if the difference between uncompressed and decompressed data exhibits a peaked and narrow PDF without tails, as it happens for near lossless techniques, whenever the user defined MAD is sufficiently smaller than the standard deviation σ_n of the background noise. Both MAD and σ_n are intended to be expressed in either physical units, for calibrated data, or as digital counts otherwise. Therefore, noise modeling and estimation from the uncompressed data becomes a major task to accomplish a virtually lossless compression [6]. The underlying assumption is that the dependence of the noise on the signal is null, or weak. However, signal independence of the noise may not strictly hold for hyperspectral images, especially for new-generation instruments. This further uncertainty in the noise model may be overcome by imposing a margin on the relationship between target MAD and RMS value of background noise.

For a DPCM scheme, the relationship between MAD and quantization step size is $\Delta = 2\text{MAD} + 1$, while the relationship between the variance of quantization noise, which is equal to MSE, and the step size is $\varepsilon^2 = \Delta^2/12$. Hence, the rationale of virtually lossless compression can be summarized as follows. Firstly, measure the background noise RMS, σ_n; if $\sigma_n < 1$, lossless compression is mandatory. If $1 \leq \sigma_n < 3$, lossless compression is recommended, but near lossless compression with MAD = 1 ($\Delta = 3$) is feasible. For $3 \leq \sigma_n < 5$, a strictly virtually lossless compression would require MAD = 1, and so on.

The signal may have been previously quantized based on different requirement; afterwards a check on the noise is made to decide whether lossless compression is really necessary, or near lossless compression could be used instead without penalty, being "de facto" *virtually lossless*.

The key to achieve a compression preserving the scientific quality of the data for remote-sensing is represented by the following twofold recommendation:

1. Absence of *tails* in the PDF of the error between uncompressed and decompressed image, in order to maximize the ratio RMSE / MAD, i.e., ε/MAD, or equivalently to minimize MAD for a given RMSE.
2. MSE lower by one order of magnitude (10 dB) than the variance of background noise σ_n^2.

Near-lossless methods are capable to fulfill such requirements, provided that the quantization step size Δ is chosen as an odd integer such that $\Delta \approx \sigma_n$. More exactly, the relationship between MAD and σ_n, also including a margin of approximately 1 dB, is:

$$\text{MAD} = \lfloor \max\{0, (\sigma_n - 1)/2\} \rfloor \qquad (6.34)$$

Depending on application context and type of data, the relationship (6.34) may be relaxed, e.g., by imposing that the ratio $\sigma_n^2 / \varepsilon^2$ is greater than, say, 3 dB, instead of the 11 dB, given by (6.34). If the data are intrinsically little noisy, the protocol may lead to the direct use of lossless compression, i.e., $\Delta = 1$.

If compression ratios greater than those of the reversible case are required, near lossless compression with MAD ≥ 1 of low-noise bands may be inadequate to preserve the scientific quality of the data, because the compression-induced MSE is not one order of magnitude lower than σ_n^2, as it would be recommended for *virtually lossless* compression. However, it may become mandatory to increase compression ratios above the values typical of the strictly virtually lossless protocol. To adjust the compression ratio, a real valued positive scale factor q can be introduced, such that the quantization step size of the nth band is given by:

$$\Delta_n = \text{round}[q \cdot \sigma_n]. \qquad (6.35)$$

where the roundoff is to the nearest *odd* integer. If $q \leq 1$ a strictly *virtually lossless* compression is achieved, since the compression-induced quadratic distortion is less than one tenth of the intrinsic noisiness of the data. Otherwise, if $q > 1$, compression is widely *virtually lossless*, even though distortion is properly allocated among the spectral bands. As an example, Fig. 6.5 shows quantization step sizes for three different values of q, if the test sequence *Cuprite Mine* of AVIRIS 1997 is near lossless compressed by means of the S-RLP algorithm. The estimation of the background noise RMS of the test sequence has been performed by considering a method based on the joint 2D PDF of the local statistics, described in [4]. To compare *virtually lossless* compression with a unique quantizer for the whole data cube, the step size of the latter, yielding the same

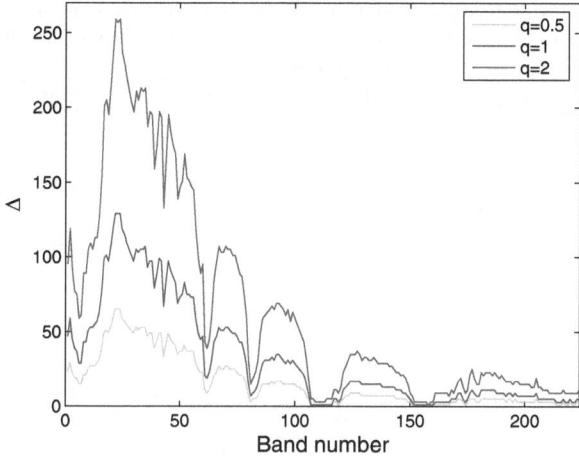

Fig. 6.5 Quantization step size varying with band number for the test sequence *Cuprite Mine*

compression ratio as the former, is the odd integer roundoff of the geometric mean of the step sizes in (6.35), at least for small distortions, in order to have an independent quantizer at each band.

8   Experimental Results and Comparisons

Experimental results are presented for the two data sets constituted by AVIRIS 1997 and AVIRIS 2006 data. Lossless compression is first considered on both data sets. Afterwards some results are presented for near-lossless and virtually lossless compression on a particular scene of the *Cuprite Mine* sequence of the AVIRIS 1997 data set. Eventually, some considerations on virtually lossless compression on discrimination of materials are introduced.

8.1   Lossless Compression Performance Comparisons

The lossless compression experiments have been performed by comparing some of the most performing schemes reported in the literature such as FL [27], SLSQ [42], M-CALIC [30], C-DPCM [33], S-RLP and S-FMP for the classical methods, and LUT-NN [32], LAIS-LUT [23], S-RLP-LUT and S-FMP-LUT for the LUT-based algorithms. Eventually, also TSP [27] is reported, an algorithm specifically designed to cope with the calibrated-induced artifacts. Some of these methods have not been made available by authors on the 2006 AVIRIS data set.

Table 6.1 Bit rates (bit/pel/band on disk) for lossless compression of AVIRIS 1997 test hyperspectral images. S-RLP-LUT and S-FMP-LUT use $N = 20$ previous bands and $M = 4$ LUT's per band. Best results are reported in bold

	Cuprite	Jasper	Lunar	Moffett	Average
FL	4.82	4.87	4.83	4.93	4.86
SLSQ	4.94	4.95	4.95	4.98	4.96
M-CALIC	4.89	4.97	4.80	4.65	4.83
C-DPCM	4.68	4.62	4.75	4.62	4.67
S-RLP	4.69	4.65	4.69	4.67	4.67
S-FMP	4.66	4.63	4.66	4.63	4.64
LUT-NN	4.66	4.95	4.71	5.05	4.84
LAIS-LUT	4.47	4.68	4.53	4.76	4.71
S-RLP-LUT	3.92	4.05	3.95	4.09	4.00
S-FMP-LUT	3.89	**4.03**	3.92	**4.05**	3.97
TSP	**3.77**	4.08	**3.81**	4.12	**3.95**

8.1.1 1997 AVIRIS Data Set

Table 6.1 lists the bit rates for several lossless compression methods applied to the standard 1997 AVIRIS images. The compared methods are FL [27], SLSQ [42], M-CALIC [30], C-DPCM [33], S-RLP, S-FMP, LUT-NN [32], LAIS-LUT [23], S-RLP-LUT, S-FMP-LUT and TSP [27].

It is apparent from the results reported in Table 6.1 that the best average performance is obtained by TSP and by the LUT-based algorithms, which are very effective even if not specifically designed for this type of data, because they are able to exploit the artificial regularities in the image histogram.

Concerning the S-RLP algorithm, Fig. 6.6 reports a comparison with 3D-RLP by varying the number S of the pixels in the prediction causal neighborhood $\mathcal{P}_S(n)$.

In this experiment, the fourth scene of the test image *Cuprite '97* has been reversibly compressed by means of S-RLP and 3D-RLP. Bit rates, reported in bits per pixel per band, include all coding overhead: predictor coefficients, block labels, and arithmetic codewords. For each λ, prediction lengths equal to 5, 14, and 20 have been considered together with 16 predictors. 3D prediction of RLP is always carried out from a couple of previous bands, except for the first band, coded in *intra* mode, i.e., by 2D DPCM, and the second band, which is predicted from one previous band only. The purely spectral 1D prediction of S-RLP is carried out from the available previous bands up the requested prediction length.

As it appears from the plots in Fig. 6.6, S-RLP outperforms 3D-RLP, especially when prediction length is lower, a case of interest for customized satellite on-board implementations. The performance of both S-RLP and 3D-RLP cannot be improved significantly by increasing the number and length of predictors, because of overhead information increasing as well.

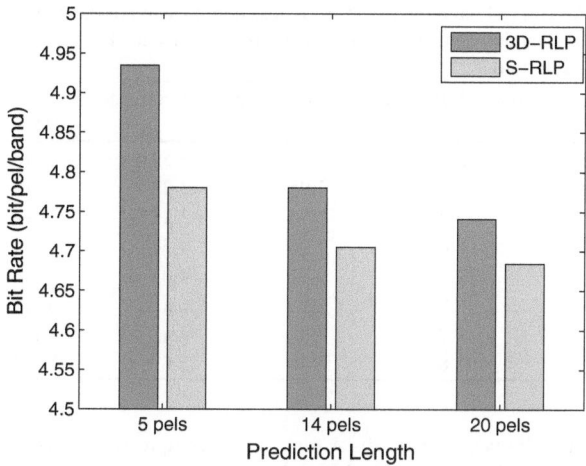

Fig. 6.6 S-RLP versus 3D-RLP varying with the prediction length. Bit rates on disk for the lossless compression of the fourth scene of AVIRIS Cuprite Mine 1997

Fig. 6.7 Band 50 of the 2006 AVIRIS Yellowstone 10 scene

8.1.2   2006 AVIRIS Data Set

A new data set of calibrated and uncalibrated AVIRIS images has been provided by Consultative Committee for Space Data Systems (CCSDS) and is now available to scientific users for compression experiments at the web site of address:(http:// compression.jpl.nasa.gov/hyperspectral). This data set consists of five calibrated and the corresponding raw 16-bit images acquired over Yellowstone, WY. Each image is composed by 224 bands and each scene (the scene numbers are 0, 3, 10, 11, 18) has 512 lines [27]. The 50th band of Yellowstone 10 is reported in Fig. 6.7.

Table 6.2 Bit rates (bit/pel/band on disk) for lossless compression of the calibrated 2006 AVIRIS data set. S-RLP-LUT and S-FMP-LUT have been utilized only for the first two scenes. The best results are reported in bold

	Yellowstone 0	Yellowstone 3	Yellowstone 10	Yellowstone 11	Yellowstone 18	Average
FL	3.91	3.79	3.37	3.59	3.90	3.71
S-RLP	3.58	3.43	2.95	3.27	3.46	3.34
S-FMP	**3.54**	**3.39**	**2.94**	**3.25**	**3.44**	**3.31**
LUT-NN	4.82	4.62	3.96	4.34	4.84	4.52
LAIS-LUT	4.48	4.31	3.71	4.02	4.48	4.20
SRLP-LUT	3.95	3.82	n.a.	n.a.	n.a.	n.a.
S-FMP-LUT	3.91	3.78	n.a.	n.a.	n.a.	n.a.
TSP	3.99	3.86	3.42	3.67	3.97	3.78

Table 6.3 Bit rates (bit/pel/band on disk) for lossless compression of the uncalibrated 2006 AVIRIS data set. S-RLP-LUT and S-FMP-LUT have been utilized only for the first two scenes. The best results are reported in bold

	Yellowstone 0	Yellowstone 3	Yellowstone 10	Yellowstone 11	Yellowstone 18	Average
FL	6.20	6.07	5.60	5.81	6.26	5.99
S-RLP	5.88	5.72	5.21	5.54	5.75	5.62
S-FMP	**5.84**	**5.67**	**5.18**	**5.48**	**5.68**	**5.57**
LUT-NN	7.14	6.91	6.26	6.69	7.20	6.84
LAIS-LUT	6.78	6.60	6.00	6.30	6.82	6.50
SRLP-LUT	6.21	6.05	n.a.	n.a.	n.a.	n.a.
S-FMP-LUT	6.17	6.01	n.a.	n.a.	n.a.	n.a.
TSP	6.27	6.13	5.64	5.88	6.32	6.05

Tables 6.2 and 6.3 report the lossless compression performances on the 2006 AVIRIS calibrated and uncalibrated data sets, respectively. The compared methods are FL [28], S-RLP, S-FMP, LUT-NN [32], LAIS-LUT [23], TSP [27], S-RLP-LUT and S-FMP-LUT. For these last algorithms, only two scenes have been compressed.

The scores show that the LUT-based methods and TSP are not able to obtain the same performances on both calibrated and uncalibrated 2006 AVIRIS data set with respect to the best performances obtained on the 1997 AVIRIS data set. S-RLP and S-FMP are the most effective and gain more than 10% and 6% over FL and 12% and 7% on TSP, on calibrated and uncalibrated data, respectively. FL algorithm is effective notwithstanding its simplicity.

8.2 Near Lossless Compression Performance Comparisons

Near lossless compression tests have been performed on the fourth scene of *Cuprite Mine*. Rate-Distortion (RD) plots are reported in Fig. 6.8a for S-RLP and 3D-RLP operating with $M = 16$ predictors and $S = 20$ coefficients per predictor. PSNR of

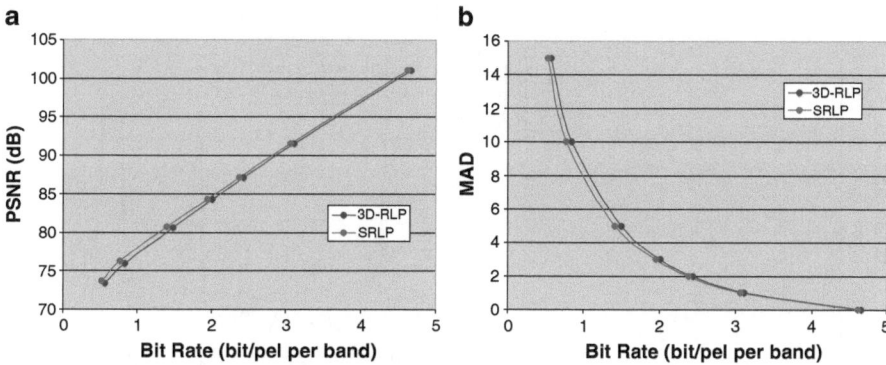

Fig. 6.8 Performance plots for near-lossless compression of AVIRIS *CupriteMine '97* test hyperspectral images: (**a**) PSNR vs. bit rate; (**b**) maximum absolute distortion (*MAD*) vs. bit rate

the whole image is calculated from the *average* MSE of the sequence of bands. Due to the sign bit, the full scale g_{fs} in (6.15) was set equal to $2^{15} - 1 = 32,767$ instead of 65,535, since small negative values, introduced by removal or dark current during calibration, are very rare and totally missing in some scenes. Hence, the PSNR attains a value of $10\log_{10}(12g_{fs}^2) \approx 102$ dB, due to integer roundoff noise only, when reversibility is reached. The correction for roundoff noise has a twofold advantage. Firstly, lossless points appear inside the plot and can be directly compared. Secondly, all PSNR-bit rate plots are straight lines with slope ≈ 6 dB/bit for bit rates larger than, say, 1 bit/pel, in agreement with RD theory [24] (with a uniform threshold quantizer). For lower bit rates, the quantization noise feedback causes an exponential drift from the theoretical straight line.

The results follow the same trends as the lossless case for S-RLP and are analogous to those of 3D-RLP, reported in [6], and of M-CALIC, reported in [30]. The near-lossless bit rate profiles are rigidly shifted downward from the lossless case by amounts proportional to the logarithm of the quantization-induced distortion. This behavior does not occur for low bit rates, because of the quantization noise feedback effect: prediction becomes poorer and poorer as it is obtained from the highly distorted reconstructed samples used by the predictor, which must be aligned to the decoder.

Interestingly, the difference in bit rate between S-RLP and 3D-RLP at a given PSNR is only two hundredths of bit/pel near the lossless point, but grows up to one tenth of bit/pel at a rate equal to 1.5 bit/pel, typical of a high-quality lossy compression. Comparisons with an up-to-date algorithm [35], implementing the state-of-the-art JPEG2000 multi-component approach, reveal that S-RLP outperforms the wavelet-based encoder by approximately 3.5 dB at 2 bit/pel. However, this difference reduces to 2.5 dB at 1 bit/pel and vanishes around 0.25 bit/pel, because of the quantization noise feedback effect, which is missing in the 3D wavelet coder. This moderate loss of performance is the price that

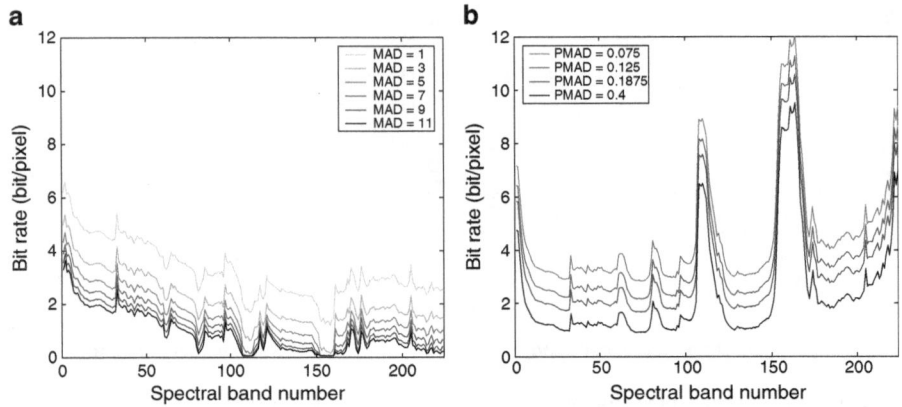

Fig. 6.9 Bit rates produced by 3D RLP on the *Cuprite* sequence of bands: (**a**) linear quantization to yield user-defined MAD values; (**b**) logarithmic quantization to yield user-defined PMAD values

embedded coders have to pay. DPCM does not allow progressive reconstruction to be achieved, but yields higher PSNR, at least for medium-high bit rates. The further advantage of DPCM is that it is near-lossless, unlike JPEG 2000, which can be made lossless, but not near-lossless, unless an extremely cumbersome quantizer is employed, with further loss in performances [15].

The near-lossless performance is shown in the MAD-bit rate plots of Fig. 6.8b. Since the average standard deviation of the noise was found to be around 10 according to [3], a *virtually-lossless* compression (maximum compression-induced absolute error lower than the standard deviation of the noise [29]) is given by S-RLP at a bit rate around 1.6 bit/pel/band, thereby yielding a compression ratio CR = 10 with respect to uncompressed data and CR $\approx$ 3 relative to lossless compression. Figure 6.8 evidences that the increment in performance of S-RLP over 3D-RLP is more relevant for such a bit rate than for higher rates.

A different experiment reported for 3D-RLP but representative of all DPCM-based methods is reported in Fig. 6.9a for linear quantization and in Fig. 6.9b for the logarithmic quantizer.

All bands have been compressed in both MAD-constrained mode (linear quantization) and PMAD constrained mode (logarithmic quantization). Bit rates varying with band number, together with the related distortion parameters are shown in Fig. 6.9. The bit rate plots follow similar trends varying with the amount of distortion, but quite different trends for the two types of distortion (i.e., either MAD or PMAD). For example, around the water vapor absorption wavelengths ($\approx$ Band 80) the MAD-bounded plots exhibit pronounced valleys, that can be explained because the intrinsic SNR of the data becomes lower; thus the linear quantizer dramatically abates the *noisy* prediction errors. On the other hand, the PMAD-bounded encoder tends to quantize the noisy residuals more finely when the signal is lower. Therefore bit rate peaks are generated instead of valleys. More generally speaking, bit rate peaks from the PMAD-bounded encoder are

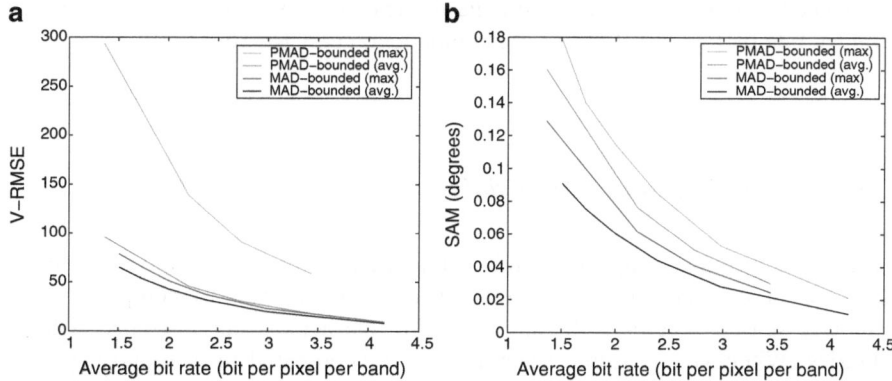

Fig. 6.10 Vector distortions vs bit rate for compressed AVIRIS *Cuprite Mine '97* data. Radiometric distortion (**a**) VRMSE; spectral distortion (**b**) SAM

associated with low responses from the spectrometer. This explains why the bit rate plots of Fig. 6.9b never fall below 1 bit/pixel per band.

Some of the radiometric distortion measures defined in Sect. 6.1 have been calculated on the distorted hyperspectral pixel vectors achieved by decompressing the bit streams generated by the near-lossless 3D-RLP encoder, both MAD- and PMAD-bounded. VRMSEs of the vector data, both *average* (6.18) and *maximum* (6.19), are plotted in Fig. 6.10a as a function of the bit rate from the encoder.

The MAD-bounded encoder obviously minimizes both the average and maximum of VRMSE, that is the Euclidean norm of the pixel error vector. A further advantage is that average and maximum VRMSE are very close to each other for all bit rates. The PMAD-bounded encoder is somewhat poorer: average VRMSE is comparable with that of the former, but peak VRMSE is far larger, due to the high-signal components that are coarsely quantized in order to minimize PMAD. Trivial results, not reported in the plots, are that MAD of the data cube (6.22) is exactly equal to the desired value, whereas the PMAD, being unconstrained, is higher. Symmetric results have been found by measuring PMAD on MAD-bounded and PMAD-bounded decoded data.

As far as *radiometric* distortion is concerned, results are not surprising. Radiometric distortions measured on vectors are straightforwardly derived from those measured on scalar pixel values. The introduction of such *spectral* measurements as SAM (6.24) and SID (6.28) may overcome the rationale of *distortion*, as established in the signal/image processing community. Figure 6.10b shows spectral distortions between original and decompressed hyperspectral pixel vectors. The PMAD-bounded algorithm yields plots that lie in the middle between the corresponding ones produced by the MAD-bounded algorithm and are very close to each other too. Since the *maximum* SAM is a better clue of spectral quality of the decoded data than the *average SAM* may be, a likely conclusion would be that PMAD-bounded compression optimizes the *spectral* quality of the data, while MAD-bounded is superior in terms of *radiometric* quality. Furthermore, the maximum SAM introduced by the P-MAD bounded logarithmic quantizer is lower than 0.2 ° for an average rate of 1 bit/pixel per vector component, i.e., CR=16.

8.3   Discrimination of Materials

Analysis procedures of hyperspectral vectors are usually performed on reflectance data, especially when the goal is identification of materials by comparing their remotely sensed spectra with sample spectra extracted from reference spectral libraries. Whenever measured spectra are to be compared to laboratory spectra, the radiance data are converted into reflectances, e.g., by means of the following simplified formula:

$$\rho(\lambda) = \frac{R(\lambda) \cdot \pi}{I(\lambda) \cdot T(\lambda)} \qquad (6.36)$$

in which $\rho(\lambda)$ is reflectance, $I(\lambda)$ is solar irradiance on ground, $T(\lambda)$ is atmospheric transmittance, and $R(\lambda)$ is at-sensor radiance, all functions of wavelength λ. Distortions introduced by compression on radiance will be amplified or attenuated depending on the values of the product $I(\lambda) \cdot T(\lambda)$. So, spectral distortion, e.g., SAM, must be measured between reflectance pixel spectra, rather than radiance pixel spectra. Extensive results of spectral discrimination from compressed hyperspectral data have demonstrated that a SAM distortion lower than 0.5 ° has negligible impact on the capability of automated classifiers of identifying spectral signatures of materials [26].

As a matter of fact, uniform distortion allocation over the entire spectrum yields minimum angle error between original and decompressed vectors. Band-variable distortion allocation following the virtually lossless protocol does not minimize error. However, when decompressed radiance data are converted into reflectance data by means of (6.36), the distribution of errors in decompressed reflectance spectra becomes approximately flat with the wavelength. In fact, quantization step sizes for virtually lossless compression follow the trend in Fig. 6.5, while the product $I(\lambda) \cdot T(\lambda)$ by which compression errors on radiance are multiplied to yield compression errors on reflectance follows an opposite trend. Eventually, virtually lossless compression is preferable to near-lossless compression at the same rate, at least for detection of materials using spectral libraries, because SAM of reflectances produced from virtually lossless data are in average lower than those coming from near lossless data.

9 Conclusions

This chapter has pointed out that quality issues are crucial for compression of radiance data produced by imaging spectrometers. Most widely used lossless compression techniques may be possibly replaced by lossy, yet error-bounded, techniques (near-lossless compression). Both lossless and near-lossless compression techniques can be implemented as adaptive DPCM encoders, with minor differences between the former and the latter. The rationale that compression-induced distortion is more tolerable, i.e., less harmful, in those bands, in which the noise is higher, and vice-versa, constitutes the *virtually lossless* paradigm, which provides operational criteria to design quantization in DPCM schemes. The quantization step size of each band should be a fraction of the measured standard deviation of instrumental noise. For typical VNIR+SWIR (400–2500 nm) spectra, virtually lossless compression exploits a band-varying quantization, with step sizes approximately decaying with increasing wavelengths. Once decompressed radiance spectra are converted to reflectance by removing the contribution of solar irradiance and atmospheric transmittance, the distribution of compression-induced errors with the wavelength is roughly equalized. Hence, angular errors introduced by compression errors on radiances will be lower for virtually lossless compression than for near-lossless compression. This feature is expected to be useful for discrimination of materials from compressed data and spectral libraries.

References

1. Abrardo, A., Alparone, L., Bartolini, F.: Encoding-interleaved hierarchical interpolation for lossless image compression. Signal Processing **56**(2), 321–328 (1997)
2. Aiazzi, B., Alba, P., Alparone, L., Baronti, S.: Lossless compression of multi/hyper-spectral imagery based on a 3-D fuzzy prediction. IEEE Trans. Geosci. Remote Sensing **37**(5), 2287–2294 (1999)
3. Aiazzi, B., Alparone, L., Barducci, A., Baronti, S., Marcoionni, P., Pippi, I., Selva, M.: Noise modelling and estimation of hyperspectral data from airborne imaging spectrometers. Annals of Geophysics **41**(1), 1–9 (2006)
4. Aiazzi, B., Alparone, L., Barducci, A., Baronti, S., Pippi, I.: Estimating noise and information of multispectral imagery. J. Optical Engin. **41**(3), 656–668 (2002)
5. Aiazzi, B., Alparone, L., Baronti, S.: A reduced Laplacian pyramid for lossless and progressive image communication. IEEE Trans. Commun. **44**(1), 18–22 (1996)
6. Aiazzi, B., Alparone, L., Baronti, S.: Near-lossless compression of 3-D optical data. IEEE Trans. Geosci. Remote Sensing **39**(11), 2547–2557 (2001)
7. Aiazzi, B., Alparone, L., Baronti, S.: Context modeling for near-lossless image coding. IEEE Signal Processing Lett. **9**(3), 77–80 (2002)
8. Aiazzi, B., Alparone, L., Baronti, S.: Fuzzy logic-based matching pursuits for lossless predictive coding of still images. IEEE Trans. Fuzzy Systems **10**(4), 473–483 (2002)
9. Aiazzi, B., Alparone, L., Baronti, S.: Near-lossless image compression by relaxation-labelled prediction. Signal Processing **82**(11), 1619–1631 (2002)
10. Aiazzi, B., Alparone, L., Baronti, S.: Lossless compression of hyperspectral images using multiband lookup tables. IEEE Signal Processing Lett. **16**(6), 481–484 (2009)

11. Aiazzi, B., Alparone, L., Baronti, S., Lastri, C.: Crisp and fuzzy adaptive spectral predictions for lossless and near-lossless compression of hyperspectral imagery. IEEE Geosci. Remote Sens. Lett. **4**(4), 532–536 (2007)

12. Aiazzi, B., Alparone, L., Baronti, S., Lotti, F.: Lossless image compression by quantization feedback in a content-driven enhanced Laplacian pyramid. IEEE Trans. Image Processing **6** (6), 831–843 (1997)

13. Aiazzi, B., Alparone, L., Baronti, S., Santurri, L.: Near-lossless compression of multi/ hyperspectral images based on a fuzzy-matching-pursuits interband prediction. In: S.B. Serpico (ed.) Image and Signal Processing for Remote Sensing VII, vol. 4541, pp. 252–263 (2002)

14. Alecu, A., Munteanu, A., Cornelis, J., Dewitte, S., Schelkens, P.: On the optimality of embedded deadzone scalar-quantizers for wavelet-based L-infinite-constrained image coding. IEEE Signal Processing Lett. **11**(3), 367–370 (2004)

15. Alecu, A., Munteanu, A., Cornelis, J., Dewitte, S., Schelkens, P.: Wavelet-based scalable L-infinity-oriented compression. IEEE Trans Image Processing **15**(9), 2499–2512 (2006)

16. Baraldi, A., Blonda, P.: A survey of fuzzy clustering algorithms for pattern recognition–Parts I and II. IEEE Trans. Syst. Man Cybern.–B **29**(6), 778–800 (1999)

17. Benazza-Benyahia, A., Pesquet, J.C., Hamdi, M.: Vector-lifting schemes for lossless coding and progressive archival of multispectral images. IEEE Trans. Geosci. Remote Sensing **40**(9), 2011–2024 (2002)

18. Bezdek, J.C.: Pattern Recognition with Fuzzy Objective Function Algorithm. Plenum Press, New York (1981)

19. Carpentieri, B., Weinberger, M.J., Seroussi, G.: Lossless compression of continuous-tone images. Proc. of the IEEE **88**(11), 1797–1809 (2000)

20. Chang, C.I.: An information-theoretic approach to spectral variability, similarity, and discrimination for hyperspectral image analysis. IEEE Trans. Inform. Theory **46**(5), 1927–1932 (2000)

21. Deng, G., Ye, H., Cahill, L.W.: Adaptive combination of linear predictors for lossless image compression. IEE Proc.-Sci. Meas. Technol. **147**(6), 414–419 (2000)

22. Golchin, F., Paliwal, K.K.: Classified adaptive prediction and entropy coding for lossless coding of images. In: Proc. IEEE Int. Conf. on Image Processing, vol. III/III, pp. 110–113 (1997)

23. Huang, B., Sriraja, Y.: Lossless compression of hyperspectral imagery via lookup tables with predictor selection. In: L. Bruzzone (ed.) Proc. of SPIE, Image and Signal Processing for Remote Sensing XII, vol. 6365, pp. 63650L.1–63650L.8 (2006)

24. Jayant, N.S., Noll, P.: Digital Coding of Waveforms: Principles and Applications to Speech and Video. Prentice Hall, Englewood Cliffs, NJ (1984)

25. Ke, L., Marcellin, M.W.: Near-lossless image compression: minimum entropy, constrained-error DPCM. IEEE Trans. Image Processing **7**(2), 225–228 (1998)

26. Keshava, N.: Distance metrics and band selection in hyperspectral processing with applications to material identification and spectral libraries. IEEE Trans. Geosci. Remote Sensing **42**(7), 1552–1565 (2004)

27. Kiely, A.B., Klimesh, M.A.: Exploiting calibration-induced artifacts in lossless compression of hyperspectral imagery. IEEE Trans. Geosci. Remote Sensing **47**(8), 2672–2678 (2009)

28. Klimesh, M.: Low-complexity adaptive lossless compression of hyperspectral imagery. In: Satellite Data Compression, Communication and Archiving II, *Proc. SPIE*, vol. 6300 pp. 63000N.1–63000N.9 (2006)

29. Lastri, C., Aiazzi, B., Alparone, L., Baronti, S.: Virtually lossless compression of astrophysical images. EURASIP Journal on Applied Signal Processing **2005**(15), 2521–2535 (2005)

30. Magli, E., Olmo, G., Quacchio, E.: Optimized onboard lossless and near-lossless compression of hyperspectral data using CALIC. IEEE Geosci. Remote Sensing Lett. **1**(1), 21–25 (2004)

31. Matsuda, I., Mori, H., Itoh, S.: Lossless coding of still images using minimum-rate predictors. In: Proc. IEEE Int. Conf. on Image Processing, vol. I/III, pp. 132–135 (2000)

32. Mielikainen, J.: Lossless compression of hyperspectral images using lookup tables. IEEE Signal Proc. Lett. **13**(3), 157–160 (2006)
33. Mielikainen, J., Toivanen, P.: Clustered DPCM for the lossless compression of hyperspectral images. IEEE Trans. Geosci. Remote Sensing **41**(12), 2943–2946 (2003)
34. Mielikainen, J., Toivanen, P., Kaarna, A.: Linear prediction in lossless compression of hyperspectral images. J. Optical Engin. **42**(4), 1013–1017 (2003)
35. Penna, B., Tillo, T., Magli, E., Olmo, G.: Progressive 3-D coding of hyperspectral images based on JPEG 2000. IEEE Geosci. Remote Sensing Lett. **3**(1), 125–129 (2006)
36. Pennebaker, W.B., Mitchell, J.L.: JPEG: Still Image Compression Standard. Van Nostrand Reinhold, New York (1993)
37. Ramabadran, T.V., Chen, K.: The use of contextual information in the reversible compression of medical images. IEEE Trans. Medical Imaging **11**(2), 185–195 (1992)
38. Rao, A.K., Bhargava, S.: Multispectral data compression using bidirectional interband prediction. IEEE Trans. Geosci. Remote Sensing **34**(2), 385–397 (1996)
39. Rao, K.K., Hwang, J.J.: Techniques and Standards for Image, Video, and Audio Coding. Prentice Hall, Engl. Cliffs, NJ (1996)
40. Reichel, J., Menegaz, G., Nadenau, M.J., Kunt, M.: Integer wavelet transform for embedded lossy to lossless image compression. IEEE Trans. Image Processing **10**(3), 383–392 (2001)
41. Rice, R.F., Plaunt, J.R.: Adaptive variable-length coding for efficient compression of space-craft television data. IEEE Trans. Commun. Technol. **COM-19**(6), 889–897 (1971)
42. Rizzo, F., Carpentieri, B., Motta, G., Storer, J.A.: Low-complexity lossless compression of hyperspectral imagery via linear prediction. IEEE Signal Processing Lett. **12**(2), 138–141 (2005)
43. Roger, R.E., Cavenor, M.C.: Lossless compression of AVIRIS images. IEEE Trans. Image Processing **5**(5), 713–719 (1996)
44. Said, A., Pearlman, W.A.: An image multiresolution representation for lossless and lossy compression. IEEE Trans. Image Processing **5**(9), 1303–1310 (1996)
45. Tate, S.R.: Band ordering in lossless compression of multispectral images. IEEE Trans. Comput. **46**(4), 477–483 (1997)
46. Taubman, D.S., Marcellin, M.W.: JPEG2000: Image compression fundamentals, standards and practice. Kluwer Academic Publishers, Dordrecht, The Netherlands (2001)
47. Wang, J., Zhang, K., Tang, S.: Spectral and spatial decorrelation of Landsat-TM data for lossless compression. IEEE Trans. Geosci. Remote Sensing **33**(5), 1277–1285 (1995)
48. Weinberger, M.J., Rissanen, J.J., Arps, R.B.: Applications of universal context modeling to lossless compression of gray-scale images. IEEE Trans. Image Processing **5**(4), 575–586 (1996)
49. Weinberger, M.J., Seroussi, G., Sapiro, G.: The LOCO-I lossless image compression algorithm: principles and standardization into JPEG-LS. IEEE Trans. Image Processing **9**(8), 1309–1324 (2000)
50. Witten, I.H., Neal, R.M., Cleary, J.G.: Arithmetic coding for data compression. Commun. ACM **30**, 520–540 (1987)
51. Wu, X., Bao, P.: L_∞ constrained high-fidelity image compression via adaptive context modeling. IEEE Trans. Image Processing **9**(4), 536–542 (2000)
52. Wu, X., Memon, N.: Context-based, adaptive, lossless image coding. IEEE Trans. Commun. **45**(4), 437–444 (1997)
53. Wu, X., Memon, N.: Context-based lossless interband compression–Extending CALIC. IEEE Trans. Image Processing **9**(6), 994–1001 (2000)

Chapter 7
Ultraspectral Sounder Data Compression by the Prediction-Based Lower Triangular Transform

Shih-Chieh Wei and Bormin Huang

Abstract The Karhunen–Loeve transform (KLT) is the optimal unitary transform that yields the maximum coding gain. The prediction-based lower triangular transform (PLT) features the same decorrelation and coding gain properties as KLT but with lower complexity. Unlike KLT, PLT has the perfect reconstruction property which allows its direct use for lossless compression. In this paper, we apply PLT to carry out lossless compression of the ultraspectral sounder data. The experiment on the standard ultraspectral test dataset of ten AIRS digital count granules shows that the PLT compression scheme compares favorably with JPEG-LS, JPEG2000, LUT, SPIHT, and CCSDS IDC 5/3.

1 Introduction

Contemporary and future ultraspectral infrared sounders such as AIRS [1], CrIS [2], IASI [3] and GIFTS [4] represent a significant technical advancements in for environmental and meteorological prediction and monitoring. Given the large 3D volume of data obtained from high spectral and spatial observations, the use of effective data compression techniques will be beneficial for data transfer and storage. When the ultraspectral sounder data is used to retrieve geophysical parameters like the vertical profiles of atmospheric temperature, moisture and trace gases, the retrieval process involves solving the radiative transfer equation which is an ill-posed inverse problem and sensitive to the noise and error in the

S.-C. Wei (✉)
Department of Information Management, Tamkang University, Tamsui, Taiwan
e-mail: seke@mail.im.tku.edu.tw

B. Huang
Space Science and Engineering Center, University of Wisconsin, Madison, USA
e-mail: bormin@ssec.wisc.edu

B. Huang (ed.), *Satellite Data Compression*, DOI 10.1007/978-1-4614-1183-3_7,
© Springer Science+Business Media, LLC 2011

data [5]. Thus, lossless or near lossless compression of the data is desired in order to avoid substantial degradation during retrieval.

Past studies on lossless compression of the ultraspectral sounder data can be categorized into clustering-based, prediction-based and transform-based methods [6]. Since a higher correlation exists in spectral dimension than in spatial dimension of ultraspectral sounder data [6], there were studies on band reordering as preprocessing before compression [7, 8]. The Karhunen–Loeve transform (KLT), a.k.a. the principal component analysis (PCA) or the Hoteling transform, is the optimal unitary transform that yields the maximum coding gain. However, considering its data-dependent computational cost involving the eigenvectors from the input covariance matrix, KLT is often only used as a benchmark for performance comparison. Phoong and Lin [9] developed the prediction-based lower triangular transform (PLT) that features the same decorrelation and coding gain properties as KLT but with a lower design and implementational cost. They showed promising results when PLT was applied to lossy compression of 2D imagery and the AR(1) process [9]. However, the original PLT by Phoong et al. requires the input vector to be a blocked version of a scalar wide sense stationary (WSS) process. Weng et al. [10] proposed a generalized triangular decomposition (GTD) which allows the input vector to be a vector WSS process. GTD has the same coding gain as KLT and it includes KLT and PLT as special cases [10].

Furthermore, unlike KLT, PLT has a perfect reconstruction (PR) property which makes it usefull for lossless compression. Since PLT provides the same coding gain as KLT, with lower complexity and PR property, we were motivated to apply PLT to carry out lossless compression of the ultraspectral sounder data. The compression method consists of using PLT for spectral prediction, followed by the arithmetic coding. The PLT compression ratio will be compared with prediction-based methods like LUT [11] and JPEG-LS [12], and wavelet-transform-based methods like JPEG2000 [13], SPIHT [14] and CCSDS IDC 5/3 [15].

The rest of the paper is arranged as follows. Section 2 describes the ultraspectral sounder data used in this study. Section 3 introduces our compression scheme. Section 4 shows the compression results on the ultraspectral sounder data. Section 5 gives the conclusions.

2 Data

The ultraspectral sounder data can be generated from either a Michelson interferometer (e.g. CrIS [2], IASI [3], GIFTS [4]) or a grating spectrometer (e.g. AIRS [1]). A standard ultraspectral sounder data set for compression is publicly available via anonymous ftp at ftp://ftp.ssec.wisc.edu/pub/bormin/Count/. It consists of ten granules, five daytime and five nighttime, selected from representative geographical regions of the Earth. Their locations, UTC times and local time adjustments are listed in Table 7.1. This standard ultraspectral sounder data set adopts the NASA EOS AIRS digital counts made on March 2, 2004. The AIRS data includes 2,378 infrared channels in the 3.74–15.4 μm region of the spectrum. A day's worth of AIRS data is divided into 240 granules, each of 6 min durations. Each granule consists of 135 scan lines containing 90 cross-track footprints per scan line; thus there are a total of $135 \times 90 = 12{,}150$ footprints per granule. More information regarding the AIRS instrument may be acquired from the NASA AIRS web site at http://www-airs.jpl.nasa.gov.

The digital count data ranges from 12 to 14 bits for different channels. Each channel is saved using its own bit depth. To make the selected data more generic to other ultraspectral sounders, 271 AIR-specific bad channels identified in the supplied AIRS infrared channel properties file are excluded. Each resulting granule is saved as a binary file, arranged as 2,107 channels, 135 scan lines, and 90 pixels for each scan line. Figure 7.1 shows the AIRS digital counts at wavenumber 800.01 cm^{-1} for some of the ten selected granules. In these granules, coast lines are depicted by solid curves and multiple clouds at various altitudes are shown as different shades of pixels.

Table 7.1 Ten selected airs granules for study of ultraspectral sounder data compression

Granule number	UTC time	Local time adjustment	Location
Granule 9	00:53:31 UTC	−12 H	Pacific Ocean, daytime
Granule 16	01:35:31 UTC	+2 H	Europe, nighttime
Granule 60	05:59:31 UTC	+7 H	Asia, daytime
Granule 82	08:11:31 UTC	−5 H	North America, nighttime
Granule 120	11:59:31 UTC	−10 H	Antarctica, nighttime
Granule 126	12:35:31 UTC	−0 H	Africa, daytime
Granule 129	12:53:31 UTC	−2 H	Arctic, daytime
Granule 151	15:05:31 UTC	+11 H	Australia, nighttime
Granule 182	18:11:31 UTC	+8 H	Asia, nighttime
Granule 193	19:17:31 UTC	−7 H	North America, daytime

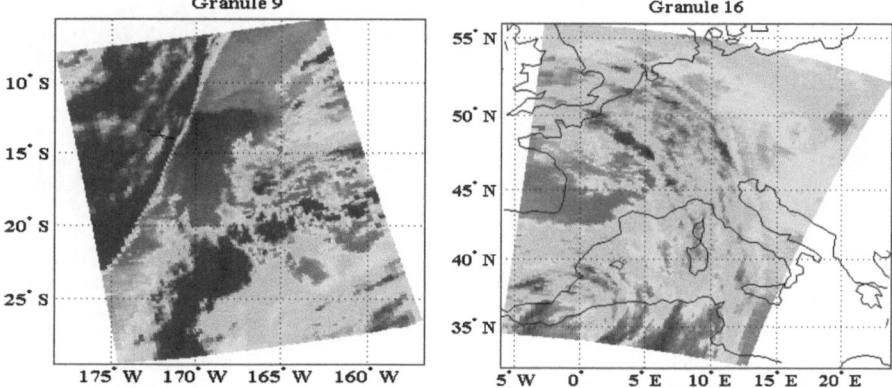

Fig. 7.1 The ten selected AIRS digital count granules at wavenumber 800.01 cm^{-1} on March 2, 2004

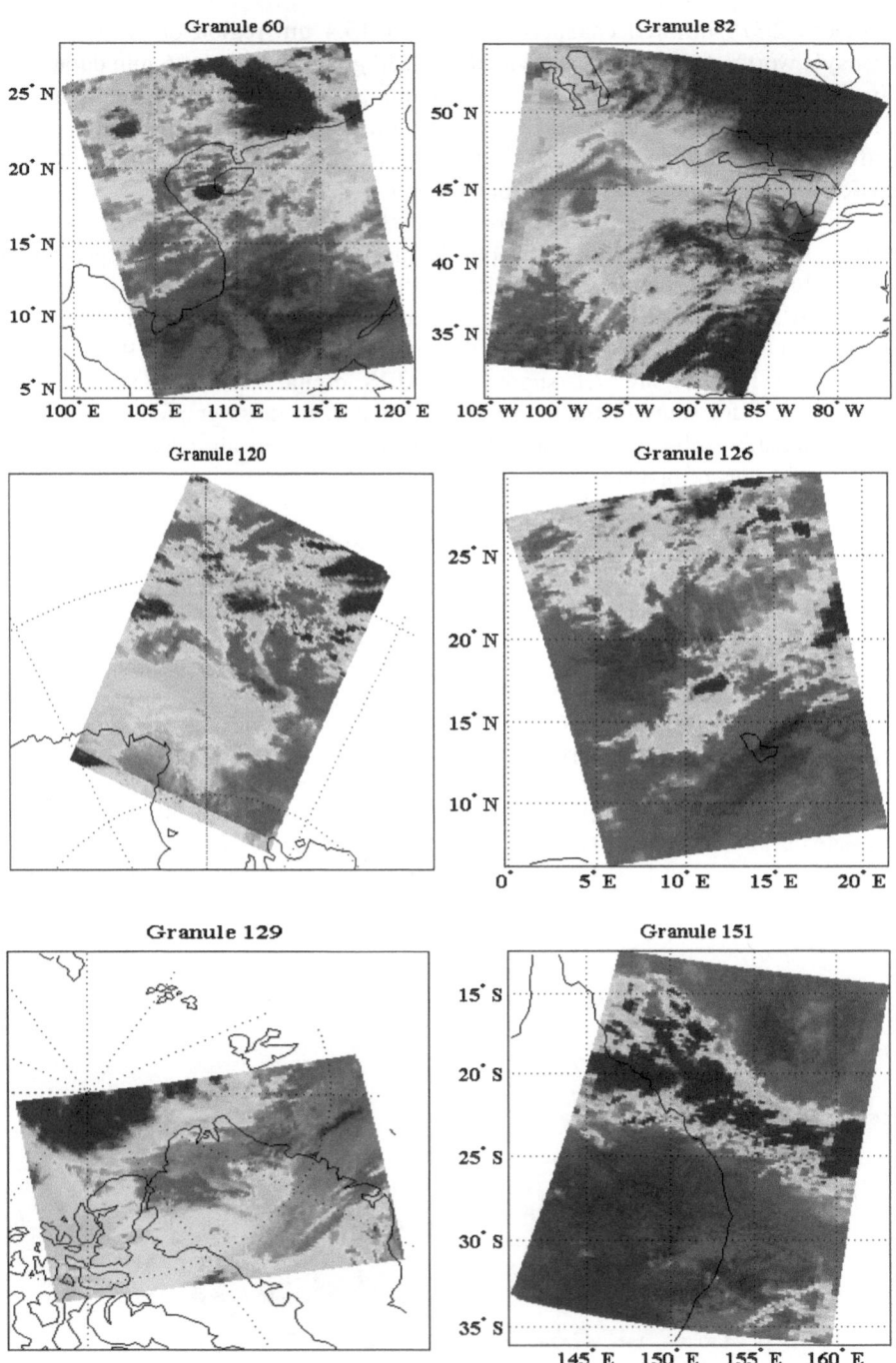

Fig. 7.1 (continued)

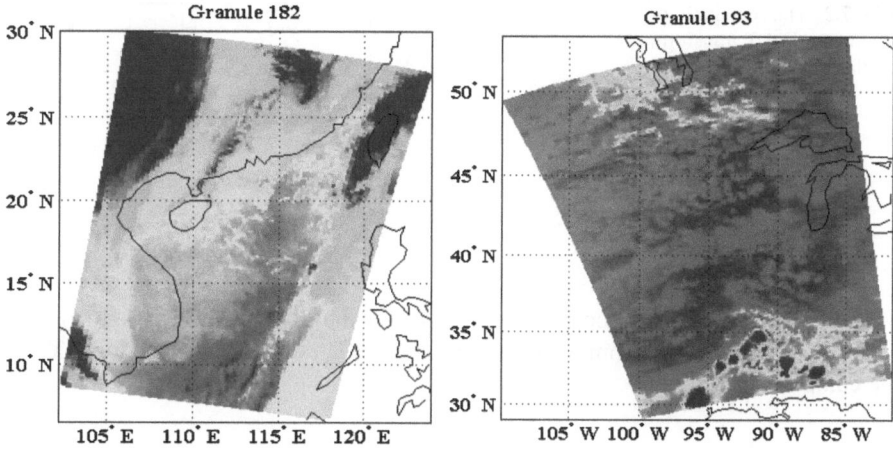

Fig. 7.1 (continued)

3 The Compression Scheme

For ultraspectral sounder data, the spectral correlation is generally much stronger than the spatial correlation [6]. To de-correlate the spectral dependency, a linear prediction with a fixed number of predictors has been used [16]. The prediction-based lower triangular transform (PLT) is also based on linear prediction but uses as many orders of predictors as possible. However, without doing linear regression on each order, PLT can be directly computed by the LDU matrix decomposition [9]. Figure 7.2 gives the schematic of the PLT transform coding scheme.

Let $x(n)$ be a sequence of scalar observation signals. In order to exploit the correlation in the sequence, M consecutive scalar signals are grouped together to form a blocked version of vector signal $\vec{x}(t)$. Each vector signal $\vec{x}(t)$ then goes through a PLT transform $\mathbf{T}$ to obtain a vector of transform coefficients $\vec{y}(t)$. The transform coefficients are often quantized and have smaller variances than the original signal for storage or transfer purposes. Finally, when necessary, an inverse PLT transform $\mathbf{T}^{-1}$ is applied to the transform coefficients to restore the original signal.

To compute the PLT transform and the inverse transform, the statistic of the source signal are required. Suppose a total of $N \times M$ samples are collected, the $N \times M$ signal matrix X can be expressed as:

$$X = \begin{pmatrix} x(1) & x(M+1) & x(2M+1) & \cdots & x((N-1)M+1) \\ x(2) & x(M+2) & x(2M+2) & \cdots & x((N-1)M+2) \\ x(3) & x(M+3) & x(2M+3) & \cdots & x((N-1)M+3) \\ \vdots & \vdots & \vdots & \ddots & \vdots \\ x(M) & x(2M) & x(3M) & \cdots & x(NM) \end{pmatrix}$$

$$= (\vec{x}(1) \quad \vec{x}(2) \quad \vec{x}(3) \quad \cdots \quad \vec{x}(N))$$

Fig. 7.2 The schematic of
the original PLT transform
coding scheme

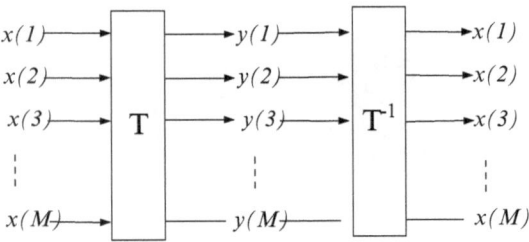

Fig. 7.2 The schematic of the original PLT transform coding scheme

Assume that R_x is the autocorrelation matrix of the observation signal $x(n)$ with order M, by the LDU decomposition of R_x, the PLT transform for the signal X can be computed as follows:

$$P = L^{-1} \text{with } R_x = LDU$$

where L is the lower triangular matrix with all diagonal entries equal to 1, D is a diagonal matrix, U is the upper triangular matrix with all diagonal entries equal to 1, and P is the desired prediction-based lower triangular (PLT) transform T which can be computed by the matrix inversion of L. Since R_x is symmetric and the LDU matrix decomposition is unique [17], we have $LDU = R_x = R_x^T = U^T D L^T$ or $U = L^T$. Let $y(n)$ be the coefficient of the transform P on $x(n)$. $y(n)$ will be the error between the original signal $x(n)$ and the prediction based on previous $M-1$ signals, i.e. $x(n-1)$ through $x(n-M+1)$. Let R_y be the autocorrelation matrix of the prediction error $y(n)$. By the transform property $R_y = P R_x(M) P^T$ and the LDU decomposition of $R_x = LDU$, R_y can be shown to be diagonal as follows [9]:

$$R_y = PR_xP^T = P(LDU)P^T = PP^{-1}D(P^T)^{-1}P^T = D$$

It means that the prediction error is de-correlated by the initial selection of $P = L^{-1}$. The entries on the diagonal of D will be the variance of the prediction error of orders 0 through $M-1$. Since the lower triangular transform P consists of all linear predictors below order M, computation of P can be carried out by the Levinson-Durbin algorithm [18] which has lower complexity than the LDU decomposition when the scalar signal satisfies the wide sense stationary (WSS) property.

For compression of the ultraspectral sounder data, each granule is considered to consist of $N = n_s$ observations at different locations with each observation containing $M = n_c$ channels of data. Specifically, we select the prediction order M as the number of channels n_c to fully exploit the spectral correlation in reducing the prediction error. That is, all previous available channels are used to predict the current channel. Figure 7.3 gives the schematic of the PLT transform coding model for our compression scheme.

Let $X = [x_1 \, x_2 \, x_3 \ldots x_{nc}]^T$ be the original mean-subtracted ultraspectral sounder data consisting of n_c channels by n_s observations, $Y = [y_1 \, y_2 \, y_3 \ldots y_{nc}]^T$ be the n_c x

Fig. 7.3 The schematic of the PLT transform coding model for our compression scheme

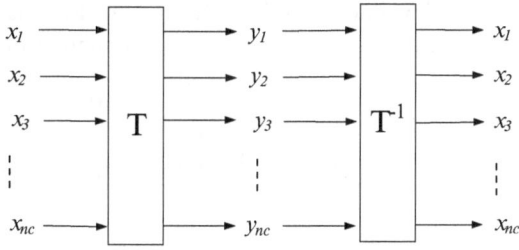

n_s transform coefficients or prediction errors, and P be the prediction-based lower triangular transform. Here X, Y, and P can be written as follows:

$$X = \begin{pmatrix} \vec{x}_1 \\ \vec{x}_2 \\ \vec{x}_3 \\ \vdots \\ \vec{x}_{n_c} \end{pmatrix} = \begin{pmatrix} x_1(1) & x_1(2) & x_1(3) & \cdots & x_1(n_s) \\ x_2(1) & x_2(2) & x_2(3) & \cdots & x_2(n_s) \\ x_3(1) & x_3(2) & x_3(3) & \cdots & x_3(n_s) \\ \vdots & \vdots & \vdots & \vdots & \ddots & \vdots \\ x_{n_c}(1) & x_{n_c}(2) & x_{n_c}(3) & \cdots & x_{n_c}(n_s) \end{pmatrix}$$

$$= (\ \vec{x}(1) \ \ \vec{x}(2) \ \ \vec{x}(3) \ \cdots \vec{x}(ns))$$

$$Y = \begin{pmatrix} \vec{y}_1 \\ \vec{y}_2 \\ \vec{y}_3 \\ \vdots \\ \vec{y}_{n_c} \end{pmatrix} = \begin{pmatrix} y_1(1) & y_1(2) & y_1(3) & \cdots & y_1(n_s) \\ y_2(1) & y_2(2) & y_2(3) & \cdots & y_2(n_s) \\ y_3(1) & y_3(2) & y_3(3) & \cdots & y_3(n_s) \\ \vdots & \vdots & \vdots & \vdots & \ddots & \vdots \\ y_{n_c}(1) & y_{n_c}(2) & y_{n_c}(3) & \cdots & y_{n_c}(n_s) \end{pmatrix}$$

$$= (\vec{y}(1) \ \ \vec{y}(2) \ \ \vec{y}(3) \ \cdots \vec{y}(ns))$$

$$P = \begin{pmatrix} 1 & 0 & 0 & \cdots & 0 \\ p_{1,0} & 1 & 0 & \cdots & 0 \\ p_{2,0} & p_{2,1} & 1 & \cdots & 0 \\ \vdots & \vdots & \vdots & \ddots & \vdots \\ p_{n_c-1,0} & p_{n_c-1,1} & p_{n_c-1,2} & \cdots & 1 \end{pmatrix}$$

Then the transform coefficient or prediction error Y can be computed by $Y = PX$ or

$$y_1 = x_1$$
$$y_2 = x_2 - \hat{x}_2 = x_2 + p_{1,0}x_1$$
$$y_3 = x_3 - \hat{x}_3 = x_3 + p_{2,0}x_1 + p_{2,1}x_2$$
$$\cdots$$

$$y_{n_c} = x_{n_c} - \hat{x}_{n_c}$$
$$= x_{n_c} + p_{n_c-1,0}x_1 + p_{n_c-1,1}x_2 + p_{n_c-1,2}x_3 + \ldots + p_{n_c-1,n_c-2}x_{n_c-1}$$

where $\hat{x}_m$ is the prediction of channel m by use of the linear combination of all previous $m-1$ channels. A similar result can be obtained for the inverse transform

$S = P^{-1}$ which is derived from L in the LDU decomposition. Let the inverse transform S, which is also a lower triangular matrix, be of the form

$$S = \begin{pmatrix} 1 & 0 & 0 & \cdots & 0 \\ s_{1,0} & 1 & 0 & \cdots & 0 \\ s_{2,0} & s_{2,1} & 1 & \cdots & 0 \\ \vdots & \vdots & \vdots & \ddots & \vdots \\ s_{n_c-1,0} & s_{n_c-1,1} & s_{n_c-1,2} & \cdots & 1 \end{pmatrix}$$

Then the original signal X can be computed by $X = SY$ or

$$x_1 = y_1$$
$$x_2 = y_2 + \hat{x}_2 = y_2 + s_{1,0}y_1$$
$$x_3 = y_3 + \hat{x}_3 = y_3 + s_{2,0}y_1 + s_{2,1}y_2$$
$$\cdots$$
$$x_{n_c} = y_{n_c} + \hat{x}_{n_c}$$
$$= y_{n_c} + s_{n_c-1,0}y_1 + s_{n_c-1,1}y_2 + s_{n_c-1,2}y_3 + \cdots + s_{n_c-1,n_c-2}y_{n_c-1}$$

From the above formulation, it can be seen that either the transform P or the inverse transform S alone can be used to compress the signal X into the prediction error Y in order to reconstruct the original X from Y. However, to minimize the amount of data in transfer, quantization on transform kernels P or S and quantization on the prediction error Y are required. Furthermore, perfect reconstruction is required for our lossless compression application. To meet these requirements, a minimum noise ladder-based structure with the perfect reconstruction property is adopted [9]. When the transform P is used for both encoding and decoding, a minimum noise ladder-based structure for encoding follows. Note that the flooring function is chosen for quantization. Both X and Y will be integers.

$$y_1 = x_1$$
$$y_2 = x_2 + floor(p_{1,0}x_1)$$
$$y_3 = x_3 + floor(p_{2,0}x_1 + p_{2,1}x_2)$$
$$\cdots$$
$$y_{n_c} = x_{n_c} + floor(p_{n_c-1,0}x_1 + p_{n_c-1,1}x_2 + p_{n_c-1,2}x_3 + \cdots + p_{n_c-1,n_c-2}x_{n_c-1})$$

A corresponding minimum noise ladder-based structure using P for decoding follows:

$$x_1 = y_1$$
$$x_2 = y_2 - floor(p_{1,0}x_1)$$
$$x_3 = y_3 - floor(p_{2,0}x_1 + p_{2,1}x_2)$$
$$\cdots$$
$$x_{n_c} = y_{n_c} - floor(p_{n_c-1,0}x_1 + p_{n_c-1,1}x_2 + p_{n_c-1,2}x_3 + \cdots + p_{n_c-1,n_c-2}x_{n_c-1}).$$

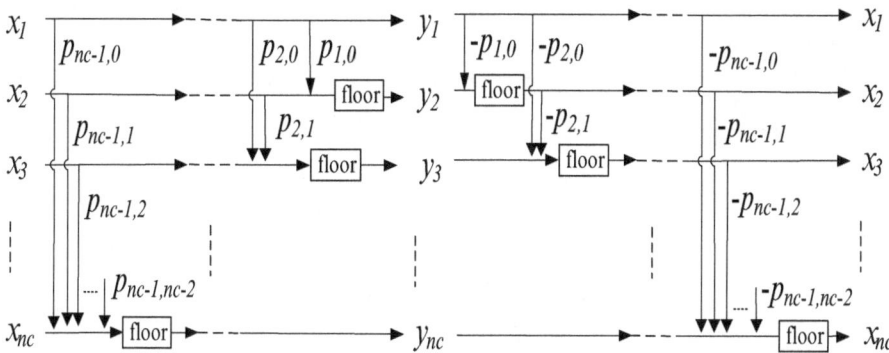

Fig. 7.4 A minimum noise ladder-based structure of PLT based on the transform P and the flooring quantizer. It has the perfect reconstruction property for lossless compression

Similarly when the inverse transform S is used for both encoding and decoding, a minimum noise ladder-based structure with the perfect reconstruction property follows. Note that the ceiling function is chosen here for quantization.

$$y_1 = x_1$$
$$y_2 = x_2 - ceil(s_{1,0}y_1)$$
$$y_3 = x_3 - ceil(s_{2,0}y_1 + s_{2,1}y_2)$$
$$\ldots$$
$$y_{n_c} = x_{n_c} - ceil(s_{n_c-1,0}y_1 + s_{n_c-1,1}y_2 + s_{n_c-1,2}y_3 + \ldots + s_{n_c-1,n_c-2}y_{n_c-1})$$

A corresponding minimum noise ladder-based structure using S for decoding follows.

$$x_1 = y_1$$
$$x_2 = y_2 + ceil(s_{1,0}y_1)$$
$$x_3 = y_3 + ceil(s_{2,0}y_1 + s_{2,1}y_2)$$
$$\ldots$$
$$x_{n_c} = y_{n_c} + ceil(s_{n_c-1,0}y_1 + s_{n_c-1,1}y_2 + s_{n_c-1,2}y_3 + \ldots + s_{n_c-1,n_c-2}y_{n_c-1})$$

Figure 7.4 shows the diagram for a minimum noise ladder-based structure of PLT for encoding and decoding where the transform P and the flooring quantizer are used. Similarly, Fig. 7.5 shows the diagram for a minimum noise ladder-based structure of PLT for encoding and decoding where the inverse transform S and the ceiling quantizer are used.

Note that in both designs, all inputs to the multipliers of the transform kernel are quantized values so that these same values can be used in reconstruction to restore the original values. To save storage space, quantization is not only applied to

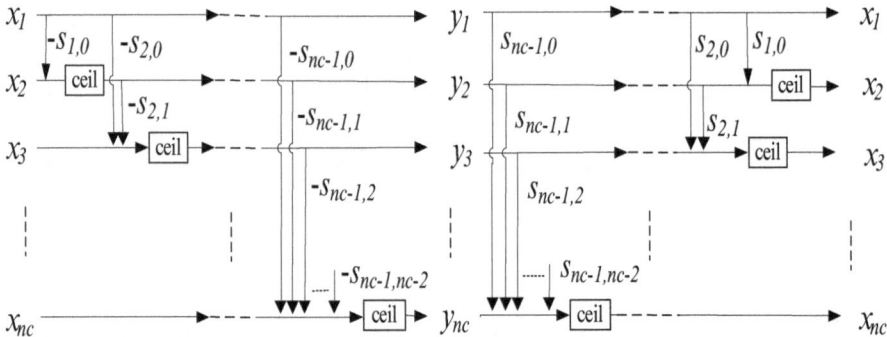

Fig. 7.5 A minimum noise ladder-based structure of PLT based on the inverse transform S and the ceiling quantizer. It also has the perfect reconstruction property for lossless compression

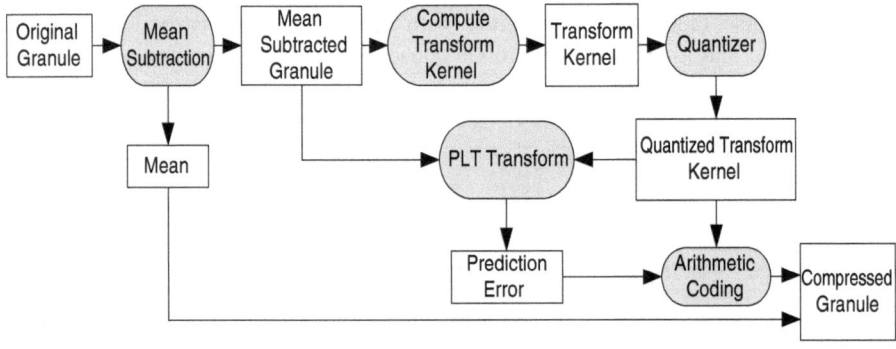

Fig. 7.6 The data flow diagram of the PLT compression scheme

the transform coefficients but also to the transform kernel. Furthermore, the traditional standard scalar or vector quantization techniques, which require transmission of the codebook, are not used. Instead, a simple ceiling or flooring on a specified number of decimal places is used for the codebook free quantization. With the minimum noise ladder-based structure being used, only the quantized transform kernel and the quantized prediction error need to be sent for restoration. Both then go through the arithmetic coder [19] in order to enhance the compression ratio.

The data flow diagrams of our PLT compression and decompression schemes are shown in Figs. 7.6 and 7.7 respectively. As the original granule in Fig. 7.6 is mean-subtracted, in addition to the compressed transform kernel and the compressed prediction error, a mean vector of $n_c \times 1$ has to be sent to the decoder in Fig. 7.7 to recover the original granule.

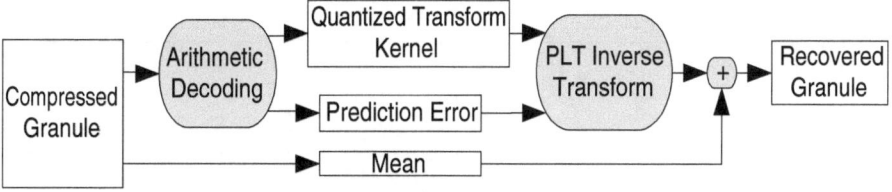

Fig. 7.7 The data flow diagram of the PLT decompression scheme

4   Results

The standard test data of ten AIRS ultraspectral digital count granules are tested. As performance index, the compression ratio and the bit rate are used. Both minimum noise ladder-based structures of PLT based on the transform P (PLT-P) and the inverse transform S (PLT-S) are tested. The quantizer used in PLT-S is tuned to keep the precision of the multipliers in transform S to two decimal places while the quantizer used in PLT-P is tuned to keep the precision of the multipliers in transform P to three decimal places. As the transformation process uses ceiling or flooring quantizers, the transform coefficient or the prediction error is an integer. A context-based arithmetic coder [20] is used to encode the transform kernel and the prediction error.

The ten plots in Fig. 7.8 show the variances of channels for the ten AIRS test granules before and after the PLT transform. The dotted curve for X is the variances of channels before the PLT transform. The solid curve for Y is the variances of channels after the PLT transform. Throughout the ten semilog plots for the ten granules, it can be seen that the variances of Y are significantly lower than those of X at most channels. The lower variances of the transform coefficients Y promise less bits in need for compression.

Figure 7.9 shows the amount of energy compaction in terms of the AM/GM ratio for the ten granules. The AM of a signal is the arithmetic mean of the variances of its subbands. The GM of a signal is the geometric mean of the variances of its subbands. It is shown that the ratio of AM to GM will always be greater than or equal to 1 [21]. AM will be equal to GM only when the variances of all subbands are equal. When a signal has a higher AM/GM ratio, there is a higher amount of energy compaction in its subband signals and the signal is good for data compression [22]. In Fig. 7.9, the AM/GM ratio is consistently much higher for the transformed coefficient Y than for the original signal X. This means that the energy compaction is much higher after the PLT transform than before the transform.

Figure 7.10 shows the coding gain of the PLT transform. The coding gain of a transform coding $\mathbf{T}$ is defined as the mean squared reconstruction error in PCM coding divided by the mean squared reconstruction error in the transform coding $\mathbf{T}$ [23]. The reconstruction error is the absolute difference between the reconstructed signal and the original signal. In fact, the coding gain of the PLT transform can be

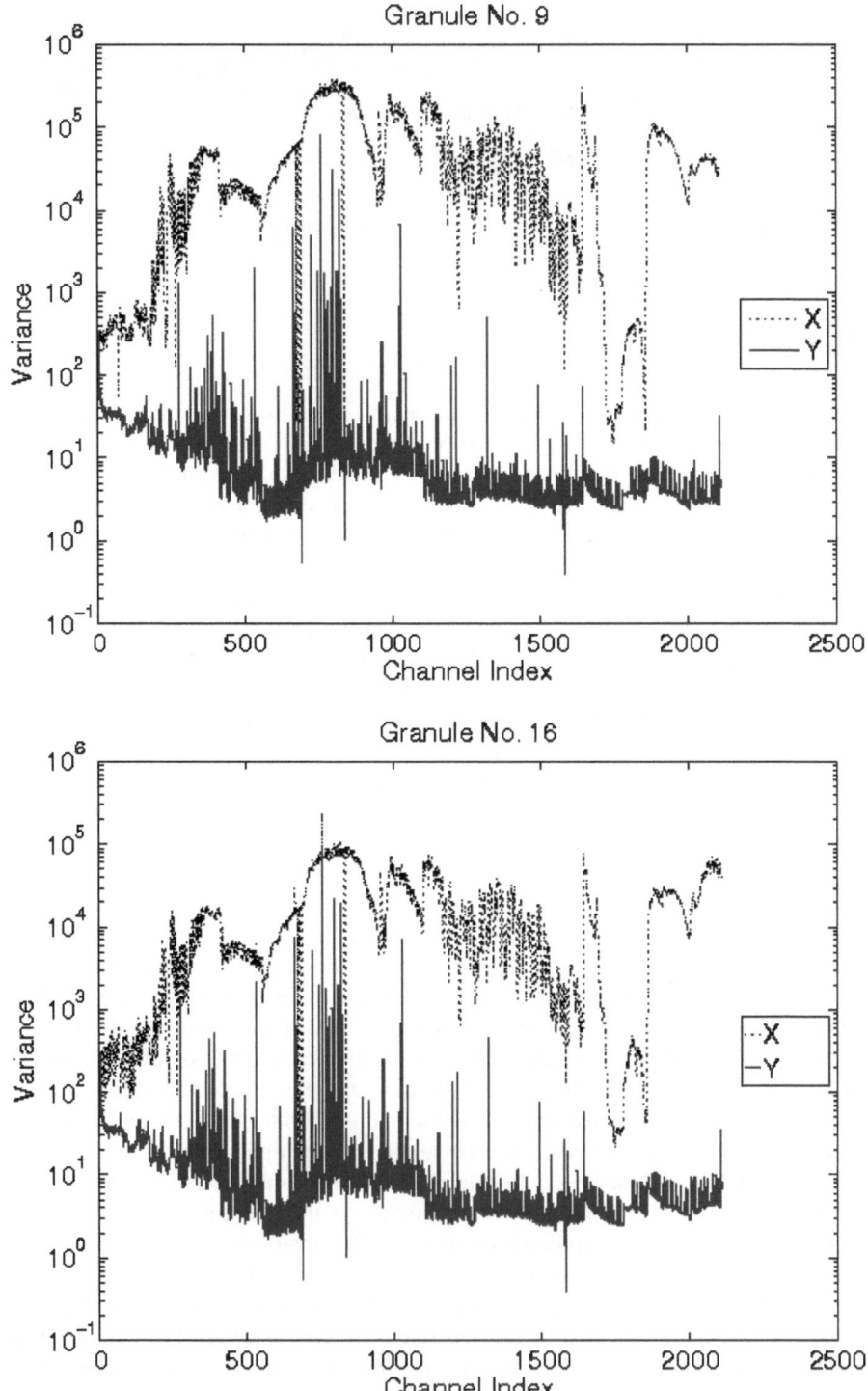

Fig. 7.8 The variances of channels X and Y before and after the PLT transform respectively for the ten AIRS granules in the dataset

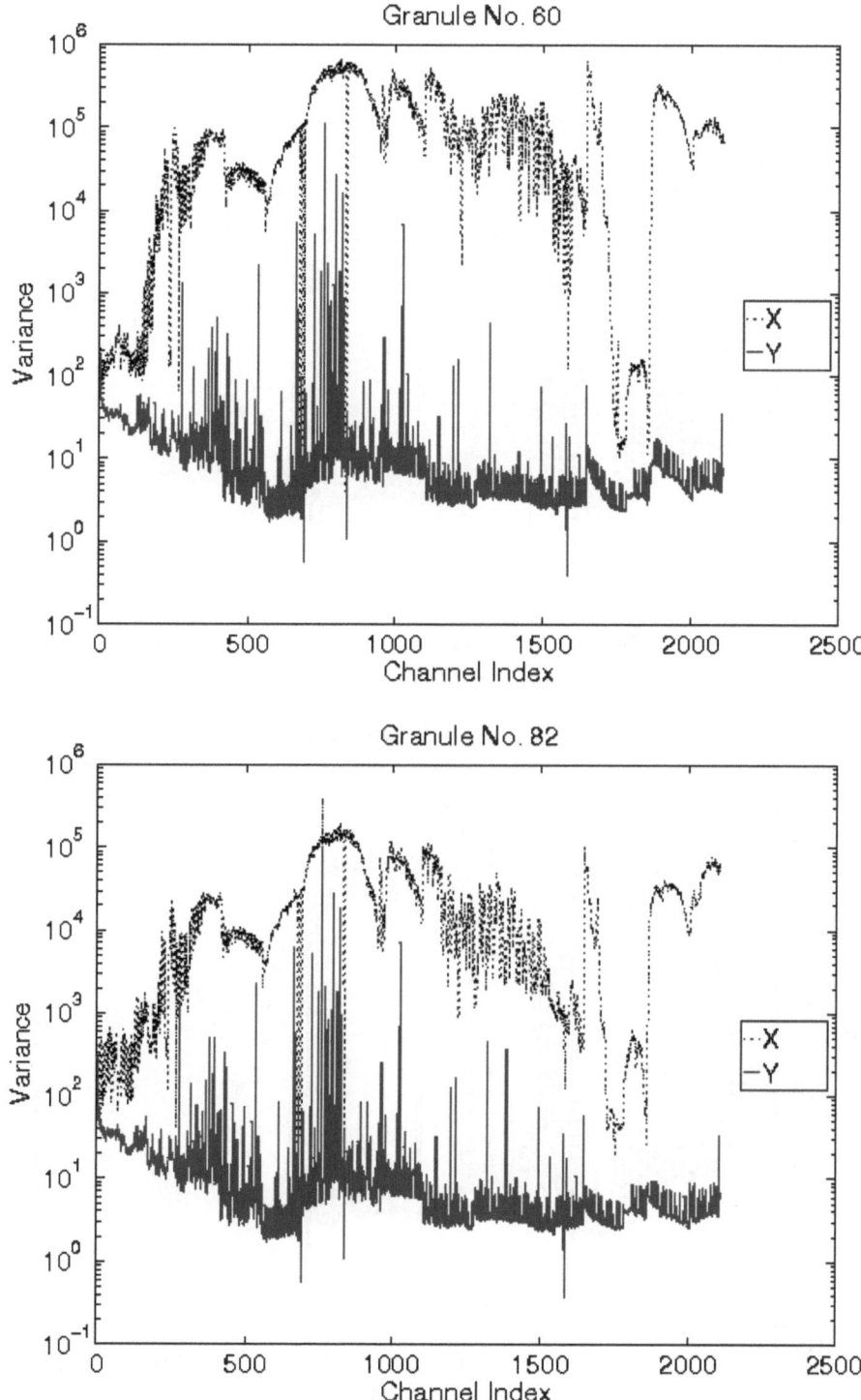

Fig. 7.8 (continued)

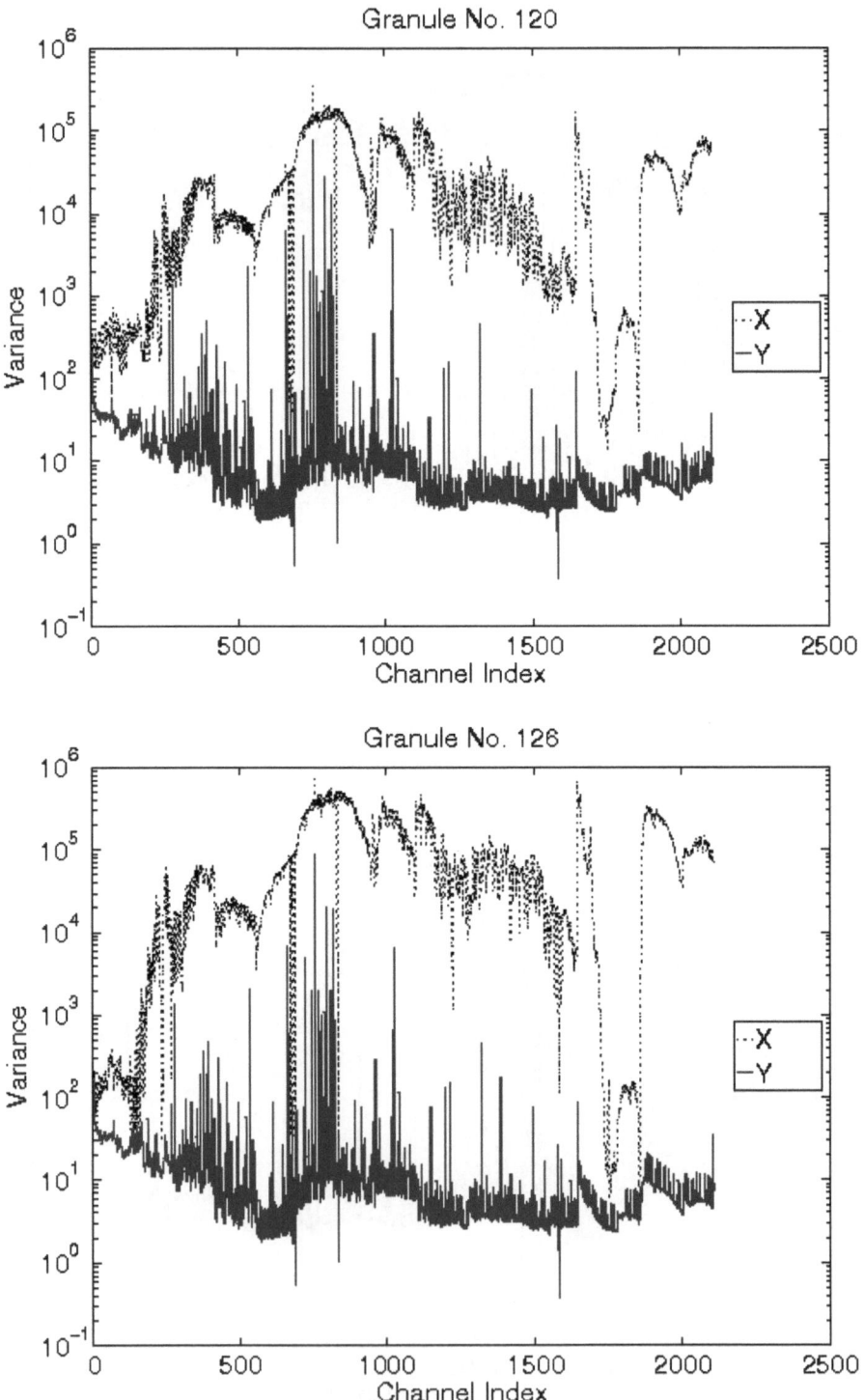

Fig. 7.8 (continued)

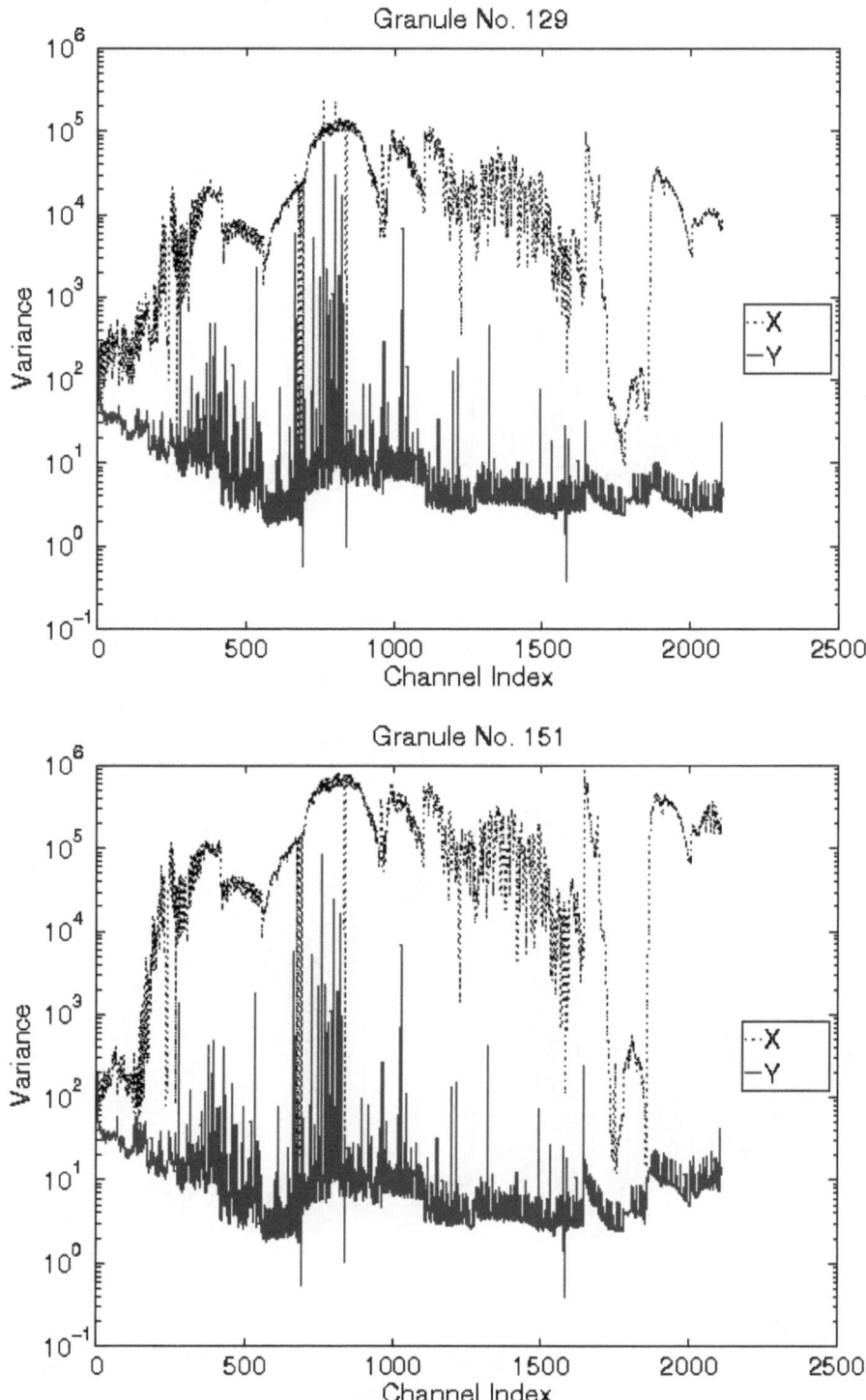

Fig. 7.8 (continued)

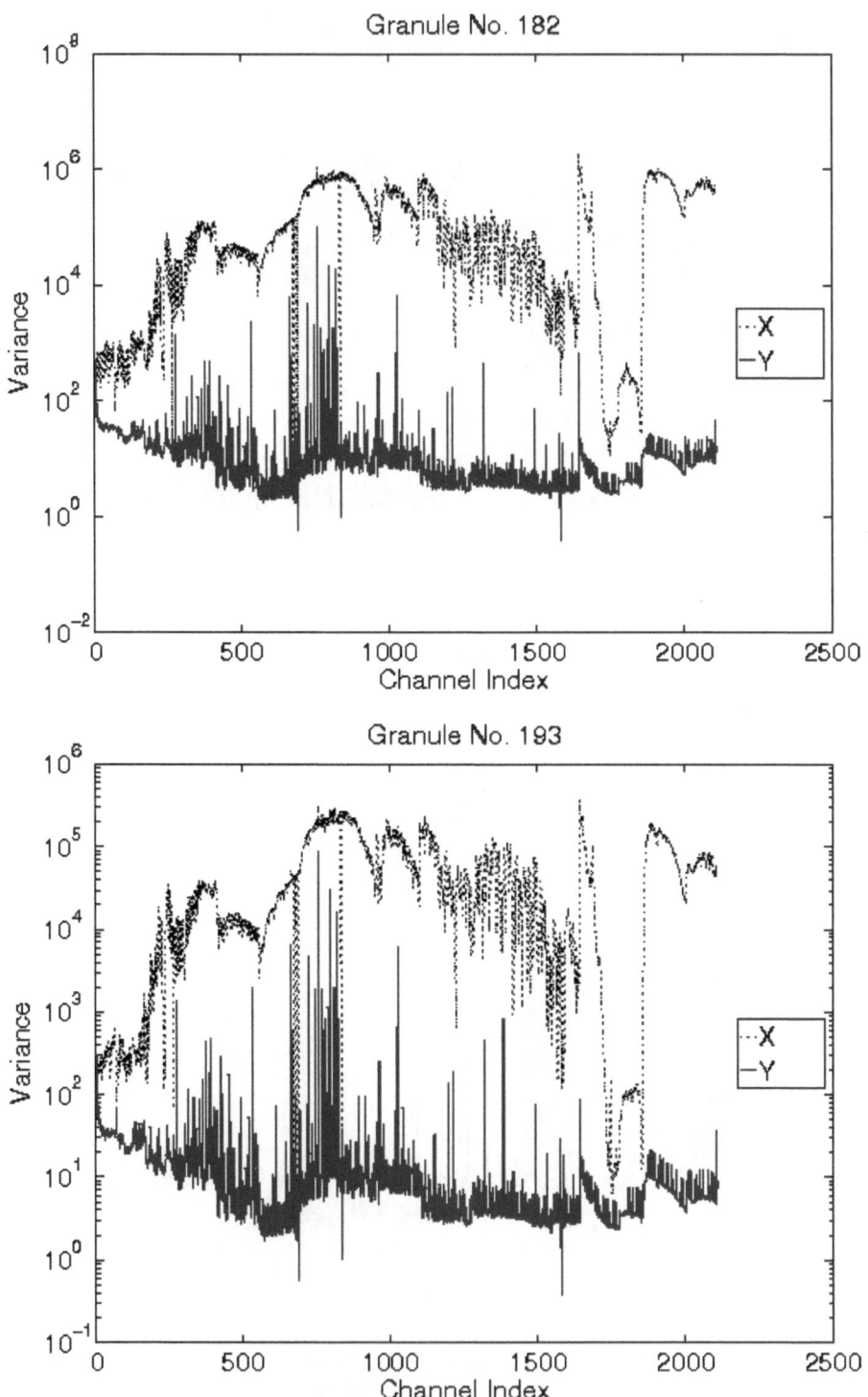

Fig. 7.8 (continued)

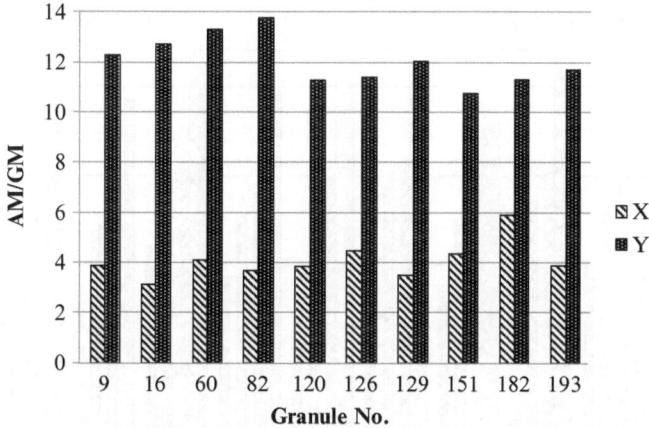

Fig. 7.9 The amount of energy compaction of the original signal X and the transformed signal Y as measured by the ratio of AM/GM for the ten AIRS granules in the dataset

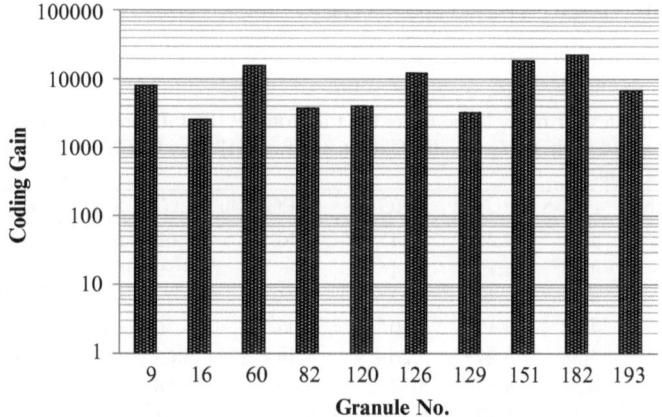

Fig. 7.10 The coding gain of the PLT transform for the ten AIRS granules in the dataset. The PLT transform has the same maximum coding gain as the KLT transform

computed by dividing the arithmetic mean of the subband variances (AM) of the original signal by the geometric mean of the subband variances (GM) of the transform coefficient [9]. In Fig. 7.10, the PLT transform has achieved the maximum coding gain, same as the KLT transform.

Figure 7.11 shows the compression ratio of the ten tested granules. The compression ratio is defined as the size of the original file divided by the size of the compressed file. For the AIRS granules, an original file containing $135 \times 90 \times 2107$ samples takes up about 41.2 MB. A higher compression ratio denotes a better compression result. The figure shows that PLT-S is slightly better than PLT-P.

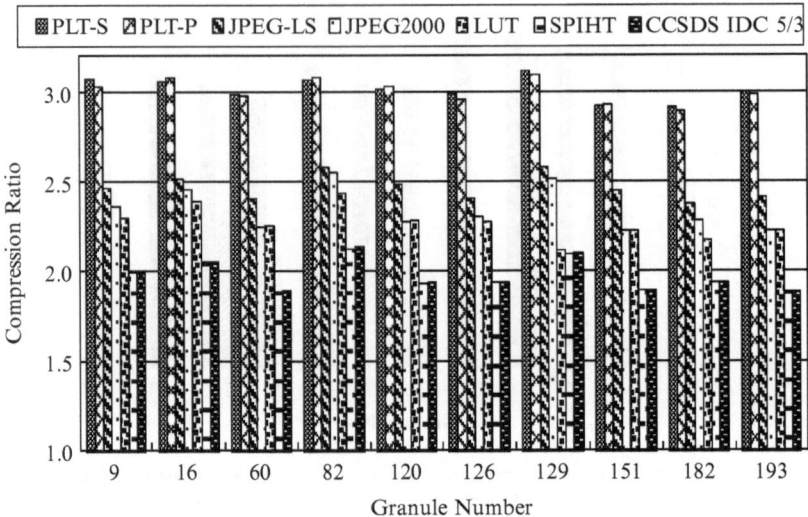

Fig. 7.11 The compression ratios of our compression scheme using the transform *P* (PLT-P) and the inverse transform *S* (PLT-S) in comparison with JPEG-LS, JPEG2000, LUT, SPIHT, and CCSDS IDC 5/3 on the ten ultraspectral granules

Table 7.2 Bit rates for our scheme using the transform *P* (PLT-P) and the inverse transform *S* (PLT-S) compared with JPEG-LS, JPEG2000, LUT, SPIHT, and CCSDS IDC 5/3 on the ten ultraspectral granules

Granule number	PLT-S	PLT-P	JPEG-LS	JPEG2000	LUT	SPIHT	CCSDS IDC 5/3
9	4.21	4.26	5.24	5.46	5.63	6.48	6.45
16	4.23	4.19	5.14	5.26	5.39	6.29	6.29
60	4.32	4.34	5.37	5.75	5.73	6.86	6.82
82	4.22	4.19	5.00	5.06	5.30	6.08	6.05
120	4.28	4.27	5.20	5.68	5.65	6.71	6.68
126	4.31	4.37	5.37	5.60	5.68	6.68	6.68
129	4.15	4.17	5.00	5.14	6.11	6.17	6.14
151	4.42	4.41	5.28	5.81	5.81	6.82	6.82
182	4.44	4.47	5.44	5.65	5.94	6.68	6.68
193	4.30	4.33	5.35	5.81	5.81	6.86	6.86

Moreover both PLT-S and PLT-P compare favorably with the prediction-based methods like LUT [11], JPEG-LS [12], the wavelet-transform-based methods like JPEG2000 [13], SPIHT [14] and the CCSDS recommendation for Image Data Compression (IDC)-5/3 scheme [15, 24] in all 10 granules [24].

The alternative compression result in terms of the bit rate is summarized in Table 7.2. The bit rate is defined as the number of bits used per sample. A lower bit rate means that fewer bits are required to encode a sample and therefore denotes better result of compression. For the AIRS granules, a sample takes up 12–14 bits depending on its channel. In average, the bit rate of the original file is 12.9 bits per samples.

5 Conclusions

The ultraspectral sounder data is characterized by its huge size and low tolerance of noise and error. Use of lossless or near lossless compression is thus desired for data transfer and storage. There have been prior works on lossless compression of ultraspectral sounder data which can be categorized into clustering-based, prediction-based, and transformation-based methods. In this work, a transformation-based method using the prediction-based lower triangular transform (PLT) is proposed. In our formulation, the PLT can use the P transform (PLT-P) or the S transform (PLT-S) for compression. To save space, a simple codebook-free quantization is applied to the transform kernel and the transform coefficient which is the prediction error for the PLT transform. Due to the minimum noise ladder structure in the PLT design, we can fully recover the granule from the quantized transform kernel and coefficients. To enhance the compression ratio, both the quantized transform kernel and the quantized transform coefficients are fed to the arithmetic coder for entropy coding.

In terms of the compression ratio, the result shows that the PLT compression scheme compares favorably with the prediction-based methods like LUT and JPEG-LS, the wavelet-transform-based methods like JPEG2000, SPIHT and CCSDS IDC 5/3 on all the ten standard AIRS granules. However, one disadvantage of the method is that intensive CPU computation is required by our PLT compression scheme. Therefore, future work includes reduction of the CPU time requirements by adopting parallel computation.

References

1. H. H. Aumann and L. Strow, "AIRS, the first hyper-spectral infrared sounder for operational weather forecasting," *Proc. of IEEE Aerosp. Conf.*, 4, pp. 1683–1692, 2001.
2. H. J. Bloom, "The Cross-track Infrared Sounder (CrIS): a sensor for operational meteorological remote sensing," *Proc. of the 2001 Int. Geosci. and Remote Sens. Symp.*, pp. 1341–1343, 2001.
3. T. Phulpin, F. Cayla, G. Chalon, D. Diebel, and D. Schlussel, "IASI onboard Metop: Project status and scientific preparation," *12th Int. TOVS Study Conf.*, Lorne, Victoria, Australia, pp. 234–243, 2002.
4. W. L. Smith, F. W. Harrison, D. E. Hinton, H. E. Revercomb, G. E. Bingham, R. Petersen, and J. C. Dodge, "GIFTS – the precursor geostationary satellite component of the future Earth Observing System," *Proc. of the 2002 Int. Geosci. and Remote Sens. Symp.*, 1, pp. 357–361, 2002.
5. B. Huang, W. L. Smith, H.-L. Huang, and H. M. Woolf, "Comparison of linear forms of the radiative transfer equation with analytic Jacobians", *Appl. Optics*, vol. 41, no. 21, pp. 4209–4219, 2002.

6. B. Huang, A. Ahuja, and H.-L. Huang, "Lossless compression of ultraspectral sounder data," *Hyperspectral Data Compression*, G. Motta and J. Storer; Eds., Springer-Verlag, pp. 75–106, 2005.

7. P. Toivanen, O. Kubasova, and J. Mielikainen, "Correlation-based band-ordering heuristic for lossless compression of hyperspectral sounder data", *IEEE Geosci. Remote Sens. Lett.*, vol. 2, no. 1, pp.50–54, 2005.

8. B. Huang, A. Ahuja, H.-L. Huang, T. J. Schmit, and R. W. Heymann, "Lossless compression of 3D hyperspectral sounding data using context-based adaptive lossless image codec with bias-adjusted reordering," *Optical Engineering*, vol. 43, no. 9, pp. 2071–2079, 2004.

9. S.-M. Phoong and Y.-P. Lin, "Prediction-based lower triangular transform," *IEEE Trans. Signal Processing*, vol. 48, no. 7, pp. 1947–1955, 2000.

10. C.-C. Weng, C.-Y. Chen and P. P. Vaidyanathan, "Generalized triangular decomposition in transform coding," *IEEE Trans. Signal Processing*, vol. 58, no. 2, pp. 566–574, 2010.

11. B. Huang, and Y. Sriraja, "Lossless compression of hyperspectral imagery via lookup tables with predictor selection," *Proc. SPIE*, vol. 6365, pp.63650L.1, 2006.

12. ISO/IEC 14495–1 and ITU Recommendation T.87, "Information Technology – lossless and near-lossless compression of continuous-tone still images," 1999.

13. D. S. Taubman and M. W. Marcellin, *JPEG2000: Image compression fundamentals, standards, and practice*, 2002.

14. A. Said, and W. A. Pearlman, "A new, fast, and efficient image codec based on set partitioning in hierarchical trees," *IEEE Trans. Circuits. Sys. Video Tech.*, vol. 6, pp. 243–250, June 1996.

15. CCSDS, "Consultative Committee for Space Data Systems," http://www.ccsds.org.

16. B. Huang, A. Ahuja, H.-L. Huang, T.J. Schmit, R.W. Heymann, "Fast precomputed VQ with optimal bit allocation for lossless compression of ultraspectral sounder data", *Proc. IEEE Data Comp. Conf.*, pp. 408–417, March 2005.

17. G. H. Golub and C. F. V. Loan, *Matrix computations*, John Hopkins University Press, 1996.

18. A. Gersho and R. M. Gray, *Vector quantization and signal compression*, Kluwer Academic Publishers, 1992.

19. I. Witten, R. Neal, and J. Cleary, "Arithmetic coding for data compression", *Comm. ACM*, vol. 30, no. 6, pp. 520–540, June 1987.

20. M. R. Nelson, "Arithmetic coding and statistical modeling", *Dr. Dobb's Journal*, pp. 16–29, February 1991.

21. Y. You, *Audio coding- theories and applications*, Springer, 2010.

22. K. Sayood, *Introduction to data compression*, 2nd Ed., Morgan Kaufmann Publishers, 2000.

23. N. S. Jayant and P. Noll, *Digital coding of waveforms- principles and applications to speech and video*, Prentice Hall, 1984.

24. J. Serra-Sagrista, F. Garcia, J. Minguillon, D. Megias, B. Huang, and A. Ahuja, "Wavelet lossless compression of ultraspectral sounder data," *Proc. Int. Geosci. Rem. Sens. Symp.*, vol. 1, pp. 148–151, July 2005.

Chapter 8
Lookup-Table Based Hyperspectral Data Compression

Jarno Mielikainen

Abstract This chapter gives an overview of the lookup table (LUT) based lossless compression methods for hyperspectral images. The LUT method searches the previous band for a pixel of equal value to the pixel co-located to the one to be coded. The pixel in the same position as the obtained pixel in the current band is used as the predictor. Lookup tables are used to speed up the search. Variants of the LUT method include predictor guided LUT method and multiband lookup tables.

1 Introduction

Hyperspectral imagers produce enormous data volumes. Thus, a lot of effort has been spent to research more efficient ways to compress hyperspectral images. Three different types of compression modalities for hyperspectral images can be defined. Lossy compression achieves the lowest bit rate among the three modalities. It does not bind the difference between each reconstructed pixel and the original pixel. Instead, the reconstructed image is required to be similar to the original image on mean-squared error sense. Near lossless compression bounds the absolute difference between each reconstructed pixel and the original pixel by a predefined constant. Lossless compression requires the exact original image to be reconstructed from the compressed data. Since lossless compression techniques involve no loss of information they are used for applications that cannot tolerate any difference between the original and reconstructed data.

In hyperspectral images the interband correlation is much stronger than the intraband correlation. Thus, interband correlation must be utilized for maximal compression performance. Transform-based and vector-quantization-based methods have not been able to achieve state-of-the-art lossless compression results

J. Mielikainen (✉)
School of Electrical and Electronic Engineering, Yonsei University, Seoul, South Korea
e-mail: mielikai@gmail.com

B. Huang (ed.), *Satellite Data Compression*, DOI 10.1007/978-1-4614-1183-3_8,
© Springer Science+Business Media, LLC 2011

for hyperspectral images. Therefore, lossless compression of hyperspectral data is performed by using prediction-based approaches. However, there have been some studies on transform-based [1–3] and vector-quantization based [4–6] methods. Vector quantization is an asymmetric compression method; compression is much more computationally intensive than decompression. On the other hand, transform-based methods have been more successful in lossy compression than lossless compression.

Prediction based methods for lossless compression of hyperspectral images can be seen as consisting of three steps:

1. Band ordering.
2. Modeling extracting information on the redundancy of the data and describing this redundancy in the form of a model.
3. Coding describes the model and how it differs from the data using a binary alphabet.

The problem of optimal band ordering for hyperspectral image compression has been solved in [7]. Optimal band reordering is achieved by computing a minimum spanning tree for a directed graph containing the sizes of the encoded residual bands. A correlation-based heuristic for estimating the optimal order was proposed in [8]. Another prediction method based on reordering was introduced in [9]. However, in this chapter, all the experiments are performed using natural ordering of the bands to facilitate comparisons to the other methods in the literature.

In this chapter, we concentrate on lookup table (LUT) based approaches to modeling and we are will gives an overview of LUT based lossless compression methods for hyperspectral images.

This chapter is organized as follows. In Sect. 2 we will present a short review of previous work in lossless compression of hyperspectral images. Section 3 presents basic LUT method. In Sect. 4 predictor guided LUT is described. Use of a quantized index in LUT method is discussed in Sect. 5. Multiband generalization of LUT method is presented in Sect. 6. Experiments results are shown in Sect. 7. Finally, conclusions are drawn in Sect. 8.

2 Lossless Compression of Hyperspectral Images

Previous approaches to lossless compression of hyperspectral images include A1, which is one of three distributed source coding algorithms proposed in [10]. It focuses on coding efficiency and the other two algorithms proposed in [10] are more focused on error-resiliency. The A1 algorithm independently encodes non-overlapped blocks of 16×16 samples in each band. This independency makes it easy to parallelize the algorithm. The first block of each band is transmitted uncompressed. The pixel values are predicted by a linear prediction that utilizes pixel value in previous bans, the average pixel values of both the current block and the co-located block in the previous band. Instead of sending prediction parameters to decoder they are guessed by the decoder. For each guess the pixels of the block are reconstructed and the Cyclic Redundancy Check (CRC) is computed. Once CRC matches the one included in the compressed file, the process terminates.

The FL algorithm [11] employs the previous band for prediction and adapts the predictor coefficients using recursive estimation. The BG block-based compression algorithm [12] employs a simple block-based predictor followed by an adaptive Golomb code. IP3 (third-order interband predictor) [13] method takes advantage of spatial data correlation and derives spectral domain predictor using Wiener filtering. They also employed a special backward pixel search (BPS) module for calibrated image data.

Clustered differential pulse code modulation (C-DPCM) [14] method partitions spectral vectors into clusters and then applies a separate least-squares optimized linear predictor to each cluster of each band. The method can be seen as an extension of the vector quantization method in [5]. However, the quantization step of [5] is omitted. In [15], another approach using clustering was presented. The causal neighborhoods of each pixel are clustered using fuzzy-c-means clustering. For each of the clusters, an optimal linear predictor is computed from the values, the membership degrees of which exceed a threshold. The final estimate is computed as a weighted sum of the predictors, where the weights are the membership degrees. The Spectral Fuzzy Matching Pursuits (S-FMP) method exploits a purely spectral prediction. In the same paper, a method called Spectral Relaxation-Labeled Prediction (S-RLP) was also proposed. The method partitions image bands into blocks, and a predictor, out of a set of predictors, is selected for prediction.

A method based on Context-Adaptive Lossless Image Coding (CALIC), which is called 3-D CALIC [28], switches between intra- and interband prediction modes based on the strength of the correlation between the consecutive bands. In multiband CALIC (M-CALIC) method [16], the prediction estimate is performed using two pixels in the previous bands in the same spatial position as the current pixel. The prediction coefficients are computed using an offline procedure on training data. An adaptive least squares optimized prediction technique called Spectrum-oriented Least SQuares (SLSQ) was presented in [17]. The prediction technique used is the same as the one in [18], but a more advanced entropy coder was used. The predictor is optimized for each pixel and each band in a causal neighborhood of the current pixel. SLSQ-HEU uses a heuristic to select between the intra- and interband compression modes. Also, an optimal method for inter-/intracoding mode selection called SLSQ-OPT was presented.

Selecting between a Correlation-based Conditional Average Prediction (CCAP) and a lossless JPEG was proposed in [19]. The selection is based on a correlation coefficient for contexts. The CCAP estimate is a sample mean of pixels corresponding to the current pixel in contexts that match the current pixel context. BH [20] is a block-based compressor. Each band of the input image is divided into square blocks. Next, the blocks are predicted based on the corresponding block in the previous band. Nonlinear Prediction for Hyperspectral Images (NPHI) [21] predicts the pixel in the current band based on the information in the causal context in the current band and pixels colocated in the reference band. NPHI was also extended into an edge-based technique, called the Edge-based Prediction for Hyperspectral Images, which classifies the pixels into edge and nonedge pixels. Each pixel is then predicted using information from pixels in the same pixel class within the context. In [23], a method called KSP, which employs a Kalman filter in the prediction stage, was proposed.

3   LUT Method

The LUT method [22] makes a prediction of the current pixel $p_{x,y,z}$ (xth row, yth column, and zth band) using all the causal pixels in the current and previous band. LUT method is based on the idea of Nearest Neighbor (NN) search. The NN procedure searches for the nearest neighbor in the previous band that has the same pixel value as the pixel located in the same spatial position as the current pixel in the previous band $p_{x,y,z-1}$. The search is performed in reverse raster-scan order. First, a pixel value equal to $p_{x,y,z-1}$ is searched. If an equal valued pixel is found at position $(x',y',z-1)$, then estimated pixel is predicted to have the same value as the pixel in the same position as obtained pixel in the current band $p_{x',y',z}$. Otherwise, the estimated pixel value is equal to the pixel value in the previous band $p_{x,y,z-1}$.

LUT method accelerates NN method by replacing time consuming search procedure with a lookup table operation, which uses the pixel co-located in the previous band as an index in the lookup table. The lookup table returns the nearest matching pixel.

An example illustrating the search process is shown in Figs. 8.1–8.3. The example uses two consecutive image bands, which have 3×8 pixels each. The previous band (band number 1) and current band (band number 2) are shown in Figs. 8.1 and 8.2, respectively. The corresponding lookup table is shown in Fig. 8.3. In the example, pixel $p_{3,8,2} = 325$ is the current pixel to be predicted in the current band. The causal pixels in the previous band are searched to find a match for the co-located pixel $p_{3,8,1} = 315$. Both current pixel and its co-located pixel have yellow background in Figs. 8.2 and 8.1, respectively. Three matches (green background) are returned. The pixel value in the current band that is present at the nearest matching location, $p_{2,6,1} = 315$, is used as the predictor for $p'_{3,8,2} = p_{2,6,2} = 332$. A time-consuming search was avoided because the lookup table directly returned the predictor value.

4   Predictor Guided LUT Method

In the LUT method the nearest matching pixel value might be not be as good of a match as many other matching pixels. In the previous example the pixels in the current band corresponding to the other two matching locations are closer to

Fig. 8.1 Previous image band. Co-located pixel has *yellow background*. Matching pixels have *green background*

336	335	314	335	314	335	319	327
316	315	317	315	328	315	325	319
322	334	329	314	329	324	317	315

Fig. 8.2 Current image band. Current pixel has *yellow background*. Pixels corresponding to the matching pixel have *green backgrounds*

328	339	323	339	328	332	331	335
335	324	325	327	320	332	327	335
330	350	339	324	333	325	333	325

Fig. 8.3 Lookup table

Index	Value
314	328
315	332
316	335
317	333

the actual pixel value 325 than the nearest matching pixel value 332. This type of behavior of LUT method motivated the development of Locally Averaged Interband Scaling (LAIS)-LUT method [29], which uses a predictor to guide the selection between two LUTs.

LAIS-LUT method works by first computing a LAIS estimate by scaling pixel co-located in the previous band. The LAIS scaling factor is an average of ratios between three neighboring causal pixels in the current and previous band:

$$\frac{1}{3}\left(\frac{P_{x-1,y,z}}{P_{x-1,y,z-1}} + \frac{P_{x,y-1,z}}{P_{x,y-1,z-1}} + \frac{P_{x-1,y-1,z}}{P_{x-1,y-1,z-1}}\right) \tag{8.1}$$

LAIS scaling factor in (8.1) is used to compute an estimate for the current pixel:

$$p''_{x,y,z} = \frac{1}{3}\left(\frac{P_{x-1,y,z}}{P_{x-1,y,z-1}} + \frac{P_{x,y-1,z}}{P_{x,y-1,z-1}} + \frac{P_{x-1,y-1,z}}{P_{x-1,y-1,z-1}}\right)P_{x,y,z-1} \tag{8.2}$$

Fig. 8.4 LAIS estimates for
LAIS-LUT

Pixel Position	Pixel Value	LAIS Estimate
(2,3)	324	320.1
(2,5)	327	321.9
(2,7)	332	316.2

Fig. 8.5 Two lookup tables
for LAIS-LUT

index	1st LUT	2nd LUT
314	328	324
315	332	327
316	335	-
317	333	325

LAIS-LUT uses two LUTs, which are similar to the one used in the LUT method. The second LUT is updated with the past entries of the first LUT. The predictor returned by the LUT that is the closest one to the LAIS estimate is chosen as the predictor for the current pixel. If the LUTs return no match then the LAIS estimate is used as the estimated pixel value.

We use the LUT example to illustrate the search process in LAIS-LUT. LAIS estimates for the three matching pixels in the previous example are shown in Fig. 8.4. Two LUTs corresponding to bands in Figs. 8.1 and 8.2 are shown in Fig. 8.5. Recall that the current pixel is $p_{3,8,2} = 325$ and the causal pixels in the previous band are searched to find a match for the co-located pixel $p_{3,8,1} = 315$. Out of the three matching pixels two are in LUTs (green background in Fig. 8.5). LAIS estimate (321.9) for 2nd LUT value 327 is closer than LAIS estimate (316.2) for the first LUT value 332. Therefore, pixel value from second LUT is used as the predictor for $p'_{3,8,2} = p_{2,5,2} = 327$.

5 Uniform Quantization of Co-Located Pixels

In [24], a quantization of indices in LUT method was proposed. In LAIS-QLUT method a uniform quantization of the co-located pixels is performed before using them for indexing the LUTs. The use of quantization reduces the size of the LUTs by an order of magnitude A quantized interband predictor is formed by uniformly quantizing the colocated pixel $p_{x,y,z-1}$ before using it as an index to the LUT. Naturally, this reduces the size of the LUTs by the factor that is used in the uniform quantization.

Except for a slightly simpler LAIS from [25] LAIS and an additional quantization step, LAIS-QLUT is the same algorithm as LAIS-LUT.

The LAIS scaling factor in LAIS-QLUT is an average of ratios between three neighboring causal pixels in current and previous band:

$$\frac{1}{3}\left(\frac{P_{x-1,y,z}+P_{x,y-1,z}+P_{x-1,y-1,z}}{P_{x-1,y,z-1}+P_{x,y-1,z-1}+P_{x-1,y-1,z-1}}\right) \tag{8.3}$$

Thus, the corresponding LAIS estimate the current pixel is the following:

$$p''_{x,y,z}=\frac{1}{3}\left(\frac{P_{x-1,y,z}+P_{x,y-1,z}+P_{x-1,y-1,z}}{P_{x-1,y,z-1}+P_{x,y-1,z-1}+P_{x-1,y-1,z-1}}\right)P_{x,y,z-1} \tag{8.4}$$

LAIS in LAIS-QLUT requires a division operation and four addition operations compared to the three division, one multiplication, and two addition operations required by LAIS in LAIS-LUT.

The search process in LAIS-QLUT will be illustrated using the same image bands are in the previous example. Quantized version of the previous image band is shown in Fig. 8.6 for a quantization factor 10. LAIS-Q estimates for two matching pixels are shown in Fig. 8.7. Two LUTs for LAIS-QLUT are shown in Fig. 8.8 for a quantization factor 10. The current pixel is $p_{3,8,2} = 325$ and the causal pixels in the previous band are searched to find a match for quantized co-located pixel $p_{3,8,1} / 10 = 32$. Two of matching pixels, which are in LUTs have LAIS-Q estimates of 328.2 for first LUT value 333 and 328.3 for second LUT value 325. The second LUT value is closer to the corresponding LAIS-Q estimate than the other one. Therefore, pixel value from the first LUT is used as the predictor for $p'_{3,8,2} = p_{3,6,2} = 324$.

34	34	31	34	31	34	32	33
32	32	32	32	33	32	33	32
32	33	33	31	33	32	32	32

Fig. 8.6 Quantized previous image band. Co-located pixel has *yellow background.* Matching pixels have *green background*

Fig. 8.7 LAIS estimates for
LAIS-QLUT

Pixel Position	Pixel Value	LAIS Estimate
(3,6)	325	328.3
(3,7)	333	328.2

Fig. 8.8 Two lookup table
for LAIS-QLUT

index	1st LUT	2nd LUT
31	324	328
32	333	325
33	339	350
34	332	339

There are two separate variants of LAIS-QLUT. The first variant, The LAIS-QLUT-OPT method selects the optimal uniform quantization factor for each band. In order to find the optimal quantization factor, an exhaustive search of all possible quantization values is performed. Thus, the quantization factor selection is based on which quantization factor achieves the best compression efficiency for that specific band. The excessive time complexity of the LAIS-QLUT-OPT method could be decreased slightly by computing entropy of the residual image instead of actually encoding residuals for the determination of the optimal quantization factor.

The second variant of LAIS-QLUT is called LAIS-QLUT-HEU and it uses constant quantization factors. The constant quantization factors are selected using a heuristic. The heuristic selects the constant quantization factors to be the bandwise mean values of the optimal quantization factors of an image set. A division operation required by the quantization represents the only increase in the time complexity of LAIS-QLUT-HEU compared to LAIS-LUT.

6 Multiband LUT

In [26], LUT and LAIS-LUT method have been generalized to a multiband and multi-LUT method. In the extended method, the prediction of the current band relies on N previous bands. LUTs are defined on each of the previous bands and each band contains *M* LUTs. Thus, there are *NM* different predictors to choose from. The decision among one of the possible prediction values is based on the closeness of the values contained in the LUTs to a reference prediction.

Two different types of purely spectral multiband prediction estimates were proposed for. One of the reference predictors is crisp and the other one is fuzzy. The first method is S-RLP [15]. The method partitions image bands into blocks, and a predictor, out of a set of predictors, is selected for prediction. In the S-FMP method [15] the causal neighborhoods of each pixel are clustered using fuzzy-c-means clustering. For each of the clusters, an optimal linear predictor is computed from the values, the membership degrees of which exceed a threshold. The final estimate is computed as a weighted sum of the predictors, where the weights are the membership degrees. The LUT based compression methods based on S-RLP and S-FMP are denoted as S-RLP-LUT and S-FMP-LUT, respectively.

7   Experimental Results

Airborne Visible/Infrared Imaging Spectrometer (AVIRIS) is an airborne hyperspectral system collecting spectral radiance in 224 contiguous spectral bands with wavelengths from 370 to 2,500 nm. The AVIRIS instrument consists of four spectrometers that view a *20-m²* spot on the ground from a flight altitude of *20 km*. This spot is simultaneously viewed in all the spectral bands. A spatial image is formed by moving the spectrometers perpendicular to the direction of the aircraft [27].

Experimental results are shown for two different AVIRIS data sets. The first data set consists of four calibrated radiance images from 1997 AVIRIS sample data product. The AVIRIS images are from the following four different areas: Cuprite, NV; Jasper Ridge, CA; Lunar Lake, NV; and Moffett Field, CA. They are the most widely used data for benchmarking hyperspectral image compression algorithms. Image features and the number of lines are listed in Table 8.1. Each image contains 614 samples/line and they are stored as 16-bit signed integers. A gray scale image of Moffett Field image can be seen in Fig. 8.9.

Newer data set was acquired on 2006. A new AVIRIS data set consists of five calibrated and uncalibrated 16-bit images from Yellowstone, WY and two 12-bit uncalibrated images one from Hawaii and one from Maine. Summary of the new Consultative Committee for Space Data Systems (CCSDS) AVIRIS data is given in Table 8.2. Each image is a 512-line scene containing 224 spectral bands. An example of a scene can be seen in Fig. 8.10 in the form of a false color image of calibrated Yellowstone scene 11.

Table 8.1 The standard 1997 AVIRIS images [11]

Site	Features	Lines
Cuprite	Geological features	2,206
Jasper Ridge	Vegetation	2,586
Lunar Lake	Calibration	1,431
Moffett Field	Vegetation, urbar, water	2,031

Fig. 8.9 Gray scale image of
Moffett Field image from
AVIRIS 1997 image set

This AVIRIS data is a part of the CCSDS data set, which is used to assess the performance of hyperspectral compression algorithms.

Table 8.3 shows results for the NN method. The first column depicts the length of the search window; 0 lines means that only the current line is searched. The following columns are bit rates in bits/pixel for the four test images and the average, respectively. When the search window's length is equal to the length of image, the method naturally predicts the same values as the LUT method. These results show

Table 8.2 AVIRIS images included in the CCSDS test set [11]

Site	Scene numbers	Year	Samples/line	Bit depth	Type
Yellowstone	0,3,10,11,18	2006	677	16	Calibrated
Yellowstone	0,3,10,11,18	2006	680	16	Uncalibrated
Hawaii	1	2001	614	12	Uncalibrated
Maine	10	2003	680	12	Uncalibrated

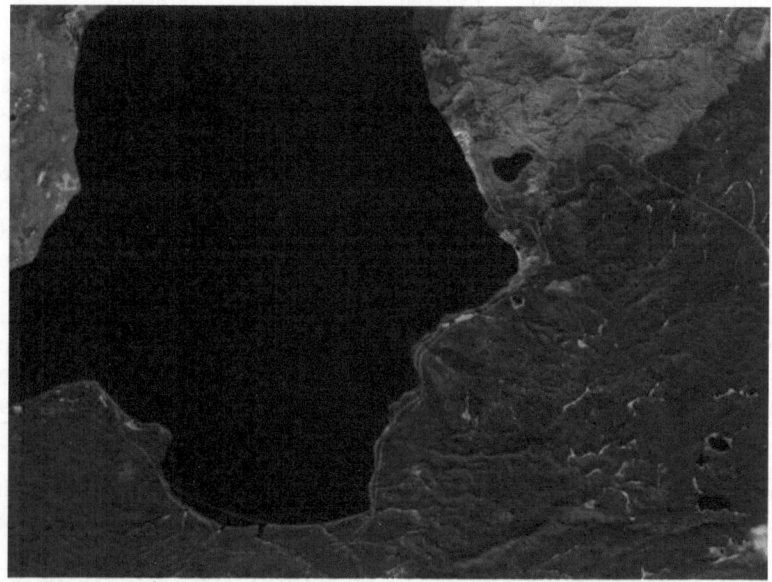

Fig. 8.10 False color image of calibrated Yellow stone 11 from CCSDS AVIRIS data set

Table 8.3 Compression results in bits/pixel for calibrated AVIRIS 1997 test images in bits per pixel

# of lines	Cuprite	Jasper ridge	Lunar lake	Moffett field
0	5.69	5.84	5.78	6.02
1	5.41	5.63	5.50	5.80
2	5.29	5.50	5.33	5.65
4	5.05	5.35	5.14	5.48
8	4.89	5.21	4.98	5.32
16	4.79	5.10	4.88	5.21
32	4.72	5.03	4.79	5.14
64	4.69	5.00	4.75	5.10
128	4.68	4.98	4.73	5.08
256	4.66	4.97	4.72	5.06
512	4.66	4.97	4.72	5.05
1,024	4.65	4.95	4.71	5.05

Table 8.4 Compression results in bits/pixel for calibrated AVIRIS 1997 test images in bits per pixel

	Cuprite	Jasper ridge	Lunar lake	Moffett field	Average
JPEG-LS	7.66	8.38	7.48	8.04	7.89
Diff. JPEG-LS	5.50	5.69	5.46	5.63	5.57
3D-CALIC	5.23/5.39	5.19/5.37	5.18/5.32	4.92/5.05	5.19/5.28
BH	–/5.11	–/5.23	–/5.11	–/5.26	–/5.18
M-CALIC	4.97/5.10	5.05/5.23	4.88/5.02	4.72/4.89	4.98/5.06
SLSQ-OPT	4.94/5.08	4.95/5.08	4.95/5.08	4.98/5.10	4.96/5.09
CCAP	–/4.92	–/4.95	–/4.97	–	–
KSP	–/4.88	–/4.95	–/4.89	–/4.92	–/4.91
FL#	4.82	4.87	4.82	4.93	4.86
NPHI	4.79	4.89	4.97	4.79	4.86
C-DPCM	–/4.68	–/4.62	–/4.75	–/4.62	–/4.67
S-RLP	4.69	4.65	4.69	4.67	4.67
S-FMP	4.66	4.63	4.66	4.63	4.64
LUT	4.66	4.95	4.71	5.05	4.84
LAIS-LUT	4.47	4.68	4.53	4.76	4.61
LAIS-QLUT-HEU	4.30	4.62	4.36	4.64	4.48
LAIS-QLUT-OPT	4.29	4.61	4.34	4.63	4.47
S-RLP-LUT	3.92	4.05	3.95	4.09	4.00
S-FMP-LUT	3.89	4.03	3.92	4.05	3.97
IP3-BPS	3.76	4.06	3.79	4.06	3.92

that limiting the search window size significantly affects the performance of the NN method compared to the full search. Thus, a large search window is necessary in order to achieve good compression ratios.

Table 8.4 shows compression results for AVIRIS 1997 data. The results are reported for the band-interleaved-by-line (BIL) and band-sequential (BSQ) formats. In the BIL format, the current line, along with the two previous lines, is available. For BSQ data, the current band and several previous bands are available for processing. The LUT family does not benefit from the BSQ data format. This is due to two factors. First, LUT and LAIS-LUT methods only utilize one previous band. Second, LAIS-LUT methods need only the data from the current and previous image lines. Those lines were already provided by the BIL data format. Most compression method exhibit identical compression results for both BIL and BSQ data. Only one bits/pixel value is shown for those methods. For the other methods both BIL and BSQ results are provided. The results for the two different data formats are separated by a forward-slash and dash denotes unavailable results. Differential JPEG-LS computes the difference between each band and the previous band before running JPEG-LS on residual data.

Experimental results show that LUT based algorithms work extremely well for calibrated AVIRIS 1997 data. Even the low time complexity LAIS-LUT and QLAIS-LUT variants have close to the state-of-the-art compression ratios. IP3-BPS method takes ten times longer than LUT and five times longer than LAIS-LUT or LAIS-QLUT-HEU to compress AVIRIS image [13].

Table 8.5 Compression results in bits/pixel for 16-bit raw CCSDS AVIRIS test images in bits per pixel

Algorithm	Scene 0	Scene 3	Scene 10	Scene 11	Scene 18	Average
JPEG-LS	9.18	8.87	7.32	8.50	9.30	8.63
BG	6.46	6.31	5.65	6.05	6.40	6.17
A1	6.92	6.78	6.10	6.53	6.92	6.65
LUT	7.13	6.91	6.25	6.69	7.20	6.84
LAIS-LUT	6.78	6.60	6.00	6.30	6.82	6.50
FL#	6.20	6.07	5.60	5.81	6.26	5.99
IP3	6.20	6.08	5.56	5.81	6.25	5.98
C-DPCM-20	5.88	5.71	5.20	5.52	5.75	5.61
C-DPCM-80	5.82	5.65	5.17	5.47	5.69	5.56

Table 8.6 Compression results in bits/pixel for 12-bit raw CCSDS AVIRIS test images in bits per pixel

Algorithm	Hawaii	Maine	Average
JPEG-LS	4.58	4.50	4.54
A1	3.49	3.65	3.57
LUT	3.27	3.44	3.36
LAIS-LUT	3.05	3.19	3.12
BG	3.03	3.17	3.10
IP3	2.55	2.68	2.62
FL#	2.58	2.63	2.61
C-DPCM-20	2.43	2.57	2.50
C-DPCM-80	2.38	2.52	2.45

The LUT method requires a full LUT for each band. Assuming 16-bit LUTs, each LUT's memory requirements are roughly equivalent to 107 lines of an AVIRIS image data. The LUT's memory requirements are independent of the spatial size of the image. Therefore, the relative size of the LUTs compared to the image gets smaller as the spatial size of the image gets larger. For our test images, the amount of the memory required by LUTs is 4–7% of the memory used by the image. The average quantization factor for LAIS-QLUT-HEU was 28. Thus, the average LUT memory requirement is roughly equivalent to four lines of AVIRIS image data compared to 107 lines of data in the original LUT method. We have also experimented with the optimization of the quantization factors for each image instead of for each band. That procedure gave a quantization factor of ten for all the test images. The average bit rate was 4.60 bits/pixel. This compares unfavorably to the 4.47 bits/pixel average bit rate of LAIS-QLUT-HEU. Therefore, separate bandwise quantization factors are worthwhile.

Tables 8.5–8.7 depict compression results for new AVIRIS data in bits per pixel for various different compression methods. C-DPCM-20 and C-DPCM-80 refer to the prediction length 20 and 80 for C-DPCM, respectively. A modified C-DPCM method uniformly quantizes coefficients to 12 bits instead of 16 bits in the original C-DPCM.

Table 8.7 Compression results in bits/pixel for calibrated CCSDS AVIRIS test images in bits per pixel

Algorithm	Scene 0	Scene 3	Scene 10	Scene 11	Scene 18	Average
JPEG-LS	6.95	6.68	5.19	6.24	7.02	6.42
A1	4.81	4.69	4.01	4.41	4.77	4.54
LUT	4.81	4.62	3.95	4.34	4.84	4.51
LAIS-LUT	4.48	4.31	3.71	4.02	4.48	4.20
BG	4.29	4.16	3.49	3.90	4.23	4.01
FL#	3.91	3.79	3.37	3.59	3.90	3.71
IP3	3.81	3.66	3.13	3.45	3.75	3.56
C-DPCM-20	3.61	3.43	2.97	3.28	3.49	3.36
C-DPCM-80	3.53	3.36	2.93	3.22	3.43	3.29

The results for uncalibrated CCSDS AVIRIS test data in Tables 8.5 and 8.6 show that LUT-based methods lose their performance advantage when applied to uncalibrated data. Moreover, the results in Table 8.7 show that LUT-based algorithms that exploit calibration artifacts in AVIRIS 1997 images have no performance advantage on the calibrated CCSDS AVIRIS images.

8 Conclusions

An overview of the lookup table (LUT) based lossless compression methods for hyperspectral images have been presented in this chapter. Experimental results on AVIRIS data showed that the LUT based algorithms work extremely well for old calibrated AVIRIS data. Even the low-complexity LAIS-LUT and QLAIS-LUT variants have close to the state-of-the-art compression ratios.

LUT-based methods exploit artificial regularities that are introduced by the conversion of raw data values to radiance units [11]. The calibration-induced artifacts are not present in the newer AVIRIS images in Consultative Committee for Space Data Systems (CCSDS) test set. Thus, LUT based method do not work as well on raw or the newer AVIRIS images in 2006, which use new calibration measures.

Acknowledgement This work was supported by the Academy of Finland.

References

1. A. Bilgin, G. Zweig, and M. Marcellin, "Three-dimensional image compression with integer wavelet transforms," Appl. Opt., vol. 39, no. 11, pp. 1799–1814, Apr. 2000.
2. B. Baizert, M. Pickering, and M. Ryan, "Compression of hyperspectral data by spatial/spectral discrete cosine transform," in Proc. Int. Geosci. Remote Sens. Symp., 2001, vol. 4, pp. 1859–1861, doi: 10.1109/IGARSS.2001.977096.

3. J. Mielikainen and A. Kaarna, "Improved back end for integer PCA and wavelet transforms for lossless compression of multispectral images," in Proc. 16th Int. Conf. Pattern Recog., Quebec City, QC, Canada, 2002, pp. 257–260, doi: 10.1109/ICPR.2002.1048287.
4. M. Ryan and J. Arnold, "The lossless compression of AVIRIS images vector quantization," IEEE Trans. Geosci. Remote Sens., vol. 35, no. 3, pp. 546–550, May 1997, doi: 10.1109/36.581964.
5. J. Mielikainen and P. Toivanen, "Improved vector quantization for lossless compression of AVIRIS images," in Proc. XI Eur. Signal Process. Conf., Toulouse, France, Sep. 2002, pp. 495–497.
6. G. Motta, F. Rizzo, and J. Storer, "Partitioned vector quantization application to lossless compression of hyperspectral images," in Proc. IEEE Int. Conf. Acoust., Speech, Signal Process., Jul. 2003, vol. 1, pp. 553–556, doi: 10.1109/ICME.2003.1220977.
7. S. Tate, "Band ordering in lossless compression of multispectral images," IEEE Trans. Comput., vol. 46, no. 4, pp. 477–483, Apr. 1997, doi: 10.1109/12.588062.
8. P. Toivanen, O. Kubasova, and J. Mielikainen, "Correlation-based bandordering heuristic for lossless compression of hyperspectral sounder data," IEEE Geosci. Remote Sens. Lett., vol. 2, no. 1, pp. 50–54, Jan. 2005, doi: 10.1109/LGRS.2004.838410.
9. J. Zhang and G. Liu, "An efficient reordering prediction based lossless compression algorithm for hyperspectral images," IEEE Geosci. Remote Sens. Lett., vol. 4, no. 2, pp. 283–287, Apr. 2007, doi: 10.1109/LGRS.2007.890546.
10. A. Abrardo, M. Barni, E. Magli, F. Nencini, "Error-Resilient and Low-Complexity On-board Lossless Compression of Hyperspectral Images by Means of Distributed Source Coding," IEEE Trans. Geosci. Remote Sens., vol. 48, no. 4, pp. 1892–1904, 2010, doi:10.1109/TGRS.2009.2033470.
11. A. B. Kiely, M. A. Klimesh, "Exploiting Calibration-Induced Artifacts in Lossless Compression of Hyperspectral Imagery," IEEE Trans. Geosci. Remote Sens., vol. 47, no. 8, pp. 2672–2678, 2009, doi:10.1109/TGRS.2009.2015291.
12. M. Slyz, L. Zhang, "A block-based inter-band lossless hyperspectral image compressor," in Proc. of IEEE Data Compression Conference, pp. 427–436, 2005, doi: 10.1109/DCC.2005.1.
13. C.-C. Lin, Y.-T. Hwang., "An Efficient Lossless Compression Scheme for Hyperspectral Images Using Two-Stage Prediction", vol. 7, no. 3, pp. 558–562, 2010, doi:10.1109/LGRS.2010.2041630.
14. J. Mielikainen, P. Toivanen, "Clustered DPCM for the Lossless Compression of Hyperspectral Images", IEEE Trans. Geosci. Remote Sens., vol. 41, no. 12, pp. 2943–2946, 2003 doi:10.1109/TGRS.2003.820885.
15. B. Aiazzi, L. Alparone, S. Baronti, and C. Lastri, "Crisp and fuzzy adaptive spectral predictions for lossless and near-lossless compression of hyperspectral imagery," IEEE Geosci. Remote Sens. Lett., vol. 4, no. 4, pp. 532–536, Oct. 2007, 10.1109/LGRS.2007.900695.
16. E. Magli, G. Olmo, and E. Quacchio, "Optimized onboard lossless and near-lossless compression of hyperspectral data using CALIC," IEEE Geosci. Remote Sens. Lett., vol. 1, no. 1, pp. 21–25, Jan. 2004, doi:10.1109/LGRS.2003.822312.
17. F. Rizzo, B. Carpentieri, G. Motta, and J. Storer, "Low-complexity lossless compression o hyperspectral imagery via linear prediction," IEEE Signal Process. Lett., vol. 12, no. 2, pp. 138–141, Feb. 2005, doi:10.1109/LSP.2004.840907.
18. J. Mielikainen and P. Toivanen, "Parallel implementation of linear prediction model for lossless compression of hyperspectral airborne visible infrared imaging spectrometer images," J. Electron. Imaging, vol. 14, no. 1, pp. 013010-1–013010-7, Jan.–Mar. 2005, doi:10.1117/1.1867998.
19. H. Wang, S. Babacan, and K. Sayood, "Lossless Hyperspectral-Image Compression Using Context-Based Conditional Average," IEEE Transactions on Geoscience and Remote Sensing, vol. 45, no. 12, pp. 4187–8193, Dec. 2007, doi:0.1109/TGRS.2007.906085.

20. M. Slyz and L. Zhang, "A block-based inter-band lossless hyperspectral image compressor," in Proc. Data Compression Conf., Snowbird, UT, 2005, pp. 427–436, doi:10.1109/DCC.2005.1.
21. S. Jain and D. Adjeroh, "Edge-based prediction for lossless compression of hyperspectral images," in Proc. Data Compression Conf., Snowbird, UT, 2007, pp. 153–162, doi:10.1109/DCC.2007.36.
22. J. Mielikainen, "Lossless compression of hyperspectral images using lookup tables," IEEE Sig. Proc. Lett., vol. 13, no. 3, pp. 157–160, 2006, doi:10.1109/LSP.2005.862604.
23. E. Magli, "Multiband lossless compression of hyperspectral images," *IEEE Transactions on Geoscience and Remote Sensing*, vol. 47, no. 4, pp. 1168–1178, Apr. 2009, doi:10.1109/TGRS.2008.2009316.
24. J. Mielikainen, P. Toivanen, "Lossless Compression of Hyperspectral Images Using a Quantized Index to Lookup Tables," vol. 5, no. 3, pp. 474–477, doi:10.1109/LGRS.2008.917598.
25. J. Mielikainen, P. Toivanen, and A. Kaarna, "Linear prediction in lossless compression of hyperspectral images," Opt. Eng., vol. 42, no. 4, pp. 1013–1017, Apr. 2003, doi:10.1117/1.1557174.
26. B. Aiazzi, S. Baronti, S., L. Alparone, "Lossless Compression of Hyperspectral Images Using Multiband Lookup Tables," IEEE Signal Processing Letters, vol. 16, no. 6, pp. 481–484. Jun. 2009, doi:10.1109/LSP.2009.2016834, 0.1109/LSP.2009.2016834.
27. W. Porter and H. Enmark, "A system overview of the Airborne Visible/Infrared Imaging Spectrometer (AVIRIS)," Proc. SPIE, vol. 834, pp. 22–31, 1997
28. X. Wu and N. Memon, "Context-based lossless interband compression—Extending CALIC," IEEE Trans. Image Process., vol. 9, no. 6, pp. 994–1001, Jun. 2000, doi:10.1109/83.846242.
29. B. Huang, Y. Sriraja, "Lossless compression of hyperspectral imagery via lookup tables with predictor selection," in Proc. SPIE, vol. 6365, pp. 63650L.1–63650L.8, 2006, doi:10.1117/12.690659.

Chapter 9
Multiplierless Reversible Integer TDLT/KLT for Lossy-to-Lossless Hyperspectral Image Compression

Jiaji Wu, Lei Wang, Yong Fang, and L.C. Jiao

1 Introduction

Hyperspectral images have wide applications nowadays such as in atmospheric detection, remote sensing and military affairs. However, the volume of a hyperspectral image is so large that a 16bit AVIRIS image with a size $512 \times 512 \times 224$ will occupy 112 M bytes. Therefore, efficient compression algorithms are required to reduce the cost of storage or bandwidth.

Lossy-to-lossless compression will be of great importance in telemedicine and satellite communications for legal reasons and research requirements. To realize scalable coding, most of the state-of-the-art compression methods adopt three dimensional discrete wavelet transform (3D-DWT) [1–3] or wavelet transform/karhunen-loeve transform (DWT/KLT) [4–6], where a 9/7 floating-point filter (9/7F filter) is always used for de-correlation in lossy compression. Lossless compression schemes include methods based on vector quantization (VQ), prediction, integer transforms and so on. Although prediction-based methods perform well, they do not have the ability to perform progressive lossy-to-lossless compression since this depends on transform-based methods [7]. Sweldens [8, 9] proposed a lifting scheme for the realization of wavelet transforms. Bilgin et al. [10] introduced a reversible integer wavelet transform method for 3D image compression. Xiong et al. [11] applied 3D integer wavelet transforms for medical image compression and pointed out that the transform has to be unitary to achieve good lossy coding performance. Some researchers have studied integer KLT for spectral decorrelation. Hao et al. [12] proposed reversible integer KLT (RKLT) and Galli et al. [13] improved it. However, in the spatial domain, integer wavelet transforms are still commonly

J. Wu (✉) • L. Wang • L.C. Jiao
School of Electronic Engineering, Xidian University, xi'an, China
e-mail: wujj@mail.xidian.edu.cn

Y. Fang
College of Information Engineering, Northwest A&F University, yangling, China

B. Huang (ed.), *Satellite Data Compression*, DOI 10.1007/978-1-4614-1183-3_9,
© Springer Science+Business Media, LLC 2011

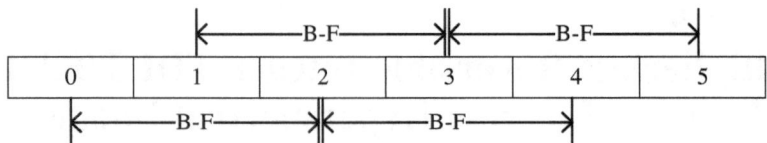

Fig. 9.1 B-F standards for basis functions of LOT; one B-F imposes on three neighboring segments but just output one segment of values

used. A drawback of the wavelet-based compression method is that 53DWT is usually applied instead of 97DWT in lossy-to-lossless compression schemes, and this will lead to performance degradation. Another disadvantage of DWT is that it cannot compete with DCT, due to the constraint of CPU performance and computer memory, especially in real-time and low-complexity applications, because the computing complexity of DWT increases exponentially with image size [14].

The computational complexity of DWT increases when the image size increases because of the global transform [14]. However, DCT has its own special advantages such as low memory cost, flexibility at block by block level and parallel processing. DCT is approximately equal to the KLT basis matrix for the first-order Markov process while image segments always satisfy this condition. Therefore, DCT performs well in image decorrelation and it is widely adopted by image/video compression standards such as JPEG, MPEG, and H.26X. Although the DCT-based coding method has been a popular method for image and video compression, a key problem of this type of coding at low bit rates is the so-called "block effect". The reason is because DCT-based coding always independently processes each block. In order to reduce the block effect of DCT compression, some deblocking methods based on filtering were proposed [15, 16], in which some low-pass filters were applied to the boundary pixels. However, filtering-based methods usually blur image content.

To resolve the problem of the block effect of DCT, lapped orthogonal transform (LOT) was proposed by Cassereau [17] and an analytical solution has been given by Malvar [18]. LOT improves DCT by designing basis functions which impose themselves on neighboring segments, as depicted in Fig. 9.1. Correlations between neighboring segments can be explored in this way, and discontinuity between reconstructed segments can be reduced [19]. In lossy compression, although LOT can reduce block effects efficiently, the lapped filtering of LOT has to follow behind the DCT. For this reason, forward LOT is difficult to make compatible with a DCT-based coding standard.

To overcome the disadvantage of traditional LOT, Tran et al. [20, 21] have designed a family of time domain lapped transforms (TDLT) by adding various pre- and post-filters in the existing block-based architecture in the time domain. Tran's algorithm can achieve competitive compression performance compared with DWT-based methods while reducing or even eliminating the block artifacts to guarantee good visual quality. TDLT, a combination of pre- and post-filters with DCT transform, can be illustrated in this way; the inputs of DCT and the outputs of the inverse DCT are processed by a pre-filter and post-filter, respectively. The pre-filter for TLDT is placed in front of the forward DCT, so the TDLT is easily made compatible with current DCT-based coding standards. The filtering process is conducted on the two neighboring blocking coefficients. The function of the pre-filter is to make the input data of each DCT block as homogenous as possible, like a flattening operation, whereas the function of the post-filter is to reduce the block artifacts. In [20], lossy compression has been realized by using different versions of TDLT based on various decompositions of the filtering matrix, and lossless compression has also been realized by using a reversible transform for lifting-based filters and multiplier-less approximations of DCT, known as binDCT in [22]. BinDCT is realized by quantizing the transform coefficients of conventional plane rotation-based factorizations of the DCT matrix, and can be implemented using only binary shift and addition operations.

Microsoft has developed a TDLT-based image coding technique called HD-Photo [23, 24], which enables reversible compression by using lifting scheme. HD-Photo has been taken as the basis technique for JPEG-XR, which is a new compression format supporting high dynamic range and promising significant improvements in image quality and performance for end-to-end digital photography. In HD-Photo, a hierarchical lapped biorthogonal transform (LBT) is adopted and the Huffman coding is performed in chunks organized as a function of resolution [23]. Both lossy and lossless compression can be realized by HD-Photo and this is one of its advantages over JPEG, which needs two coders in different applications. In HD-Photo, the LBT is realized by factorizing the core transform and overlap filtering into rotation operators and is implemented using a lifting structure to promise a reversible integer transform [25]. In addition, the new compression scheme retains advantages such as in-place calculation, amenability to parallelized implementation, flexibility and adaptivity on the block level and so on.

Although TDLT performs even better than DWT does in energy compatibility and lossy compression, it does not perform well in the lossless compression where the reversible transform is required. In fact, for hyperspectral image compression, a completely reversible transform method is often required to realize lossy-to-lossless coding.

In this chapter, we take a practical and innovative approach to replace integer DWT with integer reversible time domain lapped transform (RTDLT) in the spatial domain, and RKLT is applied in the spectral dimension. Here this RTDLT and RKLT are realized by an improved matrix factorization method. RTDLT can realize integer reversible transform and hence we have adopted a progressive lossy-to-lossless hyperspectral image compression method based on RTDLT and RKLT. Block transforming coefficients in the spatial domain are reorganized into sub-band structures so as to be coded by wavelet-based coding methods. In addition, an improved 3D embedded zero-block coding method used to code transformed coefficients is integrated in this work.

Moreover, we also extend RTDLT to 3D reversible integer lapped transform (3D-RLT), which can replace 3D integer WT and realize progressive lossy- to-lossless compression and performs better than 3D-WT in most cases. 3D-RLT is implemented at block-level and has a fixed transform basis matrix. It is therefore suitable for memory-limited systems or on board spacecraft where component constraints are significant.

Our proposed methods retain the character of scalability in reconstruction quality and spatial resolution so, at the decoder, observers can review the whole image from inferior quality to the completely reconstructed image. Experimental results show that the proposed methods perform well in both lossy and lossless compression. To reduce complexity, our proposed methods are implemented using shift and addition without any multiplier, with the help of multi-lifting.

2   Multi-lifting Scheme

We adopt a lifting scheme to realize the reversible integer transform since lifting schemes have been widely used and have many advantages such as (a) fast implementation, (b) in-place calculations, (c) immediate access to inverse transform, (d) a natural understanding of original complex transforms. Firstly, we give a simple review of our lifting scheme.

2.1   Lifting Scheme and Applications

To realize a reversible integer transform, traditional algorithms always adopt dual-lifting, as depicted in Fig. 9.2 where x_i and y_i stand for input and output signal, respectively and p and u stand for prediction and update coefficients, respectively. For instance, Daubechies and Sweldens proposed a lifting-based wavelet transform method [9], Chen et al. [26] proposed Integer DCT using the Walsh-Hadamard Transform (WHT) and lifting scheme, Abhayaratne [27] proposed an N-point integer-to-integer DCT (I2I-DCT) by applying recursive methods and lifting techniques, where N is power of 2 and Liang et al. [22] proposed two kinds of fast multiplier-less approximations of the DCT called binDCT, also with the lifting scheme. In the HD-Photo process from Microsoft, the LBT is realized by factorizing the core transform and overlap filtering into rotational operators, as depicted in Fig. 9.3; the rotation operations are also realized by a dual-lifting structure [25].

Matrix lifting has also been proposed by Li [28], and is applied in embedded audio codec. Cheng et al. have introduced the properties of the lifting matrix, based on which a new family of lapped biorthogonal transforms has been designed [29].

Lifting-based transforms can realize completely reversible integer transforms and hence can be applied in lossless compression. At the same time, lifting is lower in CPU cost and memory usage since it allows a fully in-place calculation.

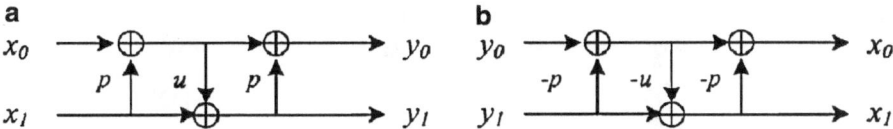

Fig. 9.2 Dual lifting structure: (**a**) Forward. (**b**) Inverse

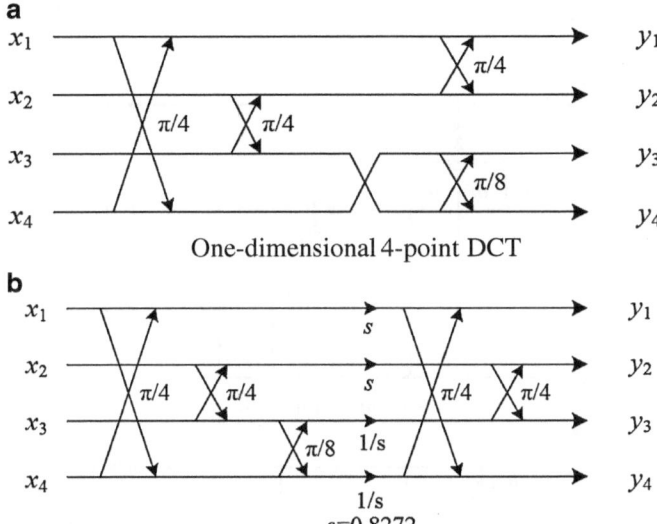

One-dimensional 4-point DCT

One-dimensional 4-point overlap filter

s=0.8272

Fig. 9.3 Transform structure of LBT in HD-Photo

2.2   *Reversible Integer Transform based on Multi-lifting Scheme*

In this chapter, we adopt a multi-lifting scheme, which is an extension of traditional dual lifting. Multi-lifting is the same concept as matrix lifting, which has been used in the JPEG2000 color transform.

If a 2×2 matrix **U** is an upper triangular matrix with the diagonal elements equal to 1, then $\mathbf{Y} = \mathbf{UX}$ can be realized as (9.1) from integer to integer; it can also be implemented by the lifting depicted in Fig. 9.4a. In addition, $\mathbf{X} = \mathbf{U}^{-1}\mathbf{Y}$ can be realized using inverse-lifting as depicted in Fig.9.4b.

$$\begin{pmatrix} y_0 \\ y_1 \end{pmatrix} = \begin{pmatrix} 1 & p \\ 0 & 1 \end{pmatrix} \begin{pmatrix} x_0 \\ x_1 \end{pmatrix} \rightarrow \begin{array}{cc} y_0 = x_0 + \lfloor px_1 \rfloor & x_1 = y_1 \\ y_1 = x_1 & x_0 = y_0 - \lfloor py_1 \rfloor \end{array} \quad (9.1)$$

In our proposed transforming scheme, 4-order, 8-order and 16-order lifting are used. For example, the 4-point reversible integer-to-integer transforms $\mathbf{Y} = \mathbf{UX}$ and its inverse transform can be realized as (9.2), where **U** is an upper triangular

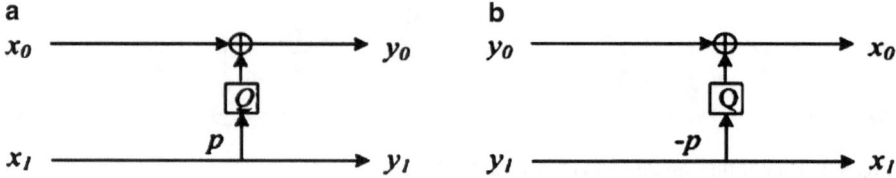

Fig. 9.4 Reversible integer to integer transform based on lifting: (**a**) Forward integer lifting.
(**b**) Inverse integer lifting

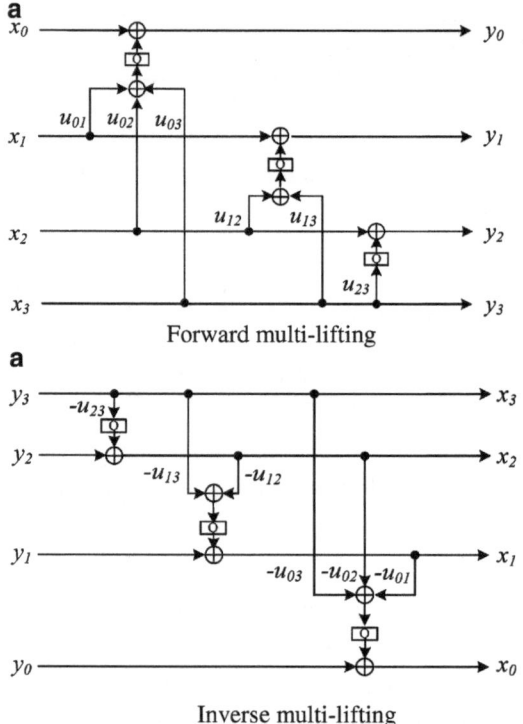

Fig. 9.5 Forward and inverse multi-lifting schemes

matrix of size 4. The multi-lifting implementation is also depicted in Fig. 9.5. What
should be noted is that the same rounding operation is implied on the sum of a set of
multiplications ($\sum_{j=i+1}^{N-1} u_{i,j}x_j$) in the forward and inverse transforms to guarantee
completely reversible integer-to-integer transforms.

$$
\begin{aligned}
y_0 &= x_0 + \lfloor u_{01}x_1 + u_{02}x_2 + u_{03}x_3 \rfloor & x_3 &= y_3, \\
y_1 &= x_1 + \lfloor u_{12}x_2 + u_{13}x_3 \rfloor & x_2 &= y_2 - \lfloor u_{23}x_3 \rfloor, \\
y_2 &= x_2 + \lfloor u_{23}x_3 \rfloor & x_1 &= y_1 - \lfloor u_{12}x_2 + u_{13}x_3 \rfloor, \\
y_3 &= x_3 & x_0 &= y_0 - \lfloor u_{01}x_1 + u_{02}x_2 + u_{03}x_3 \rfloor, \quad (9.2)
\end{aligned}
$$

3 Reversible Integer to Integer TDLT/RKLT

In this section, we will introduce the implementation of RTDLT/RKLT which is based on the matrix factorization method [30]. We will first introduce techniques of integer-to-integer transform by multi-lifting based on matrix factorization. We will then demonstrate details in the design of RTDLT/RKLT.

3.1 Matrix Factorization and Multi-lifting

Based on matrix factorization theory [30], a nonsingular matrix can be factorized into a product of at most three triangular elementary reversible matrices (TERMs). If the diagonal elements of the TERM are equal to 1, the reversible integer to integer transform can be realized by multi-lifting.

First, we review how to realize an approximation of floating-point to integer for a transform basis using matrix factorization theory.

Suppose $\mathbf{A}$ is a transform basis matrix,

$$\mathbf{A} = \begin{pmatrix} a_{1,1}^{(0)} & a_{1,2}^{(0)} & \cdots & a_{1,N}^{(0)} \\ a_{2,1}^{(0)} & a_{2,2}^{(0)} & \cdots & a_{2,N}^{(0)} \\ \vdots & \vdots & \ddots & \vdots \\ a_{N,1}^{(0)} & a_{N,2}^{(0)} & \cdots & a_{N,N}^{(0)} \end{pmatrix}. \tag{9.3}$$

There exists a permutation matrix P_1 which can make $P_{1,N}^1$ not equal to zero:

$$P_1 A = \begin{pmatrix} p_{1,1}^{(1)} & p_{1,2}^{(1)} & \cdots & p_{1,N}^{(1)} \\ p_{2,1}^{(1)} & p_{2,2}^{(1)} & \cdots & p_{2,N}^{(1)} \\ \vdots & \vdots & \ddots & \vdots \\ p_{N,1}^{(1)} & p_{N,2}^{(1)} & & p_{N,N}^{(1)} \end{pmatrix}. \tag{9.4}$$

There is an operator s_1 satisfying the following formula,

$$P_{1,1}^{(1)} - s_1 P_{1,N}^{(1)} = 1. \tag{9.5}$$

Formula (9.5) can be rewritten as,

$$s_1 = \left(P_{1,1}^{(1)} - 1 \right) / P_{1,N}^{(1)}. \tag{9.6}$$

Then

$$P_1AS_1 = P_1A \begin{bmatrix} 1 & & \\ & I & \\ -s_1 & 0 & 1 \end{bmatrix} = \begin{bmatrix} 1 & & p_{1,2}^{(1)} & \cdots & p_{1,N}^{(1)} \\ p_{2,1}^{(1)} - s_1 p_{2,N}^{(1)} & p_{2,2}^{(1)} & \cdots & p_{2,N}^{(1)} \\ \vdots & \vdots & \ddots & \vdots \\ p_{N,1}^{(1)} - s_1 p_{N,N}^{(1)} & p_{N,2}^{(1)} & \cdots & p_{N,N}^{(1)} \end{bmatrix}. \tag{9.7}$$

$$L_1 = \begin{bmatrix} 1 & & \\ s_1 p_{2,N}^{(1)} - p_{2,1}^{(1)} & 1 & \\ \vdots & \vdots & I \\ s_1 p_{N,N}^{(1)} - p_{N,1}^{(1)} & \cdots & 1 \end{bmatrix}. \tag{9.8}$$

A Gaussian matrix which satisfies formula (9.8) will produce the following result,

$$L_1 P_1 A S_1 = \begin{bmatrix} 1 & & \\ s_1 p_{2,N}^{(1)} - p_{2,1}^{(1)} & 1 & \\ \vdots & \vdots & I \\ s_1 p_{N,N}^{(1)} - p_{N,1}^{(1)} & \cdots & 1 \end{bmatrix} P_1 A S_1 = \begin{bmatrix} 1 & a_{1,2}^{(2)} & \cdots & a_{1,N}^{(2)} \\ 0 & a_{2,2}^{(2)} & \cdots & a_{2,N}^{(2)} \\ \cdots & \cdots & \cdots & \cdots \\ 0 & a_{N,2}^{(2)} & \cdots & a_{N,N}^{(2)} \end{bmatrix}. \tag{9.9}$$

Denoting $A^{(1)} = L_1 P_1 A S_1$, according to the method described above, there exists a P_2 which satisfies $P_{2,N}^{(2)} \neq 0$, together with an s_2 which satisfies formula (9.10).

$$P_{2,2}^{(2)} - s_2 P_{2,N}^{(2)} = 1. \tag{9.10}$$

Following this recursive process, $P_k, s_k, L_k \ (k = 1, 2, \cdots N)$ can be determined. Accordingly, we arrive at the following formula:

$$L_{N-1} P_{N-1} \cdots L_2 P_2 L_1 P_1 A S_1 S_2 \cdots S_{N-1} = \begin{bmatrix} 1 & a_{1,2}^{(N-1)} & \cdots & a_{1,N}^{(N-1)} \\ 0 & 1 & \cdots & a_{2,N}^{(N-1)} \\ \cdots & \cdots & \cdots & \cdots \\ 0 & 0 & \cdots & a_{N,N}^{(N-1)} \end{bmatrix} = D_R U, \tag{9.11}$$

where

$$D_R = diag(1, 1, \cdots, 1, e^{i\theta}). \tag{9.12}$$

$$U = \begin{bmatrix} 1 & a_{1,2}^{(N-1)} & \cdots & a_{1,N}^{(N-1)} \\ 0 & 1 & \cdots & a_{2,N}^{(N-1)} \\ \cdots & \cdots & \cdots & \cdots \\ 0 & 0 & \cdots & 1 \end{bmatrix}. \tag{9.13}$$

Setting

$$L^{-1} = L_{N-1}(P_{N-1}L_{N-2}P_{N-1}^T) \cdots (P_{N-1}P_{N-2} \cdots P_2 L_1 P_2^T P_3^T \cdots P_{N-2}^T P_{N-1}^T). \tag{9.14}$$

$$P^T = P_{N-1}P_{N-2} \cdots P_2 P_1. \tag{9.15}$$

$$S^{-1} = S_1 S_2 \cdots S_{N-1}. \tag{9.16}$$

We conclude that

$$L^{-1}P^T A S^{-1} = D_R U. \tag{9.17}$$

We then get A = PLUS.

What should be noted is that the factorization is not unique and different factorizations affect the error between the integer approximation transform and the original transform. This will also affect the intrinsic energy-compacting capability of the original transform, so that the error should be reduced as much as possible. Quasi-complete pivoting is suggested in the progress of matrix factorization [13]. In our experiments, we found this method to be very effective in reducing the error and enhancing the stability, so achieving better integer approximation to the original floating-point transform. The improved implementation proceeds as follows:

Create a new matrix Sc_1 using the transform basis matrix A:

$$Sc_1 = \begin{bmatrix} (a_{1,1}^{(0)}) - 1/a_{1,2}^{(0)} & \cdots & (a_{1,1}^{(0)}) - 1/a_{1,N}^{(0)} \\ \vdots & \ddots & \vdots \\ (a_{N,1}^{(0)}) - 1/a_{N,2}^{(0)} & \cdots & (a_{N,1}^{(0)}) - 1/a_{N,N}^{(0)} \end{bmatrix}. \tag{9.18}$$

Choose a parameter s_1 which has a minimum absolute value in matrix Sc_1. This method is different from the traditional method which places a nonzero element at the end of every row of the transform basis matrix A in order to perform the calculation.

$$s_1 = \min\{Sc_1\} = (a_{i,1}^{(0)} - 1)/a_{i,j}^{(0)}. \tag{9.19}$$

If i is not equal to 1, that is, s_1 is not located in the first row of A, then the i-th row must be permuted to the first row using a permutation matrix P_1.

$$P_1 A = \begin{pmatrix} q_{1,1}^{(1)} & q_{1,2}^{(1)} & \cdots & q_{1,N}^{(1)} \\ q_{2,1}^{(1)} & q_{2,2}^{(1)} & \cdots & q_{2,N}^{(1)} \\ \vdots & \vdots & \ddots & \vdots \\ q_{N,1}^{(1)} & q_{N,2}^{(1)} & & q_{N,N}^{(1)} \end{pmatrix}. \tag{9.20}$$

Create a matrix S_1 which has the following shape:

$$S_1 = \begin{bmatrix} 1 & 0 & \cdots & 0 & \cdots & 0 \\ 0 & 1 & \cdots & 0 & \cdots & 0 \\ \vdots & \vdots & \ddots & & & \vdots \\ s_1 & 0 & & 1 & & 0 \\ \vdots & \vdots & & & \ddots & \vdots \\ 0 & 0 & \cdots & 0 & \cdots & 1 \end{bmatrix}. \tag{9.21}$$

Then according to formula (9.18) to (9.21), we can get a more stable $\mathbf{A} = \mathbf{PLUS}$.

It can be proved that, based on the theory described above, matrix $\mathbf{A}$ has a unit TERM factorization of $\mathbf{A} = \mathbf{PLUS}$ if and only if $\det \mathbf{A} = \det \mathbf{P} = \pm 1$, where $\mathbf{L}$ and $\mathbf{S}$ are unit lower triangular matrices, $\mathbf{U}$ is an unit upper triangular matrix and $\mathbf{P}$ is a permutation matrix.

The type-II DCT [31] in one dimension is given by the following formula:

$$X_C(k) = \varepsilon_k \sqrt{\frac{2}{N}} \sum_{n=0}^{N-1} x(n) \cos\left((2n+1)\frac{\pi k}{2N}\right), \tag{9.22}$$

for $k = 0 \ldots N-1$, where $\varepsilon_k = 1/\sqrt{2}$ if $k = 0$, and $\varepsilon_k = 1$ otherwise. The four point DCT matrix can be calculated from the above formula

$$\mathbf{A} = \begin{pmatrix} 0.5000 & 0.5000 & 0.5000 & 0.5000 \\ 0.6533 & -0.2706 & -0.2706 & -0.6533 \\ 0.5000 & -0.5000 & -0.5000 & 0.5000 \\ 0.2706 & -0.6533 & 0.6533 & -0.2706 \end{pmatrix} \tag{9.23}$$

The permutation matrix and TERMs factorized from matrix $\mathbf{A}$ are listed in Table 9.1.

Table 9.1 Permutation matrix and TERMs factorized from 4-point DCT matrix

P				L				U				S			
0	1	0	0	1				1	−0.2929	−0.0137	−0.6533	1			
1	0	0	0	0.2346	1				1	0.3066	0.6533	0	1		
0	0	1	0	0.4142	−0.7654	1				1	0.5000	0	0	1	
0	0	0	1	0.2346	0	−0.6934	1				1	0.5307	−0.8626	0.3933	1

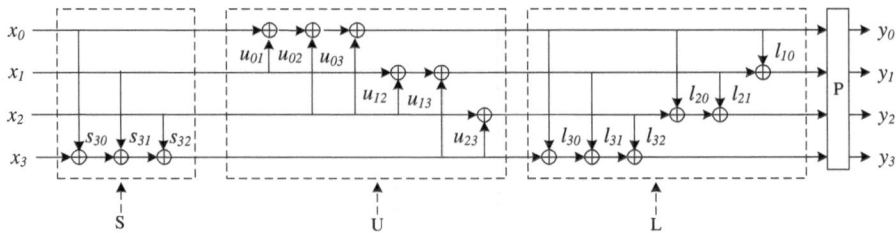

Fig. 9.6 Four-point forward DCT implemented by multi-lifting

In this way, the matrix has been decomposed into TERMs, and diagonal elements of triangular matrices are equal to 1, so the 4-point DCT can be realized by multi-lifting as depicted in Fig. 9.6.

From left to right, the input signals pass through S, U, L, and P in turn. If round the floating-point multiplication results to integers, then integer inputs will be transformed into integers. So in the inverse transform, original integers can be perfectly reconstructed as long as we subtract what was added as depicted in Fig. 9.5b.

3.2 Design of Reversible Integer to Integer TDLT/RKLT

In the proposed transform scheme, we use the modified matrix factorizing method, introduced above, to decompose the filtering matrix and DCT matrix into TERMs to realize reversible integer to integer transforms using the multi-lifting scheme.

TDLT consists of pre- and post-filters with intact DCT between them. The pre- and post-filters are exact inverses of each other. The framework of TDLT [21] can be illustrated as Fig. 9.7.

The general formula for the pre-filter [21] can be defined as:

$$\mathbf{F} = \frac{1}{2} \begin{bmatrix} \mathbf{I} & \mathbf{J} \\ \mathbf{J} & -\mathbf{I} \end{bmatrix} \begin{bmatrix} \mathbf{I} & \mathbf{0} \\ \mathbf{0} & \mathbf{V} \end{bmatrix} \begin{bmatrix} \mathbf{I} & \mathbf{J} \\ \mathbf{J} & -\mathbf{I} \end{bmatrix}, \tag{9.24}$$

where $\mathbf{I}$ and $\mathbf{J}$ are identity matrix and reversal identity matrix, respectively. Different types of TDLT can be derived with different matrices $\mathbf{V}$. Two basic types of TDLT include time domain lapped orthogonal transform (TDLOT) and time domain lapped biorthogonal transform (TDLBT). The free-control matrix $\mathbf{V}$ is defined by the following two equations:

$$\mathbf{V}_{LOT} = \mathbf{J}(\mathbf{C}_{M/2}^{II})^T \mathbf{C}_{M/2}^{IV} \mathbf{J}, \tag{9.25}$$

$$\mathbf{V}_{LBT} = \mathbf{J}(\mathbf{C}_{M/2}^{II})^T \mathbf{D}_S \mathbf{C}_{M/2}^{IV} \mathbf{J}, \tag{9.26}$$

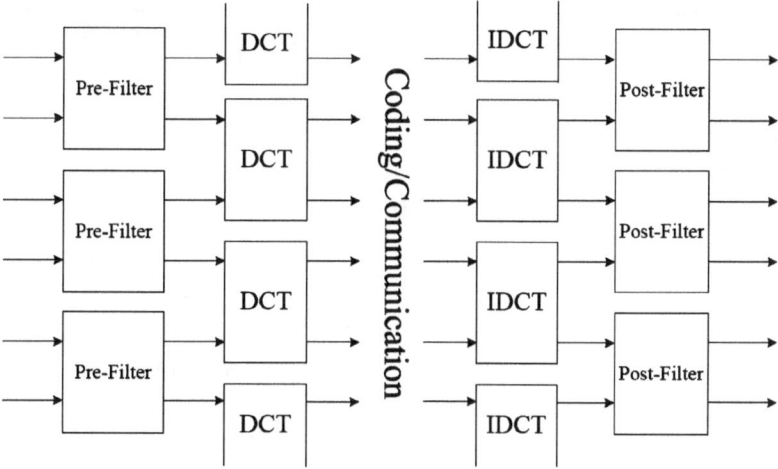

Fig. 9.7 Forward and inverse TDLT

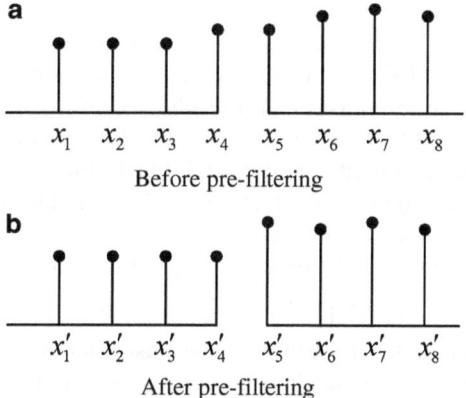

Fig. 9.8 Pre-filtering effect

where $\mathbf{C}_{M/2}^{II}$ and $\mathbf{C}_{M/2}^{IV}$ stand for the $M/2$ point type-II and type-IV DCT matrix respectively, $\mathbf{D}_S = diag\{s, 1, \cdots, 1\}$ is a diagonal matrix where s is a scaling factor and we set $s = \sqrt{2}$ in our experiments. Now we take 2×2 filters to illustrate how they work. In this case, sub-matrices of $\mathbf{F}$ degenerate into matrices of one single element with $\mathbf{I} = \mathbf{J} = \mathbf{C}_{M/2}^{II} = \mathbf{C}_{M/2}^{IV} = 1$ and $\mathbf{D}_S = s$. Let $\{x_i\}$ and $\{x'_i\}$ represent the input and output of the pre-filter, respectively, as depicted in Fig. 9.8.

The 2×2 pre-filter operates on the two adjacent elements of neighboring blocks, so just x_4 and x_5 have been modified. The relationship between $\{x_4, x_5\}$ and $\{x'_4, x'_5\}$ can be obtained as follows.

$$\begin{pmatrix} x'_4 \\ x'_5 \end{pmatrix} = \frac{1}{2} \begin{pmatrix} 1 & 1 \\ 1 & -1 \end{pmatrix} \begin{pmatrix} 1 & 0 \\ 0 & s \end{pmatrix} \begin{pmatrix} 1 & 1 \\ 1 & -1 \end{pmatrix} \begin{pmatrix} x_4 \\ x_5 \end{pmatrix} \tag{9.27}$$

Fig. 9.9 Post-filtering effect

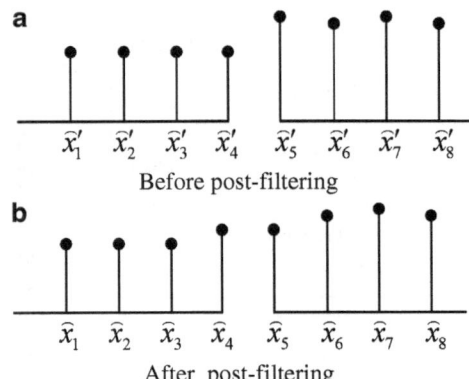

$$x'_4 = \frac{1}{2}[x_4 + x_5 - s \cdot (x_5 - x_4)] = x_4 - \frac{s-1}{2}(x_5 - x_4) \tag{9.28}$$

$$x'_5 = \frac{1}{2}[x_4 + x_5 + s \cdot (x_5 - x_4)] = x_5 + \frac{s-1}{2}(x_5 - x_4) \tag{9.29}$$

From (9.28) and (9.29) we can see that adjacent elements are modified by the pre-filter by enlarging their difference, aiming to lower the smaller-value pixel while increasing the larger-value pixel. Pre-filters with long-length basis not only enlarge the difference at boundaries from one block to another, but also make the pixels as homogenous as possible within each block. From another point of view, correlations between neighboring blocks have been reduced while correlations within one block have been increased.

The post-filter is the exact inverse of the pre-filter, and its effect on the one-dimensional vector has been depicted in Fig. 9.9, where $\{\hat{x}'_i\}$ and $\{\hat{x}_i\}$ represent the input and output of the post-filters. The post-filter decreases the difference between adjacent elements in neighboring blocks, aiming to increase the smaller-value pixel while lowering the larger-value pixel. For two-dimensional images, the post-filter aims to decrease blocking artifacts. The visual quality of post-filtered images will be shown in the experimental section, from which we will see that the post-filter performs very well.

After pre-filtering, the input data will be transformed by DCT so that most energy can be concentrated in the low frequency region which will be beneficial to coding. Type-II DCT has been adopted in this paper.

In the proposed method, we use the modified matrix factorization method [13] to decompose the filtering and DCT matrix into TERMs. This means that we do not need to search for other complex decomposition formations of the filtering matrix or DCT matrix, but simply factorize both of them to realize reversible integer-to-integer transform.

TDLOT has largely reduced the blocking artifacts of DCT and greatly improved image compression but it cannot eliminate the blocking artifacts. TDLBT is constructed to modify TDLOT by inserting a diagonal matrix **S** in **V** acting as a scaling factor in order to further reduce the block artifacts. It should be pointed out that the determinant of the filter matrix **F** does not equal 1 and should be modified to satisfy $\det \mathbf{F} = \pm 1$ before factorizing TDLBT. In our method we normalize **F** in the following way:

$$\mathbf{F}^* = \frac{\mathbf{F}}{\sqrt[M]{|\det \mathbf{F}|}}, \tag{9.30}$$

where **F** is the original M × M filter matrix, and $\mathbf{F}^*$ is the normalized matrix.

As soon as the filtering matrix and DCT matrix have been factorized into TERMs, RTDLT can be obtained, which includes reversible time domain lapped orthogonal transform (RTDLOT) and reversible time domain lapped biorthogonal transform (RTDLBT). If $\mathbf{A}_F$ and $\mathbf{A}_{DCT}$ denote the filtering matrix and DCT matrix, respectively, their factorizations will have the same format:

$$\mathbf{A}_F = \mathbf{P}_F \mathbf{L}_F \mathbf{U}_F \mathbf{S}_F \tag{9.31}$$

$$\mathbf{A}_{DCT} = \mathbf{P}_{DCT} \mathbf{L}_{DCT} \mathbf{U}_{DCT} \mathbf{S}_{DCT}. \tag{9.32}$$

Therefore, we can obtain the multi-lifting structure of RTDLT if we replace the filters and DCT in Fig. 9.7 with the structure as in Fig. 9.6.

$\mathbf{P}_F$, $\mathbf{L}_F$, $\mathbf{U}_F$, $\mathbf{S}_F$ and $\mathbf{P}_{DCT}$, $\mathbf{L}_{DCT}$, $\mathbf{U}_{DCT}$, $\mathbf{S}_{DCT}$ are TERMs decomposed from the matrices of Pre-filter and DCT, respectively. For RTDLT, because the PLUS factorizations only depend on the transform basis, we can calculate them in advance.

Obviously, the best compression method should reduce redundancies in both the spatial and spectral dimensions for hyperspectral images. In our scheme, a spatial transform (RTDLT) is first applied. Then the spectral components of each spatial frequency band are decorrelated using a Reversible Integer Karhunen-Loeve Transform (RKLT).The dataset of hyperspectral images can be represented as below:

$$X = \{X_1, X_2, X_3, \cdots, X_n\}^T, \tag{9.33}$$

where the subscript n denotes the number of bands and X_n represent the sequence of the different spectral images. The W after RTDLT is represented as

$$X \xrightarrow{\;\;RTDLT\;\;} W = \{W_1, W_2, W_3, \cdots, W_n\}^T \tag{9.34}$$

The covariance matrix C_w is defined as

$$\begin{aligned}
C_w &= E\left[\left(W - m_w\right)(W - m_w)^T\right] \\
&= \frac{1}{M} \sum_{i=0}^{M-1} (W_i - m_w)(W_i - m_w)^T
\end{aligned} \tag{9.35}$$

$$m_w = E\{W\} = \frac{1}{M}\sum_{i=0}^{M-1}W_i \qquad (9.36)$$

where M represents the number of the RTDLT coefficient vectors. Let λ_i be the eigenvalues of matrix C_w, with e_i as the corresponding eigenvectors. When the eigenvalues λ_i are arranged in descending order so that $\lambda_1 \geq \lambda_2 \geq \lambda_3 \geq \cdots \geq \lambda_l$, the auto-correlations of the transformed signal vector are arranged in descending order. The transformation matrix T_{KLT} is represented as $T_{KLT} = \{e_1, e_2, e_3, \cdots, e_n\}^T$. The matrix Y, a result of the KLT, is represented as $Y = T_{KLT}W$.

According to our proposal, the KLT matrix can be also realized by multiplierless multi-lift, based on matrix factorization. Integer approximation of the floating-point transform can be achieved if rounding operations are added, just as Eq. (9.2). Reversible integer KLT and TDLT are realized in the same way.

The RKLT concentrates much of the energy into a single band, improving overall coding efficiency.

KLT is the most efficient linear transform in the sense of energy compaction. The transform matrix can be obtained by calculating the eigenvectors of the covariance of the input data. To reduce the high computation complexity, low-complexity-KLT has been proposed by Penna et al. [32].

In our proposed method, integer reversible low-complexity-KLT (Low-RKLT) is designed for decorrelation in the spectral direction. The evaluation of the covariance matrix is simplified by sampling the input signal vectors.

If using downsampling in KLT, then formula (9.36) should be rewritten as below:

$$m'_w = \frac{1}{M'}\sum_{i=0}^{M'-1}W_i. \qquad (9.37)$$

M' denotes the number of the downsampled coefficients.

In our experiments, 100:1 scaling is applied in the sampling process. As illustrated in [32], the performance of Low-RKLT is very similar to the full-complexity RKLT, but reduced the computation complexity significantly. The computational comparison will be given in Sect. 4.

Reversible integer KLT and TDLT are realized in the same way. The integer transform can be realized by shifting and adding, only without any multiplier if the floating-point lifting coefficients are replaced by fractions, whose dominators are power of 2. For example, $15/64 = 1/8 + 1/16 + 1/32 + 1/64$ while $1/8, 1/16, 1/32$ and $1/64$ can be realized by shifting only. The multi-lifting coefficients of matrixces L, U, S decomposed from 4-point DCT have been tabulated in Table 9.2. Experimental results show that the multiplier-less DCT based on multi-lifting approximates the floating-point DCT very well. Further experiments to study the efficiency of multiplier-less RTDLT based on multi-lifting applied in lossy-to-lossless image compression will be discussed in the next section.

Table 9.2 Triangular TERM matrices with dyadic coefficients

L				U				S			
1				1	$-75/256$	$-1/64$	$-167/256$	1			
15/64	1				1	39/128	167/256	0	1		
53/128	$-49/64$	1				1	1/2	0	0	1	
15/64	0	$-89/128$	1				1	17/32	$-221/256$	101/256	1

4   Experimental Results and Discussion of RTDLT/RKLT-based Hyperspectral Image Compression

A progressive hyperspectral image compression method is designed, based on RKLT and RTDLT, with higher compression ratio and better rate distortion (RD) performance compared with the 3D-DWT-based method. The flow graph is depicted in Fig. 9.10. In the spatial domain, 2-level RTDLT and 5-level DWT are adopted separately. In addition, as shown in Figs. 9.11 and 9.12, RTDLT transform coefficients are reorganized into a tree-structure [33] so as to be coded by wavelet-based methods. In the coding method, 3DSPECK algorithm [34, 35] is applied, and JPEG2000-MC [36] has been applied in lossless.

AVIRIS hyperspectral images [37], "Jasper Ridge" (scene 1), "Cuprite" (scene 3), "Lunar Lake" (scene 2), "Low Altitude" (scene 3) and "Moffett" (scene 1) (spatially cropped to $512 \times 512, 224$ bands) are used in our experiments to test the performance of different algorithms. The test images are coded with 16 bands in a group, so the entire image of 224 bands are divided into 14 groups.

Figure 9.13 shows Jasper Ridge, Cuprite, Lunar Lake, Low Altitude and Moffett, respectively.

Lossless compression performance is compared in Table 9.3, where two coding methods, 3DSPECK and JPEG2000-MC have been adopted, and transforms include asymmetric 3D-5/3DWT (3D-53DWT), 53DWT+RKLT and RTDLT+RKLT. Based on the same codec – 3DSPECK, we can see that our proposed RTDLT/RKLT performs 7.35 ~ 8.6% better than 3D-DWT and is comparable to 53DWT+RKLT.

Lossy compression performance is given in Table 9.4. In our experiments, the performance of five types of transforms combined with 3D SPECK is compared at different bit rates, where 3D SPECK is carried with QccPack Version 0.56 [38]. It can be seen that our proposed RTDLT/RKLT consistently outperforms asymmetric 3D-97DWT by a large margin (up to 5 dB), combined with the same coding method. It also gives a gain of 0.38 ~ 0.69 dB compared with 53DWT+RKLT. Although the proposed method performs not as well as 97DWT+FloatKLT, it is capable of complete reversible transform while the latter cannot.

Figure 9.14 depicts the rate distortion (RD) performance of different transform schemes using the same codec – 3D SPECK.

From the above experimental results, we can conclude that the proposed compression method performs well in both lossy and lossless hyperspectral image compression. Among conventional transforms based on integer wavelet

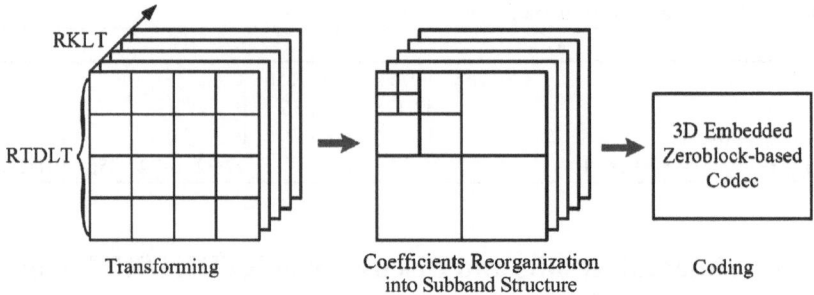

Fig. 9.10 Flow graph of the Proposed RTDLT/RKLT Compression Scheme

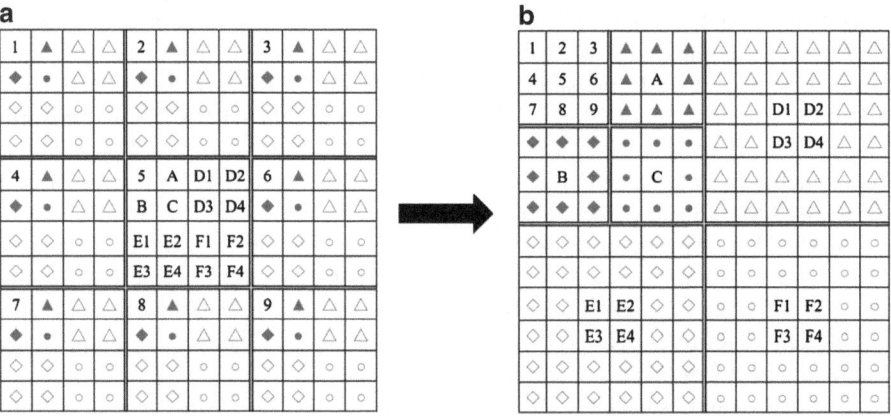

Fig. 9.11 Spatial coefficients reorganization for each band: (**a**) Original spatial coefficients distribution after RTDLT. (**b**) Reorganized coefficients as wavelet-liked subband structure

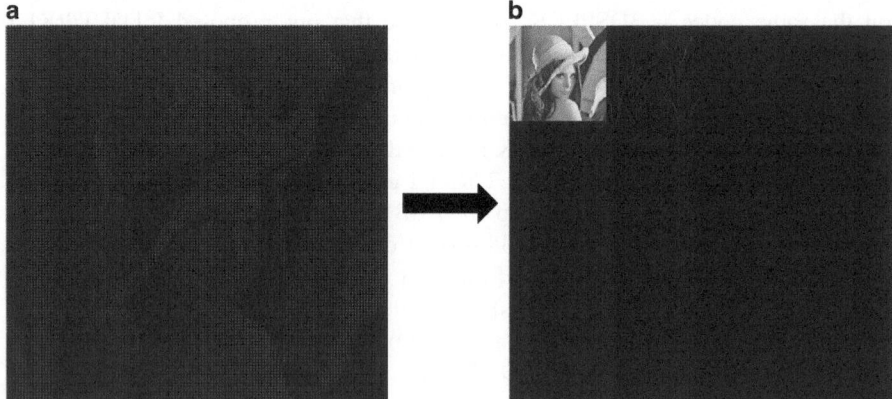

Fig. 9.12 Representations of block transform coefficients for Lena image: (**a**) Block representation after TDLT. (**b**) Wavelet-liked subband structure after coefficients reorganization

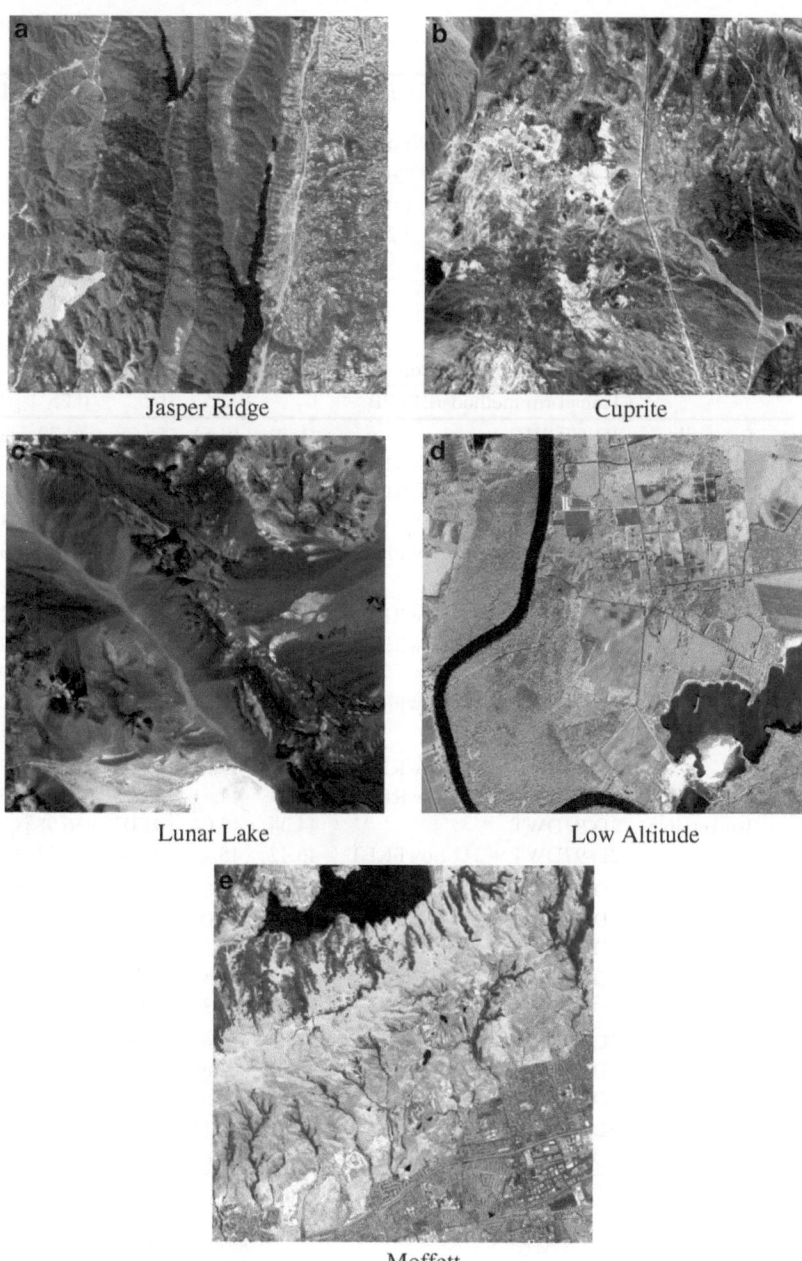

Fig. 9.13 Original AVRIS images

(i.e. 5/3 DWT), the non-unitary transform gives the best lossless compression performance but will decrease the performance of lossy compression. Therefore, from a single lossless codestream, the decoder cannot obtain such good lossy results shown in Table 9.4. However, we don't need to consider unitary problem by using

Table 9.3 Lossless compression performance comparison (in bits per pixel per band, BPPPB)

Codec	3D SPECK			JPEG2000-MC	
Transform	3D-53DWT	53DWT + Low-RKLT	RTDLT + low-RKLT	3D-53DWT	53DWT + Low-RKLT
Cuprite	5.32	4.97	4.95	5.44	5.07
Jasper	5.52	4.99	5.01	5.65	5.11
Lunar	5.29	4.97	4.96	5.42	5.08
Low	5.75	5.22	5.23	5.90	5.35
Moffett	6.67	6.11	6.11	6.86	6.32

Table 9.4 Lossy compression, SNR comparison (in dB)

		Transform methods/BPPPB	1	0.75	0.5	0.25	0.125
Cuprite	Reversible	3D-53DWT	41.79	40.64	38.78	34.72	30.73
		2D-53DWT + 1D-LowRKLT	43.77	43.07	42.21	39.29	34.33
		2D-RTDLT + 1D-LowRKLT	44.61	43.91	43.03	40.13	35.16
	Irreversible	3D-97DWT	43.24	41.93	39.89	35.77	31.62
		2D-97DWT + 1D-LowFKLT	45.65	44.65	43.54	40.46	35.38
Jasper	Reversible	3D-53DWT	34.80	33.01	30.35	25.52	21.06
		2D-53DWT + 1D-LowRKLT	38.63	37.57	35.75	30.7	25.14
		2D-RTDLT + 1D-LowRKLT	39.45	38.43	36.52	31.45	25.93
	Irreversible	3D-97DWT	36.27	34.19	31.39	26.49	22.07
		2D-97DWT + 1D-LowFKLT	40.54	39.28	37.05	31.77	26.06
Lunar	Reversible	3D-53DWT	42.75	41.68	39.89	35.84	31.63
		2D-53DWT + 1D-LowRKLT	44.55	43.85	43.05	40.42	35.61
		2D-RTDLT + 1D-LowRKLT	45.34	44.61	43.75	41.01	36.10
	Irreversible	3D-97DWT	44.34	43.04	41.07	36.98	32.65
		2D-97DWT + 1D-LowFKLT	46.42	45.44	44.36	41.51	36.67
Low	Reversible	3D-53DWT	34.36	32.82	30.51	26.31	22.43
		2D-53DWT + 1D-LowRKLT	38.04	37.06	35.65	31.84	26.89
		2D-RTDLT + 1D-LowRKLT	38.68	37.66	36.19	32.22	27.33
	Irreversible	3D-97DWT	35.65	33.87	31.46	27.21	23.28
		2D-97DWT + 1D-LowFKLT	39.49	38.34	36.72	32.64	27.61
Moffett	Reversible	3D-53DWT	41.24	39.61	37.04	31.55	26.28
		2D-53DWT + 1D-LowRKLT	43.36	41.69	39.27	33.49	27.64
		2D-RTDLT + 1D-LowRKLT	43.99	42.36	39.83	34.15	28.34
	Irreversible	3D-97DWT	41.83	40.18	37.64	32.57	27.32
		2D-97DWT + 1D-LowFKLT	44.22	42.62	40.03	34.21	28.37

uniform RTDLT framework to obtain embedded and high efficient lossy-to-lossless coding performance.

In Table 9.5, we give a computation comparison between the low-complexity KLT and full-complexity KLT. From the table we can see that the low-complexity KLT has significantly reduced the computational time of full-complexity KLT.

Fig. 9.14 RD performance of different transform schemes combined with the same codec—3DSPECK

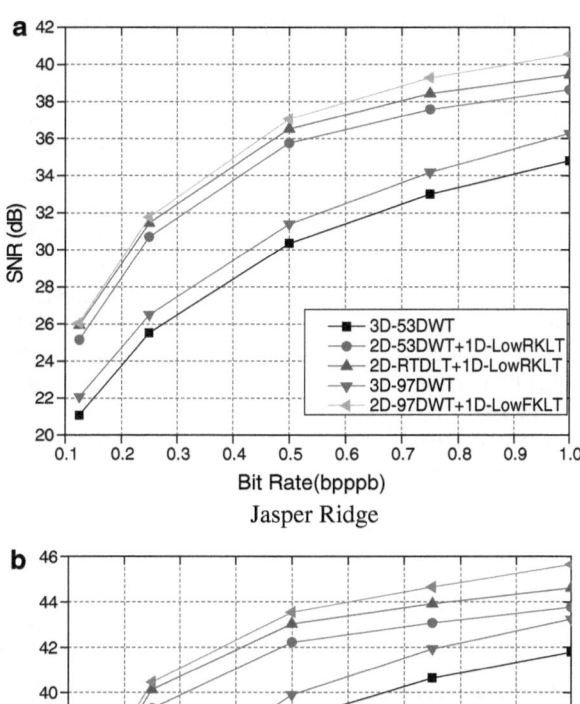

Jasper Ridge

Cuprite

Table 9.5 Computational time comparison (in seconds)

Different KLT[a]	Full-FKLT	Low-FKLT	Full-RKLT	Low-RKLT
Covariance computation	39.4375	0.2344	39.9844	0.2344
Spectral transform	71.8125	43.2813	98.2813	41.2031

[a]*Full-FKLT* full-complexity float KLT, *Low-FKLT* low-complexity float KLT, *Full-RKLT* full-complexity reversible KLT, *Low-RKLT* Low-complexity reversible KLT. The running times are measured on a Pentium IV PC at 3.20 GHz using C# Language

Table 9.6 The performance comparison of various sampled KLT(SNR, in dB)

Transform methods, sampling rates/BPPPB	1	0.75	0.5	0.25	0.125
	Cuprite (512 × 512 × 224)				
Full-FKLT	43.81	43.06	42.22	39.37	34.43
Low-FKLT, 1/10	43.78	43.05	42.22	39.38	34.42
Low-FKLT, 1/100	43.76	43.02	42.19	39.33	34.37
Full-RKLT	43.83	43.13	42.26	39.34	34.35
Low-RKLT, 1/10	43.82	43.11	42.26	39.35	34.36
Low-RKLT, 1/100	43.77	43.07	42.21	39.29	34.33
	Jasper (512 × 512 × 224)				
Full-FKLT	38.67	37.57	35.78	30.75	25.22
Low-FKLT, 1/10	38.67	37.57	35.77	30.73	25.20
Low-FKLT, 1/100	38.68	37.57	35.79	30.71	25.19
Full-RKLT	38.56	37.48	35.66	30.68	25.05
Low-RKLT, 1/10	38.61	37.55	35.71	30.69	25.06
Low-RKLT, 1/100	38.63	37.57	35.75	30.70	25.14
	Lunar (512 × 512 × 224)				
Full-FKLT	44.53	43.81	43.01	40.49	35.74
Low-FKLT, 1/10	44.53	43.81	43.02	40.49	35.73
Low-FKLT, 1/100	44.53	43.81	43.02	40.48	35.69
Full-RKLT	44.55	43.84	43.05	40.44	35.62
Low-RKLT, 1/10	44.55	43.84	43.05	40.46	35.63
Low-RKLT, 1/100	44.55	43.85	43.05	40.42	35.61

However, there are also disadvantages, since the coding gain can be accompanied by lower speed. The complexity of RTDLT/RKLT is higher than that of 3D integer wavelet transforms.

In Table 9.6, the performance comparison of various sampled 2D-53DWT+KLT is shown. The KLT compression performance degradation is negligible, when the scaling is 10:1~100:1 in the sampling process.

5   3D-RLT-based Hyperspectral Image Compression

Following on from RTDLT/RKLT-based hyperspectral image compression, some readers might think intuitively whether RTDLT could be extended to apply to a spectral domain. Certainly, the RTDLT can be extended to 3D reversible integer lapped transform (3D-RLT), which is realized by cascading three 1D RLT along spatial and spectral dimensions.

RLT is derived from reversible integer time-domain lapped transform (RTDLT) which was introduced in Sect. 3. Compared with RTDLT/RKLT, the new 3D block transform – 3D-RLT is much less complex.

We have extended RLT to three dimensions, and designed a 3D-RLT-based compression algorithm for hyperspectral images as follows.

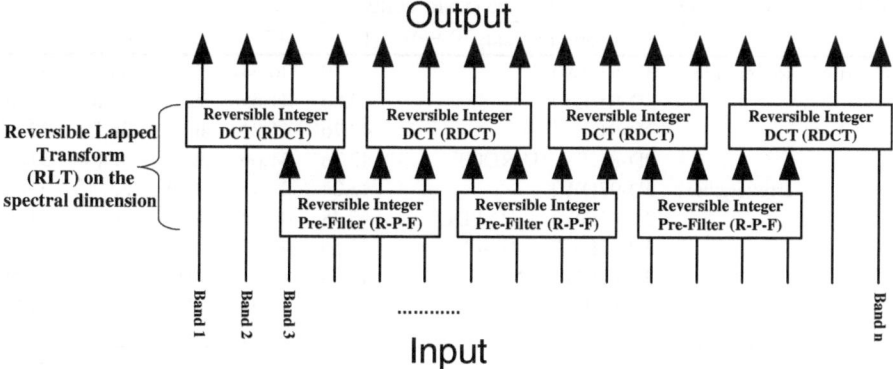

Fig. 9.15 The diagram of reversible lapped transform (*RLT*) on the spectral dimension

It is necessary to perform 2D RLT on each band of the entire image along the horizontal and vertical dimensions, followed by 1D RLT along the spectral dimension, as shown in Fig. 9.15. RLT is realized by performing a reversible integer pre-filter (R-P-F) and reversible integer DCT (RDCT) successively. In the spatial domain, RLT is performed on the unit of a block. Here, we adopt blocks with size 8×8. The test images are coded with 16 bands in a group in the spectral dimension. R-P-F can be turned on or off depending on the performance requirement and complexity limitation. If R-P-F is turned off, RLT converts to RDCT.

6 Experimental Results of 3D-RLT-based Hyperspectral Image Compression

To demonstrate the validity of the proposed 3D-RLT coding scheme, we again conducted experiments on the AVIRIS hyperspectral images [37]. 3D floating-point 9/7-tap biorthogonal WT (3D-97WT), 3D integer 5/3-tap WT (3D-53WT), 3D floating-point DCT (3D-FDCT), 3D-RDCT, 3D floating-point TDLT (3D-FLT) and 3D-RLT have been tested. It has been proved that the asymmetric 3D-WT (anisotropic wavelet) performs better than the symmetric 3D-WT (isotropic wavelet) [39]. In our experiment, we adopt asymmetric 3D-WT. RLT coefficients are approximated by hardware-friendly dyadic values which realize operations in the transforming process be realized by only shifts and additions. All the transform methods are combined with the same coding method, namely the 3D set partitioned embedded block (3D-SPECK) codec [35], to ensure a fair comparison.

Lossy compression results, based on the criterion of signal-to-noise ratio (SNR), are given in Table 9.7, from which we can see that at most bit-rates 3D-RLT performs better than all the other reversible transform methods except the floating-point transforms. Although the floating-point transforms produce the better results, 3D-RLT can be applied in progressive lossy-to-lossless compression.

Table 9.7 Lossy compression performance (SNR, in dB)

		Transform methods/BPPPB	1	0.75	0.5	0.25	0.125
Cuprite	Reversible	3D-53DWT	41.79	40.64	38.78	34.72	30.73
		3D-RDCT	42.86	41.65	39.77	35.57	31.49
		3D-RLT	42.76	41.81	40.25	36.71	32.59
		2D-RLT + 1D-RDCT	42.76	41.66	39.92	35.87	31.89
	Irreversible	3D-97DWT	43.24	41.93	39.89	35.77	31.62
		2D-97DWT + 1D-FDCT	43.48	42.23	40.24	36.11	31.95
		2D-97DWT + 1D-FLT	43.56	42.39	40.64	36.92	32.68
		3D-FDCT	43.34	42.02	40.01	35.73	31.59
		3D-FLT	43.53	42.35	40.67	36.98	32.83
Jasper	Reversible	3D-53DWT	34.80	33.01	30.35	25.52	21.06
		3D-RDCT	35.78	33.92	31.29	26.48	22.10
		3D-RLT	35.73	34.07	31.80	27.39	23.09
		2D-RLT + 1D-RDCT	35.79	34.03	31.47	26.79	22.51
	Irreversible	3D-97DWT	36.27	34.19	31.39	26.49	22.07
		2D-97DWT + 1D-FDCT	36.32	34.41	31.64	26.81	22.43
		2D-97DWT + 1D-FLT	36.35	34.52	32.00	27.45	23.01
		3D-FDCT	36.09	34.13	31.39	26.53	22.06
		3D-FLT	36.24	34.49	32.02	27.43	23.03
Lunar	Reversible	3D-53DWT	42.75	41.68	39.89	35.84	31.63
		3D-RDCT	43.77	42.59	40.65	36.44	32.16
		3D-RLT	43.66	42.72	41.14	37.59	33.38
		2D-RLT + 1D-RDCT	43.68	42.62	40.85	36.87	32.67
	Irreversible	3D-97DWT	44.34	43.04	41.07	36.98	32.65
		2D-97DWT + 1D-FDCT	44.44	43.20	41.31	37.28	33.01
		2D-97DWT + 1D-FLT	44.51	43.36	41.66	38.01	33.71
		3D-FDCT	44.26	42.96	40.91	36.69	32.33
		3D-FLT	44.45	43.31	41.59	37.92	33.68
Low	Reversible	3D-53DWT	34.36	32.82	30.51	26.31	22.43
		3D-RDCT	35.04	33.48	31.13	26.92	23.01
		3D-RLT	35.07	33.62	31.65	27.72	23.97
	Irreversible	3D-97DWT	35.65	33.87	31.46	27.21	23.28
		2D-97DWT + 1D-FDCT	35.55	33.92	31.62	27.46	23.56
		2D-97DWT + 1D-FLT	35.66	34.16	32.03	28.04	24.23
		3D-FDCT	35.31	33.65	31.22	26.94	22.95
		3D-FLT	35.51	33.92	31.76	27.79	23.91
Moffett	Reversible	3D-53DWT	41.24	39.61	37.04	31.55	26.28
		3D-RDCT	40.72	39.23	36.91	31.97	26.73
		3D-RLT	37.82	36.69	34.88	31.06	26.59
	Irreversible	3D-97DWT	41.83	40.18	37.64	32.57	27.32
		2D-97DWT + 1D-FDCT	40.91	39.46	37.26	32.41	27.18
		2D-97DWT + 1D-FLT	37.96	36.85	35.13	31.37	26.93
		3D-FDCT	40.77	39.26	36.96	32.03	26.71
		3D-FLT	37.83	36.75	34.98	31.24	26.79

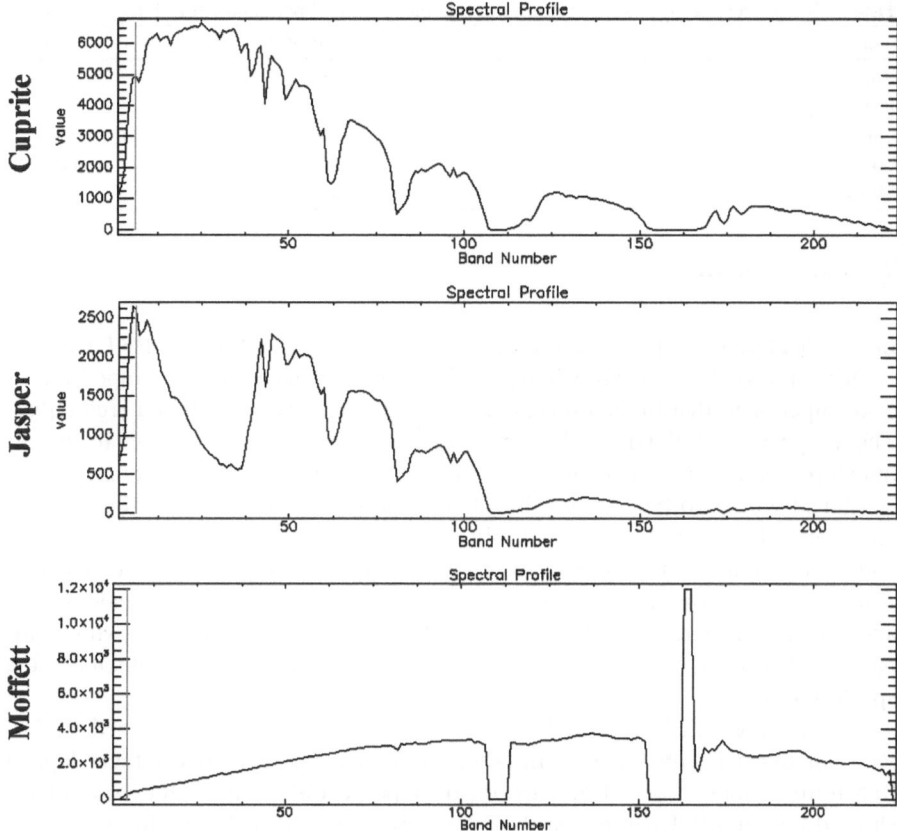

Fig. 9.16   Spectral profiles of various images

3D-RLT performs better than 3D-53DWT by about 0.71~2.03 dB, and better than 3D-97DWT in most cases. However, for the Moffett image, it is clear in comparison with 3D-DWT the performance of 3D-RLT is degraded significantly. This is because Moffett's spectral profile is much smoother than that of other images, as shown in Fig. 9.16. It is well-known, either DCT or lapped transform can be considered to be a kind of multi-band filters, so DCT and lapped transform are not suitable in the spectral dimension of this kind of images which has simple frequency characteristic.

The lossless compression performance of reversible integer transforms combined with 3DSPECK is compared in Table 9.8 based on the criterion of bit per pixel per band (bpppb). From Table 9.8 we can see that 3D-RLT is competitive with 3D-53DWT and better than 3D-RDCT in most cases.

Table 9.8 Lossless compression performance (in bits per pixel per band, BPPPB)

Images	3D-53DWT	3D-RDCT	3D-RLT
Cuprite	5.32	5.38	5.33
Jasper	5.52	5.65	5.63
Lunar	5.29	5.36	5.32
Low	5.75	5.87	5.83
Moffett	6.67	7.31	7.61

7 Conclusions

In this chapter, we first presented a new transform scheme of RTDLT/RKLT for the compression of hyperspectral images with a performance that is competitive or even superior to that of state-of-the-art conventional transform-based techniques. The proposed RTDLT/RKLT scheme can realize reversible integer-to-integer transform, and so it can be applied in progressive lossy-to-lossless compression combined with zero-block-based bit-plane coders. The new transforming scheme has some advantages such as multiplierless computing, flexibility in processing, and parallel implementation and so on. In addition, it also preserves the desirable features of the lifting scheme, such as low memory request, in-place calculation, and perfect reconstruction. While high performance means high complexity, compared with 1D wavelet transform, RKLT still has much higher complexity in the spectral dimension.

We also presented a simple but efficient compression algorithm based on 3D-RLT for hyperspectral images. In other words, the 1D reversible integer lapped transform is applied on all dimensions of hypespectral images. Our experiments showed that 3D-RLT can defeat the integer 5/3 wavelet easily in a lossy application. At the same time, the lossless performance of 3D-RLT is near that of 5/3 WT. In addition, the 3D-RLT-based method can be simplified if much lower complexity is required. When the pre/post filters are cancelled, the RLT can be converted into RDCT.

At this point, it should be noted that in our experiments 5/3 WT is unitary in lossy mode in order to obtain better performance, so the experimental results of 5/3 WT do not support completely progressive transforms. If, in order to support progressive transforms completely and, at the same time, to ensure the best lossless performance, 5/3 WT should be non-unitary, although that will degenerate the lossy performance. In contrast, the unitary wavelet transform will cause the degeneration of lossless compression. However, our proposed methods do not suffer this problem, and both RTDLT/RKLT and 3D-RLT supports progressive coding completely.

No transform provides a perfect result and is only one of many topics relating to image compression. Although our proposed RTDLT/RKLT and 3D-RLT methods are suitable for the lossy-to-lossless compression of hyperspectral images (actually, RTDLT is also suitable for the compression of natural images [40]), they are not suitable to transform 3D medical images. The reason is because DCT-based lapped transform is not suitable for medical images with a smooth texture. In this case, the wavelet transform is rather good.

In addition, if a given application is only for 3D lossless compression of hyperspectral images, it would be better to use the lookup table (LUT) coding method [41] which has low complexity and high lossless performance.

References

1. J. Xu, Z. Xiong, S. Li, and Y. Zhang, "3-D embedded subband coding with optimal truncation (3-D ESCOT)," Applied and Computational Harmonic Analysis, vol.10, pp.290–315, May 2001.
2. J. E. Fowler and D. N. Fox, "Embedded wavelet-based coding of three dimensional oceano-graphic images with land masses," IEEE Transactions on Geoscience and Remote Sensing, vol.39, no.2, pp.284–290, February 2001.
3. X. Tang, W. A. Pearlman, and J. W. Modestino, "Hyperspectral image compression using three-dimensional wavelet coding," in proceedings SPIE, vol.5022, pp.1037–1047, 2003.
4. Q. Du and J. E. Fowler, "Hyperspectral Image Compression Using JPEG2000 and Principal Component Analysis," IEEE Geoscience and Remote sensing letters, vol.4, pp.201–205, April 2007.
5. P. L. Dragotti, G. Poggi, and A. R. P. Ragozini, "Compression of multispectral images by three-dimensional SPIHT algorithm," IEEE Geoscience and Remote sensing letters, vol.38, no. 1, pp. 416–428, January 2000.
6. B. Penna, T. Tillo, E. Magli, and G. Olmo, "Transform Coding Techniques for Lossy Hyperspectral Data Compression," IEEE Geoscience and Remote sensing letters, vol.45, no.5, pp.1408–1421, May 2007.
7. B. Penna, T. Tillo, E. Magli, and G. Olmo, "Progressive 3-D coding of hyperspectral images based on JPEG 2000," IEEE Geoscience and Remote sensing letters, vol.3, no.1, pp.125–129, January 2006.
8. W. Sweldens, "The lifting scheme: A construction of second generation wavelet," in SIAM Journal on Mathematical Analysis, vol.29, pp.511–546, 1997.
9. I. Daubechies and W. Sweldens, "Factoring wavelet transforms into lifting steps," The Journal of Fourier Analysis and Applications, vol.4, pp.247–269, 1998.
10. A.Bilgin, G. Zweig, and M. W. Marcellin, "Three-dimensional image compression using integer wavelet transforms," Applied Optics, vol.39, pp.1799–1814, April 2000.
11. Z. Xiong, X. Wu, S. Cheng, and J. Hua, "Lossy-to-Lossless Compression of Medical Volu-metric Data Using Three-Dimensional Integer Wavelet Transforms," IEEE Transactions on Medical Imaging, vol.22, no.3, pp.459–470, March 2003.
12. P. Hao and Q. Shi, "Reversible integer KLT for progressive-to-lossless compression of multiple component images," in Proceedings IEEE International Conference Image Processing (ICIP'03), Barcelona, Spain, pp.I-633–I-636, 2003.
13. L. Galli and S. Salzo, "Lossless hyperspectral compression using KLT," IEEE International Geoscience and Remote Sensing Symposium, (IGARSS2004), vol.1, pp.313–316, September 2004.
14. C. Kwan, B. Li, R. Xu, X. Li, T. Tran, and T. Nguyen, "A Complete Image Compression Method Based on Overlapped Block Transform with Post-Processing," EURASIP Journal on Applied Signal Processing, pp.1–15, January 2006.
15. P. List, A. Joch, J. Lainema, G. Bjontegaard, M. Karczewicz, "Adaptive deblocking filter," IEEE Transactions on Circuits and Systems for Video Technology, vol.13, no.7, pp.614–619, July 2003.

16. Xiong ZX, Orchard MT, Zhang YQ, "A deblocking algorithm for JPEG compressed images using overcomplete wavelet representations," IEEE Transactions on Circuits and Systems for Video Technology, vol.7, no.2, pp.433–437, April 1997.
17. P. Cassereau, "A New Class of Optimal Unitary Transforms for Image Processing", Master's Thesis, Massachusetts Institute of Technology, Cambridge, MA, May 1985.
18. H. S. Malvar, "Lapped transforms for efficient transform/subband coding", IEEE Transactions on Acoustics, Speech, and Signal Processing, pp.969–978, ASSP-38. 1990.
19. C.W. Lee and H. Ko, "Arbitrary resizing of images in DCT domain using lapped transforms", Electronics Letters, vol.41, pp.1319–1320, November 2005.
20. T. D. Tran, J. Liang, and C. Tu, "Lapped transform via time-domain pre- and post-processing," IEEE Transactions on Signal Processing, vol.51, no.6, pp.1557–1571, January 2003.
21. Chengjie Tu, and Trac D. Tran, "Context-based entropy coding of block transform coefficients for image compression," IEEE Transactions on Image Processing, vol.11, no.11, pp. 1271–1283, January 2002.
22. Jie Liang, Trac D. Tran, "Fast Multiplierless Approximations of the DCT with the Lifting Scheme," IEEE Transactions on Signal Processing, vol.49, no.12, pp.3032–3044, December 2001.
23. http://www.microsoft.com/whdc/xps/hdphotodpk.mspx
24. S. Srinivasan, C. Tu, S. L. Regunathan, and G. J. Sullivan, "HD Photo: a new image coding technology for digital photography," in Proceedings SPIE Applications of Digital Image Processing XXX, San Diego, vol.6696, pp.66960A, August 2007.
25. C. Tu, S. Srinivasan, G. J. Sullivan, S. Regunathan, and H. S. Malvar, "Low-complexity hierarchical lapped transform for lossy-to-lossless image coding in JPEG XR//HD Photo," in Proceedings SPIE Applications of Digital Image Processing XXXI, San Diego,vol.7073, pp.70730 C1-12,August 2008.
26. Y. Chen, S. Oraintara, and T. Nguyen, "Integer discrete cosine transform (IntDCT)," in Proceedings 2nd International Conference Information and Communication of Signal Processing, December 1999.
27. G.C.K. Abhayaratne, "Reversible integer-to-integer mapping of N-point orthonormal block transforms," Signal Processing, vol.87, no.5, pp.950–969, 2007.
28. J. Li, "Reversible FFT and MDCT via matrix lifting," in Proceedings IEEE International Conference on Acoustics, Speech, and Signal Processing, vol. 4, pp. iv-173–iv-176, May 2004.
29. L.Z. Cheng, G.J. Zhong, and J.S. Luo, "New family of lapped biorthogonal transform via lifting steps," IEE Proceedings -Vision, Image and Signal Processing, Vol. 149, no. 2, pp. 91–96, April 2002.
30. P. Hao and Q. Shi, "Matrix Factorizations for Reversible Integer Mapping," IEEE Transactions on Signal Processing, vol.42, no.10, pp. 2314–2324,October 2001.
31. K. R. Rao and P. Yip, Discrete Cosine Transform: Algorithms, Advantages, Applications. New York: Academic, 1990.
32. B. Penna, T. Tillo, E. Magli, and G. Olmo, "Transform Coding Techniques for Lossy Hyperspectral Data Compression," IEEE Geosciense and Remote Sensing, vol. 45, no. 5, pp. 1408–1421, May 2007.
33. Z. Xiong, O. Guleryuz, and M. T. Orchard, "A DCT-based embedded image coder". IEEE Signal Processing Letters, vol. 3, pp. 289–290, 1996.
34. X. Tang and W. A. Pearlman, "Three-Dimensional Wavelet-Based Compression of Hyperspectral Images," in Hyperspectral Data Compression, pp.273–308, 2006.
35. Jiaji Wu, Zhensen Wu, Chengke Wu, "Lossly to lossless compression of hyperspectral images using three-dimensional set partitioning algorithm," Optical Engineering, vol.45, no.2, pp.1–8, February 2006.
36. Information Technology—JPEG 2000 Image Coding System—Part 2: Extensions, ISO/IEC 15444–2, 2004.
37. http://aviris.jpl.nasa.gov/html/aviris.freedata.html
38. http://qccpack.sourceforge.net/

39. E. Christophe, C. Mailhes and P. Duhamel, "Hyperspectral image compression: adapting SPIHT and EZW to Anisotropic 3-D Wavelet Coding", IEEE Transactions on Image Processing, vol.17, no.12, pp.2334–2346, 2008.
40. Lei Wang, Jiaji Wu, Licheng Jiao, Li Zhang and Guangming Shi, "Lossy to Lossless Image Compression Based on Reversible Integer DCT. IEEE International Conference on Image Processing 2008 (ICIP2008), pp.1037–1040, 2008.
41. J. Mielikainen, "Lossless compression of hyperspectral images using lookup tables," IEEE Signal Processing Letters, vol.13, no.3, pp.157–160, 2006.

Chapter 10
Divide-and-Conquer Decorrelation for Hyperspectral Data Compression

Ian Blanes, Joan Serra-Sagristà, and Peter Schelkens

Abstract Recent advances in the development of modern satellite sensors have increased the need for image coding, because of the huge volume of such collected data. It is well-known that the Karhunen-Loêve transform provides the best spectral decorrelation. However, it entails some drawbacks like high computational cost, high memory requirements, its lack of component scalability, and its difficult practical implementation. In this contributed chapter we revise some of the recent proposals that have been published to mitigate some of these drawbacks, in particular, those proposals based on a divide-and-conquer decorrelation strategy. In addition, we provide a comparison among the coding performance, the computational cost, and the component scalability of these different strategies, for lossy, for progressive lossy-to-lossless, and for lossless remote-sensing image coding.

1 Introduction

When coding a hyperspectral image in a lossy or progressive lossy-to-lossless way, it is common to employ a spectral decorrelating transform followed by a traditional transform coder. In this regard, the Karhunen-Loêve Transform (KLT) and its derivatives are the transforms that provide some of the best results [20]. Yet, the KLT has a very high computational cost that hinders its adoption in many situations.

I. Blanes (✉) • J. Serra-Sagristà
Universitat Autònoma de Barcelona, E-08290 Cerdanyola del Vallès (Barcelona), Spain
e-mail: ian.blanes@uab.es; joan.serra@uab.es

P. Schelkens
Department of Electronics and Informatics, Vrije Universiteit Brussel,
Pleinlaan 2, B-1050 Brussels, Belgium

Interdisciplinary Institute for Broadband Technology, Gaston Crommenlaan 8,
b102, B-9050 Ghent, Belgium
e-mail: peter.schelkens@vub.ac.be

B. Huang (ed.), *Satellite Data Compression*, DOI 10.1007/978-1-4614-1183-3_10,   215
© Springer Science+Business Media, LLC 2011

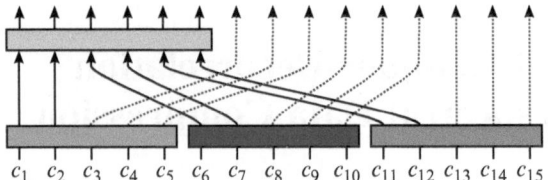

c_1 c_2 c_3 c_4 c_5 c_6 c_7 c_8 c_9 c_{10} c_{11} c_{12} c_{13} c_{14} c_{15}

Fig. 10.1 Example of divide-and-conquer strategy for 15 spectral components

Recently, a variety of divide-and-conquer strategies have been proposed to ease the computational requirements of the KLT. These strategies rely on the fact that the KLT has a quadratic cost $O(n^2)$, but that if the full transform is approximated by a set of smaller ones, then the computational cost becomes a fraction of the original cost.

The rationale behind these strategies is that only spectral components with high covariances are worth decorrelating, since in the other cases, coding gains are negligible. Hence, in these strategies, a transform is decomposed into a set of smaller transforms that provide decorrelation where it is needed and the other regions are neglected.

One of such transform decompositions is shown in Fig. 10.1. In this example, first, local decorrelation is provided by three KLT clusters, and then, some outputs of each clusters are processed again together to achieve global decorrelation. Note that due to the properties of the KLT, output components are by convention arranged in descending order according to their variance, therefore, components with low variances are discarded by not selecting the last outputs from each cluster.

This chapter reviews divide-and-conquer strategies for the KLT, and provides some insights into the details found in their practical implementation. Additionally, it includes an experimental evaluation and comparison of the various strategies in the context of the compression of hyperspectral remote-sensing images.

We note that some other recent approaches to alleviate some of the issues of the KLT are not addressed here. In particular, the reader should be aware that coding gains can be improved by not assuming Gaussian sources, and finding the Optimal Spectral Transform with an Independent Component Analysis (ICA)-based algorithm [3, 4], which could also be pre-trained [2, 5]; or that, to overcome the high computational cost of the KLT, the Discrete Cosine Transform (DCT) was proposed [1] – it assumes a Toeplitz matrix as data covariance matrix. However, the DCT has a poor performance as spectral decorrelator [20]. Similar approaches to reduce the computational cost are the fast Approximate Karhunen-Loève Transform (AKLT) [19] and the $AKLT_2$ [21], which extend the DCT with first and second order perturbations.

The rest of this chapter is organized as follows: Sect. 10.2 provides an overview of the Karhunen-Loève Transform. In Sect. 10.3 we present the different variants of divide-and-conquer strategies for spectral decorrelation. Section 10.4 provides some experimental results for compression of satellite data using the different decorrelation strategies. Finally, Sect. 10.5 contains some conclusions.

2 The Karhunen-Loêve Transform

This section describes the KLT and its practical application when used as part of a divide-and-conquer strategy. For an image of N spectral components, each one with a mean required to be zero, the KLT is defined as follows: let X be a matrix that has N rows, one for each source, and M columns, one for each spatial location. Then, Y, the outcome after applying the transform, is computed as

$$Y = \mathrm{KLT}_{\Sigma_X}(X) = Q^T X, \tag{10.1}$$

where $\Sigma_X = (1/M)XX^T$ is the covariance matrix of X, and Q is the orthogonal matrix obtained from the Eigenvalue Decomposition (ED) of Σ_X, i.e., $\Sigma_X = Q\Lambda Q^{-1}$, with $\Lambda = diag(\lambda_1, \ldots, \lambda_N)$, $|\lambda_1| \geq |\lambda_2| \geq \ldots \geq |\lambda_N|$. Being Σ_X an Hermitian matrix, according to the spectral theorem, such a decomposition always exists.

The covariance matrix of Y is the diagonal matrix Λ (i.e., $\Sigma_Y = (1/M)YY^T = (1/M)Q^TXX^TQ = Q^T\Sigma_XQ = \Lambda$), and $\lambda_1, \ldots, \lambda_N$ are the variances of each component after the transform.

The ED of the covariance matrix is usually performed using an iterative procedure that converges to the solution, such as the algorithm based on a preprocessing tridiagonalization by Householder transformations, followed by iterative QR decompositions [13]. Other diagonalization procedures with lower complexities exist [10], at the expense of higher implementation difficulties. It is highly recommended to rely on one of the existing libraries for this purpose, as several numerical stability issues appear in the process that may lead to a non-converging algorithm.

Note that this transform is dependent on Σ_X and hence different for each input. For this reason, the specific transform matrix Q^T used has to be signaled as side information, so that the decoder is able to invert the transformation.

2.1 Centering and Covariance Particularities

As noted in its definition, the KLT has to be applied on image components of zero mean. It is rare to find image components with zero mean, thus, a variable change is usually applied to address this issue, i.e.,

$$X' = X - \begin{pmatrix} \frac{1}{M}\sum_{j=1}^{M} x_{(1,j)} \\ \vdots \\ \frac{1}{M}\sum_{j=1}^{M} x_{(N,j)} \end{pmatrix} (1, \ldots, 1). \tag{10.2}$$

As the KLT does not affect the mean of components, this change of variable is only required once per image, regardless of whether a sequence of transforms or just one transform is applied.

Some particularities of the covariance calculation are also worth noting. With large enough spatial sizes, image subsampling in the KLT has been known to substantially reduce the training cost of the KLT [20]. The aim is to use only a small sample of the image spatial locations to compute the covariance matrix. A subsampling factor of $\rho = 0.01$ (1%) is known to provide almost the same compression performance results, and effectively reduce transforms training cost, leaving the application of the transform matrix – the $Q^T X$ matrix multiplication – as the main source of computational cost.

To select sampling locations, some sort of Pseudo-Random Number Generator (PRNG) is required. High quality PRNG have high cost, but, in this case, with a poor generator, results stay similar. A very fast Park-Miller PRNG (10.3) can be used, only taking four operations for each random number.

$$Y_n = M \cdot X_n \text{ div } (2^{32} - 5)$$
$$X_n = (279470273 \cdot X_{n-1}) \text{ mod } (2^{32} - 5) \tag{10.3}$$

Another particularity of the covariance matrix calculation appears when the KLT is used in a divide-and-conquer strategy, where one or more components are transformed more than once. When a component has been previously transformed and has to be transformed again, its variance does not have to be computed, as it is already available as one of the eigenvalues λ_i from the previous transform.

2.2 Spatial Subdivision

Spatial subdivision – also known as segmentation or blocking – might be used in this context for several reasons, e.g., to ease memory requirements or to limit the impact of data loss. Some issues are here discussed in relation to its use.

If a transform is applied in multiple spatial blocks, or an image spatial size is small enough, then the computational cost for the ED step stops being negligible and might mean an important fraction of the total computational cost. To address this issue, improved methods of ED can be used [10, 14].

It is also worth noting that the size of the transform side information is only dependent on the spectral size, thus with multiple spatial divisions or small spatial sizes, the size of the transform side information becomes more relevant, as the total amount of data per transform becomes smaller. Covariance subsampling is also affected by spatial subdivision for the same reasons; as the spatial size decreases, the number of sampled pixels also decreases. Hence, subsampling factors have to be increased –i.e., a larger number of samples has to be used–, to still produce good estimations. When spatial sizes are very small, covariance subsampling might even be more costly than a regular covariance calculation because of the small overhead of sample selection becoming noticeable.

Finally, in case of multiple spatial blocks, blocking artifacts might also appear on the spatial boundaries and have to be taken into consideration when evaluating specific transforms.

2.3 The Reversible KLT

Given an almost exactly-reversible floating-point KLT matrix, like the one obtained in the previous steps, one can adapt it to be used in a lossless way. This adaptation is performed factorizing the transform matrix into an equivalent sequence of Elementary Reversible Matrices (ERMs) [9]. Each of these ERMs can later be applied in an approximate way with respect to the original multiplication, which maps integers to integers, and is fully reversible. In addition to the reversibility, as the transformed coefficients are integers instead of floating-point values, the number of bitplanes to be encoded later is reduced.

A factorization known as the Reversible KLT (RKLT) with quasi-complete pivoting [12] is described now. The factorization is based on the one introduced in [16], and improves it by minimizing the adaptation differences with the original lossy transform.

The factorization is as follows: given an orthogonal KLT matrix Q^T, an iterative process over A_i is applied, starting with $A_1 = Q^T$.

1. A permutation matrix P_i is selected such that $(P_i A_i)_{(i, N)} \neq 0$.
2. The matrix S_i is computed as $S_i = I - s_i e_k e_i^T$, where e_m is the m-th vector of the standard basis, and $s_i = \frac{(P_i A_i)_{(j,i)} - 1}{(P_i A_i)_{(j,k)}}$. Indices j and k are selected so that s_i is minimal over $i \leq j \leq N, i + 1 \leq k \leq N$.
3. Gaussian elimination is applied to $P_i A_i S_i$, obtaining the Gaussian elimination matrix L_i that guarantees $(L_i P_i A_i S_i)_{(k, i)} = 0$ for all $k > i$.
4. A_{i+1} is set to the product $L_i P_i A_i S_i$, and a new iteration is started.

After $N - 1$ iterations, A_N is an upper triangular matrix, from now on called U, where all diagonal elements are 1 except the last one which might be ± 1. Then all the partial results obtained in the previous step are merged,

$$S^{-1} = \prod_{k=1}^{N-1} S_k, \tag{10.4}$$

$$L^{-1} = L_{N-1} \cdot (P_{N-1} L_{N-2} P_{N-1}^T) \cdot \cdots \cdot (P_2 L_1 P_2^T), \tag{10.5}$$

$$P^T = \prod_{k=1}^{N-1} P_{N-k}, \tag{10.6}$$

and finally the factorization is reached:

$$A = PLUS, \tag{10.7}$$

where P is a permutation matrix, and L and S are lower triangular matrices with ones in the diagonal.

Once the matrix is factorized, the transform is applied in the forward and inverse directions using interleaved rounding operations within each specially crafted matrix multiplication. The rounding operator —denoted by $[\cdot]$— is defined as the function that rounds its argument to the closest integer value, with half-way values rounded to the nearest even value. For lower triangular matrices, the multiplication – for instance for $Y = LX$– is as follows:

$$Y = \begin{pmatrix} y_1 \\ \vdots \\ y_i \\ \vdots \\ y_N \end{pmatrix} = \begin{pmatrix} x_1 \\ \vdots \\ x_i + \left[\sum_{1 \le j < i} l_{(i,j)} x_j\right] \\ \vdots \\ x_N + \left[\sum_{1 \le j < N} l_{(N,j)} x_j\right] \end{pmatrix} = [LX], \tag{10.8}$$

and for the upper triangular matrix the operation follows the same principle, but takes into account that, as $(U)_{(N,N)} = \pm 1$, $y_N = (U)_{(N,N)} x_N$.

The inverse operation is applied by undoing the previous steps in a similar way as they are applied. In this case, for the lower triangular matrices the operation is

$$X = \begin{pmatrix} x_1 \\ \vdots \\ x_i \\ \vdots \\ x_N \end{pmatrix} = \begin{pmatrix} y_1 \\ \vdots \\ y_i - \left[\sum_{1 \le j < i} l_{(i,j)} x_j\right] \\ \vdots \\ y_N - \left[\sum_{1 \le j \le N} l_{(N,j)} x_j\right] \end{pmatrix} = [L^{-1}Y]. \tag{10.9}$$

Special care has to be taken of the order of the operations. Between rounding steps, operations must be performed exactly in the same order in the forward and the inverse transforms, while blocks of operations delimited by rounding steps have to be undone in the opposite order.

On a related note, if the use of integer multiplications is to be avoided, for example because of a slow multiplication operation on embedded hardware, a further extension of the RKLT can be used where integer multiplications are factorized to a sequence of shifts and additions using the same principles as the RKLT [27, 28],

3   Divide-and-Conquer Strategies

Divide-and-conquer strategies have been known for a long time. For example, binary search, which divides a list of items to search into two smaller lists at each step, dates back at least from year 200 B.C. in Babylonia [18, p. 420]. In this sense, applying a well known strategy to the spectral decorrelation problem might seem straightforward, but it is not.

The issues with a divide-and-conquer strategy appear once the problem of how to organize such divisions is taken into account, as in this case, the possible divisions are many and optimal methods to perform such divisions other than exhaustive search are not known. The problem is further convoluted by the fact that the optimality criterion is not unique. One might want a decorrelating transform that yields high coding gains, but also that has low computational cost, and further providing good component scalability. Nonetheless, several suitable divisions strategies have been found for this problem and have recently been published in the literature. This section describes each division strategy, its heuristics, and its optimality trade-offs.

3.1   Plain Clustering

Of the divide-and-conquer strategies, plain clustering is the most simple. It consists of dividing a hyperspectral image into sets or clusters of spectral components, and then applying a KLT on each set of components independently [6, 11]. See Fig. 10.2 for an example of such structure.

Such a strategy trades global decorrelation – as correlation between components pertaining to different sets is not removed – for lower computational costs, and hence moderate coding performances are achieved. Usually, as nearby components have higher correlations, sets are performed partitioning the image into slices of continuous spectral components. Sets might be of regular size if no additional

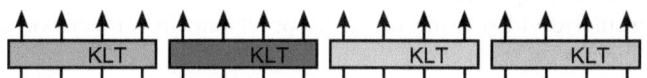

Fig. 10.2 Plain clustering applied to a 16-component image. Each vertical arrow depicts a spectral component as it is transformed by a KLT

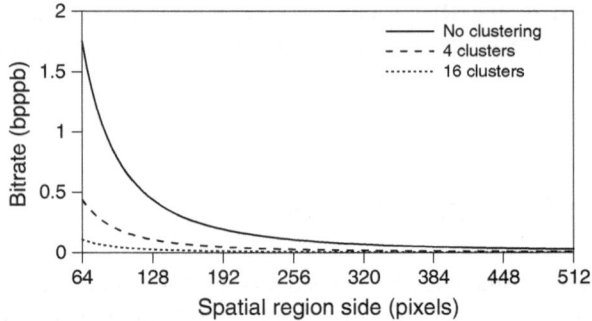

Fig. 10.3 Side information cost for a transform on an image of 224 spectral components given a square region of variable size

information of the image to be decorrelated is known a priory, or might be carefully picked so that the set boundaries fall in the border of two uncorrelated zones of an image.

This strategy might also be used when, due to the small spatial size of an image, or due to the use of small spatial subdivisions, the amount of side information required becomes significant and impacts the coding performance. In Fig. 10.3, the cost of side information is plotted in relation to the spatial size of an image.

For images as small as 100×100 pixels the amount of side information is around 0. 7 bpppb, which, for low bitrates, has a very negative impact on coding performance. In [11] it is shown that for low bitrates and small image sizes, the reduction of side information provided by plain clustering is more beneficial in coding performance terms than the penalty introduced by the lack of global decorrelation.

3.2 Recursive Structure

A recursive division of the KLT is proposed in [29, 30]. While originally proposed only for pure lossless compression of electroencephalograms and MRI data, it can also be used in lossy and progressive lossy-to-lossless coders for remote-sensing imagery compression.

The idea is to replace a full transform by three half-size blocks. The first two half-size blocks decorrelate one half of the input each, while the third decorrelates the first half of the output of the first two blocks. Then each of the three half-size blocks are further divided using the same procedure in a recursive fashion until a minimum block size is reached.

An example of such a recursive structure is provided in Fig. 10.4, where an image of eight components is decorrelated by a recursive structure divided in blocks of two components or, said otherwise, with two levels of recursion.

Fig. 10.4 An example of a recursive structure applied to an image of eight spectral components, with transforms of size two

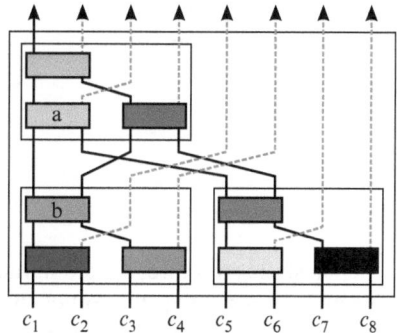

c_1 c_2 c_3 c_4 c_5 c_6 c_7 c_8

When dividing a full transform, it is important that, at each level of recursion, an interleaving permutation between the outputs of the first two half-size transforms is added before processing the inputs of the third half-size transform. If such permutation is not added, then groups of components from the first two transforms that have already been decorrelated together are again decorrelated by the third transform with no additional benefit. In Fig. 10.4, that would mean that transforms marked as a and b would perform the same task twice.

Recursive transforms provide a very comprehensive decorrelation, but at the expenses of many superfluous operations which substantially increase their computational cost, and reduce component scalability.

3.3 Two-Level Structures

Another kind of divide-and-conquer structures are the ones introduced in [22, 23]. The proposed structures are composed by two levels of decorrelation; the first level provides local decorrelation, while the second provides global decorrelation. At the first level, spectral components are decorrelated as in a clustered transform, and the most significant components of each of the clusters of the first level are forwarded to the second level for a second decorrelation, while the remaining components proceed to further coding stages as is. The most significant component of each cluster of the first level is decorrelated together with the most significant component of the other first-level clusters. The second most significant component with the second most significant components and so on and so forth until all the most significant components of each cluster are decorrelated.

An example of a two-level structure is shown in Fig. 10.5. In this particular case, the structure selects only the two most significant components of each of the three first-level clusters for further decorrelation.

In order to select the structure parameters, i.e., the number of clusters and how many components are significant at each first level cluster, two approaches are provided by the authors: a static approach and a dynamic approach.

Fig. 10.5 Example of a
two-level static structure

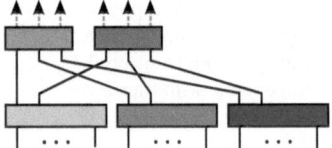

On a static structure, a fixed number of most significant components are selected for all clusters. On the other hand, on dynamic structures, the amount of components is selected on the training stage by a process similar to pruning a static structure. The pruning consists in discarding, from entering the second stage, components that have small influence on the first components of the output of the second stage, and the influence is determined by the relative weights set to each input component by the transform matrix of the second stage of an equivalent static structure.

Both structures provided a good trade-off between computational cost and coding performance. Due to the second stage of static transforms decorrelating together one component from each cluster, the static variant provides low component scalability.

3.4 Multilevel Structures

In parallel with two-level structures, multilevel structures where proposed in [6–8]. There are four variations, each of them with its own trade-offs, but all of them share the same idea, multiple levels of clustering are applied, each of them further decorrelating the most significant components of previous levels.

The first variation is a simple multilevel structure proposed in [6], where a plain clustering with clusters of fixed size is applied and then the most significant half of each cluster is selected to be further decorrelated in a next level, until only one cluster remains in the last level. No component interleaving is applied in any multilevel structure. This first kind of multilevel structures is superseded by the ones proposed in [7], where instead of using a naive regular structure, eigen-thresholding methods – methods that determine the amount of significant components after a KLT – are used to determine the amount of components to be selected from each cluster.

Two approaches are proposed to select a good structure among all the possible variations of cluster size. The first approach produces static structures, by setting a cluster size regularity constrain in each level, and performing exhaustive search on the cluster sizes for a training corpus. In this case, eigen-thresholding methods are used to estimate the best number of components to be selected for each cluster, but a single amount is fixed for all clusters of one level during the search process to

a   **b**

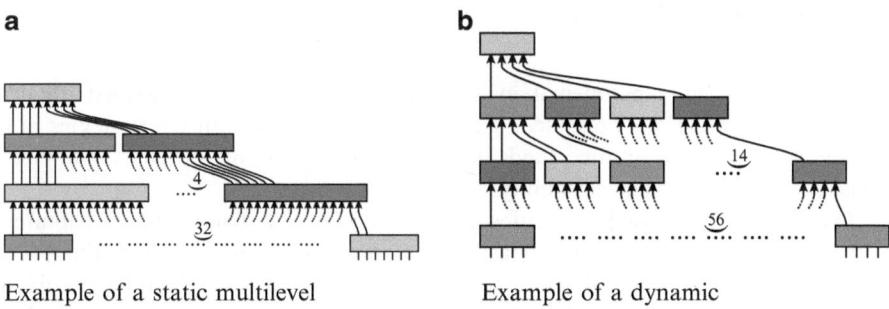

Example of a static multilevel   Example of a dynamic
structure. This particular structure   multilevel structure.
is the one that was found to be
better for the hyperspectral
AVIRIS sensor.

Fig. 10.6 Examples of multilevel structures

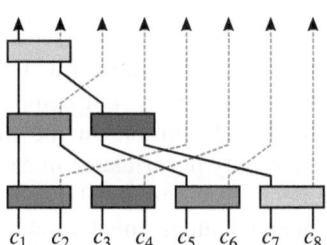

Fig. 10.7 An example of the pairwise orthogonal transform for an image of eight spectral components

minimize the combinatorial explosion. The second approach produces structures with a single cluster size but allows that the number of selected components of each cluster is determined at runtime, thus producing a dynamic structure.

Examples of both approaches are provided in Fig. 10.6 a,b. In the first case, cluster sizes are selected to be 7, 16, 12, 8 respectively at each level, and the amount of selected components for each cluster of the first level is two, six on the second level, and four on the third. In the second case, the cluster size is fixed to four for the whole structure, and an eigenthresholding algorithm selects the appropriate amount of components from each cluster, until only one cluster remains at the last level.

The fourth variation of multilevel structures is the Pairwise Orthogonal Transform (POT) introduced in [8], which is a regular structure of clusters of two components where one component is always selected from each cluster. An example structure for an image of eight spectral components is provided in Fig. 10.7. The POT provides a minimalistic fixed structure that provides moderate decorrelation with a very low cost, and is suitable for real-time processing, and power or memory-constrained environments like on-board sensors.

4   Experimental Results

In this section, an experimental evaluation of the previously described divide-and-conquer strategies is presented. This evaluation compares all the strategies in a common scenario and in three fundamental aspects: coding performance, computational cost, and component scalability.

The divide-and-conquer strategies are evaluated over radiance images captured with the Hyperion sensor on board the Earth Observing One satellite [17]. The images used are the radiance corrected version or level 1, which have a fixed width of 256 columns, a variable height, and 242 spectral components covering wavelengths from 357 to 2,576 nm. The last 18 components are uncalibrated and have been discarded so that images have 224 spectral components.

To evaluate coding performance, spectral transforms have to be paired with a complete coding system. In this case, JPEG2000 [26, 24] has been used for this purpose using the software implementation Kakadu [25]. As for the spectral transforms, an open source implementation is provided by the authors of this Chapter [15].

The following are the exact transforms that have been tested: Four levels of recursion have been used in the recursive transform. The two-level static structure has 28 clusters in the first level and selects four components of each cluster. The two-level dynamic structure has 28 clusters in the first level and selects four components of each cluster but only decorrelates in each second level cluster the four most significant components. The multilevel dynamic structure has clusters of size four, and selects components with eigenvalues above average. The multilevel static structure has four levels with 32, 8, 2, and 1 clusters in each, and selects three, five, and seven components from each cluster from the first, second and third levels respectively. The POT is not applied in a line-by-line basis, as initially described in its original report, to provide a fair comparison with the other transforms, which cannot use a line-by-line application because of too much side information costs.

The described spectral transforms are compared against a traditional KLT, and against wavelets, using a Cohen-Doubechies-Feauveau (CDF) 9/7 for lossy and a CDF 5/3 for Lossy-to-Lossless (PLL) and lossless, in both cases with five levels of transform.

Coding performance is reported in Fig. 10.8 for lossy performance, in Fig. 10.9 for PLL, and in Table 10.1 for pure lossless. For the sake of clarity, lossy and PLL results are reported using the difference of Signal-to-Noise Ratio (SNR) performance as compared with the performance of the full KLT. In this case, SNR is defined as $\mathrm{SNR} = 10\log_{10}(\sigma^2/\mathrm{MSE})$, where σ^2 is the variance of the original image. Lossless results are reported with the real bitrate required for lossless.

In Figs. 10.8 and 10.9, the first plot reports results for a plain clustering strategy and for a recursive structure, the second plot reports results for two-level structures, and the third plot reports results for the multi-level structures.

For all the experiments, evaluating either the lossy performance, the PLL performance, or the lossless performance, results for the KLT transform are usually

Table 10.1 Bitrate required for lossless coding of Hyperion Radiance images. Bitrate is reported in bpppb

	RKLT	16 clusters	Recursive	Two-level static	Two-level dynamic	ML dynamic	ML static	POT	IWT 5/3
Erta Ale	5.95	5.95	5.90	5.99	5.96	6.00	5.89	6.05	6.24
Lake Monona	6.11	6.09	6.07	6.16	6.10	6.16	6.03	6.23	6.37
Mt. St. Helens	6.00	6.06	5.97	6.04	6.06	6.11	5.93	6.23	6.38

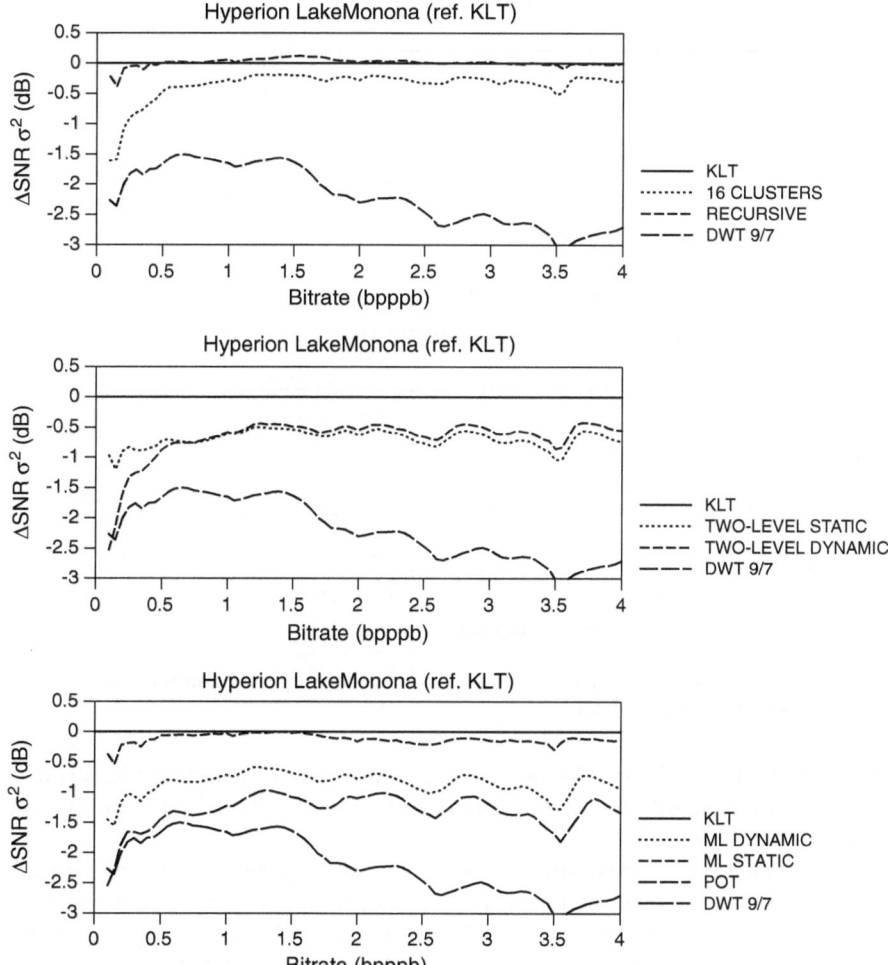

Fig. 10.8 Lossy coding performance of the divide-and-conquer spectral transforms in relation to the performance of a regular KLT

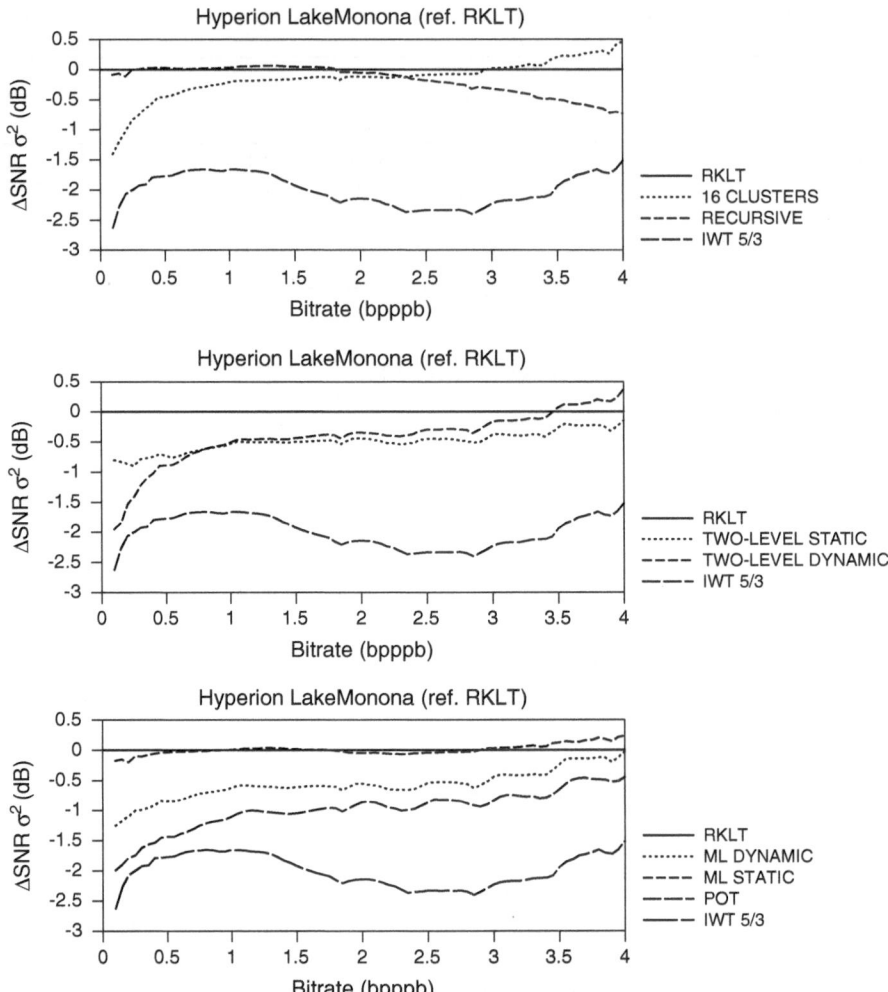

Fig. 10.9 PLL coding performance of the divide-and-conquer spectral transforms in relation to the performance of a regular KLT

the highest ones, while results for a wavelet transform are always the lowest ones, and results for the different divide-and-conquer decorrelation strategies lie somewhere in between.

As expected, the recursive structure provides very competitive coding performance results, reaching those of the classical KLT, although its computational cost is too demanding. Interestingly, the multi-level static structure also yields very good coding results, with a lower computational cost. Results for the two-level structures, either static or dynamic, are quite acceptable, with moderate penalty, similar to that achieved by a plain clustering strategy of 16 clusters. Among all the divide-and-conquer

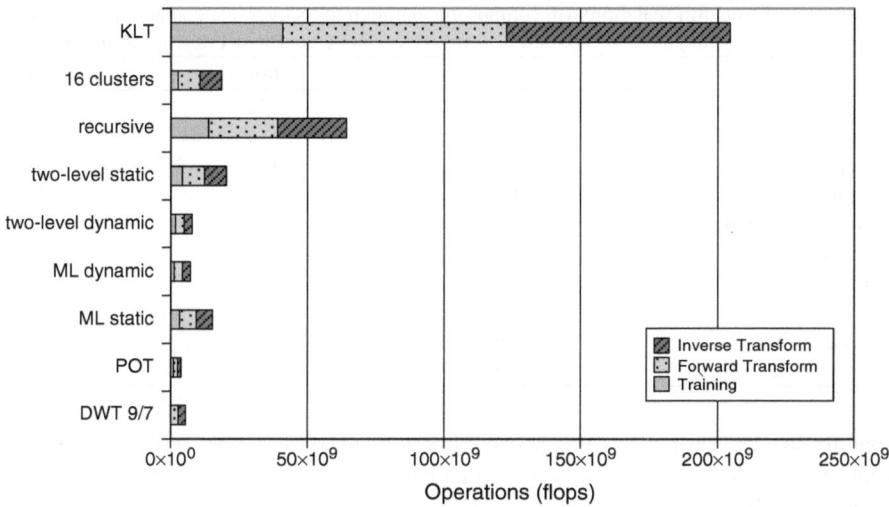

Fig. 10.10 Computational cost of lossy transforms when applied to the image Hyperion lake Monona

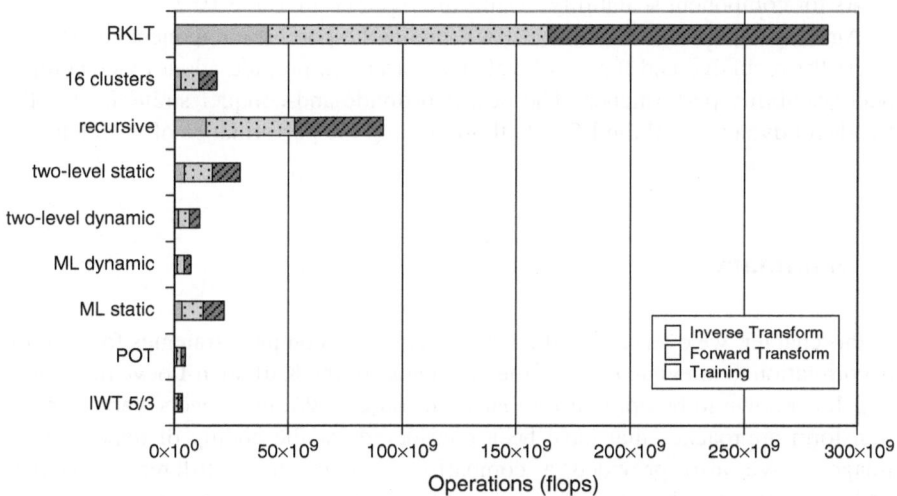

Fig. 10.11 Computational cost of lossless transforms when applied to the image Hyperion lake Monona

strategies, the POT produces the worst coding performance results, but its computational cost and its component scalability is the best one.

Computational cost results are reported in Fig. 10.10 for lossy transforms and in Fig. 10.11 for pure lossless and PLL transforms. The main difference between them is

Table 10.2 Transform scalability (in components required to recover one component). Reported wavelet scalability may be reduced on transform edges due to coefficient mirroring (up to a half)

	Avg.	Min.	Max.
Full KLT	224.0	224	224
16 clusters	14.0	14	14
Recursive	119.0	119	119
Two-level static	116.0	116	116
Two-level dynamic	9.7	8	14
ML dynamic	13.1	12	15
ML static	38.0	38	38
POT	8.9	8	9
Wavelet CDF 9/7	36.0	32	38
Wavelet CDF 5/3	16.0	11	17

that, for lossless results, a lifting scheme has to be used to ensure reversibility, which requires a few extra operations on the application and removal of transforms.

Figures 10.10 and 10.11 clearly show that the classical KLT is extremely expensive, with the recursive structure having about one fourth its computational cost. All other divide-and-conquer strategies have a much lower complexity, with static structures being more costly than dynamic structures. POT strategy has a similar complexity than wavelets.

As for component scalability, results are reported in Table 10.2.

Again, we see that the classical KLT entails the poorest component scalability, while the recursive and the two-level static structure provide about twice component scalability performance. The best two divide-and-conquer strategies are the two-level dynamic and the POT, both improving the performance of wavelets.

5 Summary

In this chapter, we overviewed the major divide-and-conquer strategies for spectral decorrelation to alleviate some of the drawbacks of the Karhunen-Loêve transform, which is known to be optimal for Gaussian sources. We reviewed several popular transform approaches that have been considered for the coding of hyperspectral imagery. We also provided a comparison among these different strategies, addressing the drawbacks of the classical KLT, in particular, and for satellite data coding, we evaluated the coding performance, the computational cost, and the component scalability.

As for coding performance, and for lossy, progressive lossy-to-lossless, and lossless image compression, we found that all divide-and-conquer strategies yield results in between the performance of the KLT and the performance of wavelets. The recursive structure produces competitive coding results, closely followed by the multi-level static structure. POT structure shows the lowest coding performance, although still above that of wavelets.

As for computational complexity, all divide-and-conquer strategies –but the recursive structure–, provide a significant saving as compared to KLT. POT structure matches the cost of wavelets.

As for component scalability, the recursive and the two-level static structures have a poor performance, the multi-level static has a medium performance, similar to that of lossy CDF 9/7 wavelets, and all the other strategies have a competitive performance, matching that of lossless CDF 5/3 wavelets, with POT yielding the best component scalability among all.

In conclusion, a POT divide-and-conquer strategy retains all the good properties of wavelet spectral transforms, and yet provides superior coding performance, becoming a convenient replacement of wavelets where the KLT is not suitable.

Acknowledgements This work was supported in part by the Spanish Government, by the Catalan Government, and by FEDER under grants TIN2009-14426-C02-01, TIN2009-05737-E/TIN, SGR2009-1224, and FPU2008. Computational resources used in this work were partially provided by the Oliba Project of the Universitat Autònoma de Barcelona. This work was also supported by the Belgian Government via the Fund for Scientific Research Flanders (postdoctoral fellowship Peter Schelkens). The authors would like to thank NASA and USGS for providing the Hyperion images.

References

1. N. Ahmed, T. Natarajan, and K.R. Rao. Discrete cosine transfom. *IEEE Trans. Comput.*, C-23(1):90–93, Jan. 1974.
2. Isidore Paul Akam Bita, Michel Barret, Florio Dalla Vedova, and Jean-Louis Gutzwiller. Lossy compression of MERIS superspectral images with exogenous quasi optimal coding transforms. In Bormin Huang, Antonio J. Plaza, and Raffaele Vitulli, editors, *Satellite Data Compression, Communication, and Processing V*, volume 7455, page 74550U. SPIE, 2009.
3. Isidore Paul Akam Bita, Michel Barret, and Dinh-Tuan Pham. On optimal orthogonal transforms at high bit-rates using only second order statistics in multicomponent image coding with JPEG2000. *Elsevier Signal Processing*, 90(3):753 – 758, 2010.
4. Isidore Paul Akam Bita, Michel Barret, and Dinh-Tuan Pham. On optimal transforms in lossy compression of multicomponent images with JPEG2000. *Elsevier Signal Processing*, 90(3):759–773, 2010.
5. Michel Barret, Jean-Louis Gutzwiller, Isidore Paul Akam Bita, and Florio Dalla Vedova. Lossy hyperspectral images coding with exogenous quasi optimal transforms. In *Data Compression Conf. 2009 (DCC 2009)*, pages 411–419. IEEE Press, Mar. 2009.
6. I. Blanes and J. Serra-Sagristà. Clustered reversible-KLT for progressive lossy-to-lossless 3d image coding. In *Data Compression Conf. 2009 (DCC 2009)*, pages 233–242. IEEE Press, Mar. 2009.
7. I. Blanes and J. Serra-Sagristà. Cost and scalability improvements to the Karhunen-Loêve transform for remote-sensing image coding. *IEEE Trans. Geosci. Remote Sens.*, 48(7):2854–2863, Jul. 2010.
8. I. Blanes and J. Serra-Sagristà. Pairwise orthogonal transform for spectral image coding. *IEEE Trans. Geosci. Remote Sens.*, 49(3):961–972, Mar. 2011.
9. F. A. M. L. Bruekers and A. W. M. van den Enden. New networks for perfect inversion and perfect reconstruction. *IEEE J. Sel. Areas Commun.*, 10(1):130–137, 1992.

10. Inderjit Singh Dhillon. *A New $O(n^2)$ Algorithm for the Symmetric Tridiagonal Eigenvalue/ Eigenvector Problem*. PhD thesis, EECS Department, University of California, Berkeley, Oct 1997.

11. Q. Du, W. Zhu, H. Yang, and J. E. Fowler. Segmented principal component analysis for parallel compression of hyperspectral imagery. *IEEE Geosci. Remote Sens. Lett.*, 6 (4):713–717, Oct. 2009.

12. L. Galli and S. Salzo. Lossless hyperspectral compression using KLT. *IEEE Int'l Geosci. and Remote Sens. Symp. Proc. (IGARSS 2004)*, 1–7:313–316, 2004.

13. G.H. Golub and C.F. van Loan. *Matrix Computations*. The Johns Hopkins University Press, Oct. 1996.

14. Ian R. Greenshields and Joel A. Rosiene. A fast wavelet-based Karhunen-Loeve transform. *Pattern Recognition*, 31(7):839–845, Jul. 1998.

15. Group on Interactive Coding of Images. Spectral transform software. http://gici.uab.cat/, 2010.

16. P. W. Hao and Q. Y. Shi. Matrix factorizations for reversible integer mapping. *IEEE Trans. Signal Process.*, 49(10):2314–2324, 2001.

17. Jet Propulsion Laboratory, NASA. Hyperspectral image compression website. http://compression.jpl.nasa.gov/hyperspectral/.

18. Donald E. Knuth. *Sorting and Searching*, volume 3 of *The Art of Computer Programming*. Addison-Wesley, second edition, 1998.

19. L.S. Lan and I.S. Reed. Fast approximate Karhunen-Loeve transform (AKLT) with applications to digital image-coding. *Elsevier Visual Commun. and Image Process.* 93, 2094:444–455, 1993.

20. Barbara Penna, Tammam Tillo, Enrico Magli, and Gabriella Olmo. Transform coding techniques for lossy hyperspectral data compression. *IEEE Trans. Geosci. Remote Sens.*, 45(5):1408–1421, May 2007.

21. A.D. Pirooz and I.S. Reed. A new approximate Karhunen-Loeve transform for data compression. *Conference Record of the Thirty-Second Asilomar Conference on Signals, Systems and Computers*, 1–2:1471–1475, 1998.

22. John A. Saghri and Seton Schroeder. An adaptive two-stage KLT scheme for spectral decorrelation in hyperspectral bandwidth compression. *Proc. SPIE*, 7443:744313, Sept. 2009.

23. John A. Saghri, Seton Schroeder, and Andrew G. Tescher. Adaptive two-stage Karhunen-Loeve-transform scheme for spectral decorrelation in hyperspectral bandwidth compression. *SPIE Optical Engineering*, 49:057001, May 2010.

24. Peter Schelkens, Athanassios Skodras, and Touradj Ebrahimi, editors. *The JPEG 2000 Suite*. Wiley, Sept. 2009.

25. D.S. Taubman. Kakadu software. http://www.kakadusoftware.com/, 2000.

26. D.S. Taubman and M.W. Marcellin. *JPEG2000: Image Compression Fundamentals, Standards, and Practice*, volume 642. Kluwer International Series in Engineering and Computer Science, 2002.

27. Lei Wang, Jiaji Wu, Licheng Jiao, and Guangming Shi. 3D medical image compression based on multiplierless low-complexity RKLT and shape-adaptive wavelet transform. *ICIP 2009. Proceedings of 2009 International Conference on Image Processing*, pages 2521–2524, Nov. 2009.

28. Lei Wang, Jiaji Wu, Licheng Jiao, and Guangming Shi. Lossy-to-lossless hyperspectral image compression based on multiplierless reversible integer TDLT/KLT. *IEEE Geosci. Remote Sens. Lett.*, 6(3):587–591, July 2009.

29. Y. Wongsawat. Lossless compression for 3-D MRI data using reversible KLT. *Int'l Conf. on Audio, Language and Image Processing, 2008. (ICALIP 2008).*, pages 1560–1564, July 2008.

30. Y. Wongsawat, S. Oraintara, and K. R. Rao. Integer sub-optimal Karhunen-Loève transform for multi-channel lossless EEG compression. *European Signal Processing Conference*, 2006.

Chapter 11
Hyperspectral Image Compression Using Segmented Principal Component Analysis

Wei Zhu, Qian Du, and James E. Fowler

Abstract Principal component analysis (PCA) is the most efficient spectral decorrelation approach for hyperspectral image compression. In conjunction with JPEG2000-based spatial coding, the resulting PCA+JPEG2000 can yield superior rate-distortion performance. However, the involved overhead bits consumed by the large operation matrix for principal component transform may affect compression performance at low bitrates, particularly when the spatial size of an image patch to be compressed is relatively small compared to the spectral dimension. In our previous research, we proposed to apply the segmented principal component analysis (SPCA) to mitigate this effect, and the resulting compression algorithm, denoted as SPCA +JPEG2000, can improve the rate-distortion performance even when PCA +JPEG2000 is applicable. In this chapter, we investigate the quality of reconstructed data after SPCA+JPEG2000 compression based on the performance in spectral fidelity, classification, linear unmixing, and anomaly detection. The experimental results show that SPCA+JPEG2000 can outperform in terms of preserving more useful data information, in addition to offer excellent rate-distortion performance. Since the spectral partition in SPCA relies on the calculation of a data-dependent spectral correlation coefficient matrix, we investigate a sensor-dependent suboptimal partition approach, which can accelerate the compression process with no much distortion.

1 Introduction

Data compression is a frequently applied technique to reduce the vast data volume of a hyperspectral image. It has been shown that principal component analysis (PCA) in conjunction with JPEG2000 [1, 2] provides prominent rate-distortion

W. Zhu • Q. Du (✉) • J.E. Fowler
Department of Electrical and Computer Engineering, Mississippi State University,
Starkville, USA
e-mail: du@ece.msstate.edu

B. Huang (ed.), *Satellite Data Compression*, DOI 10.1007/978-1-4614-1183-3_11,
© Springer Science+Business Media, LLC 2011

Table 11.1 SNR (dB) at 1.0 bpppb with different spectral decorrelation approaches for the compression of AVIRIS Jasper Ridge radiance data of size 512 × 512 with 224 bands

	None	DWT	PCA
Cuprite	38.3	51.0	54.1
Jasper Ridge	29.8	44.8	50.3
Moffett	30.6	45.5	50.9

performance for hyperspectral image compression, where PCA is for spectral coding and JPEG2000 is for spatial coding of principal component (PC) images (referred to as PCA+JPEG2000); in particular, PCA+JPEG2000 outperforms its discrete wavelet transform (DWT) counterpart, DWT+JPEG2000, where DWT is applied for spectral coding [3–7]. PCA+JPEG2000 also outperforms other DWT-based algorithms, such as 3-dimensional (3D) set partitioning in hierarchical trees (SPIHT) [8] and 3D set partitioned embedded block (SPECK) [9]. When only a set of principal components (PCs) are used for compression, the resulting SubPCA +JPEG2000 can further improve the rate-distortion performance [6].

Table 11.1 lists the signal-to-noise ratio (SNR) for PCA+JPEG2000 and DWT +JPEG2000 at 1.0 bpppb (bit per pixel per band); also shown is the performance when JPEG2000 is used with no spectral decorrelation. Here, SNR is defined as the ratio between signal variance and reconstruction error variance. We see that in all cases, although DWT-based spectral decorrelation improves SNR by around 15 dB with respect to no spectral decorrelation, PCA-based spectral decorrelation results in a further 5-dB increase. From a statistical perspective, PCA offers optimal decorrelation while highly structured correlation is known to exist between DWT coefficients, both with subbands and across subbands. While JPEG2000 exploits this DWT correlation structure spatially, no attempt is made to exploit residual correlation across components, i.e., spectrally. As a consequence, a spectral DWT leaves a significant degree of correlation present in the spectral direction; the spectral PCA, with its optimal decorrelation, thus performs better.

Figure 11.1 shows the rate-distortion curves of the DWT group (i.e., 3D SPIHT, 3D SPIHT, DWT+JPEG2000) and PCA group (PCA+JPEG2000, SubPCA +JPEG2000). Obviously, the PCA group significantly outperforms the DWT group. The rate distortion performance evaluates the data point fidelity. In order to evaluate the fidelity of pixel spectrum after compression, spectral angle mapper (SAM) is used to calculate the spectral angles before and after compression, and the angle difference is averaged and shown in Fig. 11.2. It demonstrates that the PCA group can do a better job in preserving pixel spectral signatures than the DWT group. PCA+JPEG2000 and SubPCA+JPEG2000 have been modified for anomaly detection and multi-temporal image compression [10–12]. Since SubPCA +JPEG2000 requires the estimation of an optimal number of PCs to be used at difference bitrates, we limit the discussion about PCA+JPEG2000 only in this chapter. Moreover, only DWT+JPEG2000 is used for comparison purpose hereafter, because it is the best in the DWT group.

It is worth mentioning that the Consultative Committee for Space Data Systems (CCSDS) has created the Multispectral Hyperspectral Data Compression (MHDC)

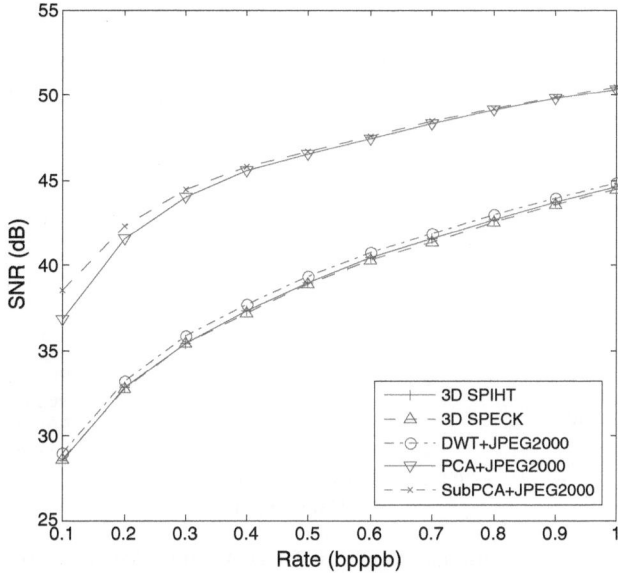

Fig. 11.1 Rate-distortion performance of PCA- and DWT-based compression algorithms for AVIRIS Jasper Ridge radiance data of size 512 × 512 with 224 bands

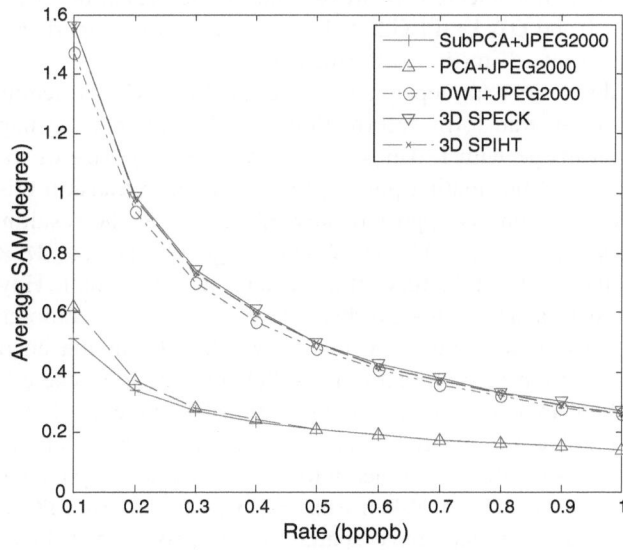

Fig. 11.2 Spectral distoration of PCA- and DWT-based compression algorithms for AVIRIS Jasper Ridge radiance data of size 512 × 512 with 224 bands

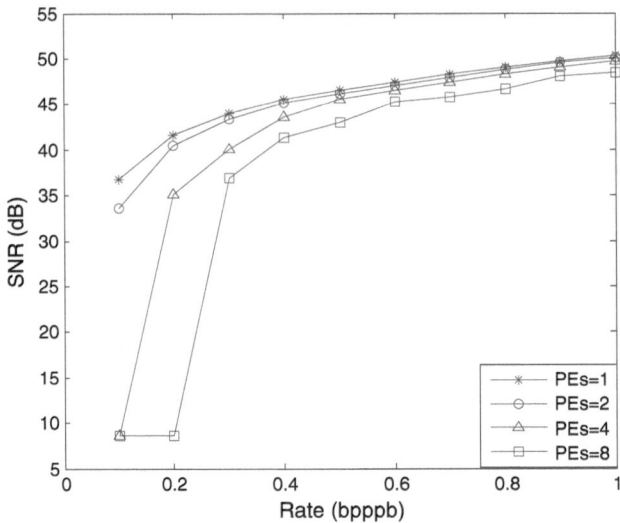

Fig. 11.3 Rate-distortion performance of parallel PCA+JPEG2000 with spatial partition for AVIRIS Jasper Ridge radiance data of size 512 × 512 with 224 bands

Working group to recommend the standard for multispectral and hyperspectral image compression. Since it targets onboard compression, where a careful tradeoff between compression performance and computational complexity has to be considered, JPEG2000 is not selected. However, during the design of CCSDS standard, PCA+JPEG2000 and DWT+JPEG2000 have been used as comparison baseline due to their excellent rate-distortion performance [13].

However, due to its data-dependent nature, PCA+JPEG2000 requires overhead bits carrying PCA transform information, which may not be negligible. For instance, for an image with L bands and $M{\times}N$ pixels, the size of transformation matrix is $L{\times}L$; if 32-bit floating point values are used to transmit this matrix, the number of overhead bits is approximately $32{\times}L{\times}L$ and the resulting change in bitrate in bpppb is $(32{\times}L)/(M{\times}N)$. When image spatial size $M \times N$ is large compared to the value of L, this bitrate change is very small. However, when image spatial size is small, this overhead severely degrades compression performance due to insufficient bits being used for data information encoding at low bitrates. This is an intrinsic problem in parallel compression, where image spatial partition is often applied to reduce computational complexity [14]. Figure 11.3 illustrates this problem in parallel compression when the image of size 512 × 512 with 224 bands is spatially partitioned into p segments, where p equals the number of processing elements (PEs). When PEs = 1, it is the original PCA+JPEG2000. When PEs = 4 (i.e., the subimage spatial size is 256 × 256), PCA fails at 0.1 bpppb; when PEs = 8 (i.e., the subimage spatial size is 128 × 128), PCA cannot work properly at 0.1 and 0.2 bpppb.

To overcome such a practical application problem of PCA+JPEG2000, we have proposed a segmented PCA (SPCA)-based compression method to improve the performance of PCA+JPEG2000 when the impact of overhead bits cannot be ignored [15]. The basic idea is to conduct spectral partition based on spectral correlation coefficient, and then PCA is applied to each spectral segment. The resulting compression method is referred to as SPCA+JPEG2000. In [15], we have shown that SPCA+JPEG2000 can not only solve the problem of the original PCA+JPEG2000 but also improve the rate-distortion performance even when PCA +JPEG2000 is applicable. In this chapter, we will investigate the quality of reconstructed data after SPCA+JPEG2000 compression by evaluating its spectral fidelity and the discrepancy on data applications in classification, anomaly detection, and linear unmixing results. Due to the lack of ground truth in practice, all the data analysis algorithms used in the evaluation are unsupervised. Since SPCA calculates the correlation coefficient matrix for spectral partition, which is data dependent, we propose a sensor-dependent suboptimal segment approach to accelerate the compression process. Its performance in rate-distortion, spectral fidelity, and data applications are also evaluated.

This chapter is organized as follows. In Sect. 2, the spectral partition for SPCA is briefly introduced. In Sect. 3, the compression scheme of SPCA+JPEG2000 is reviewed, and its performance in data analysis is thoroughly investigated in Sect. 4. In Sect. 5, the sensor-specific suboptimal SPCA is proposed to expedite the compression process and its performance in data analysis is also presented. The conclusion remarks are given in Sect. 6.

2   Segmented Principal Component Analysis

Band partition can be conducted uniformly (denoted as SPCA-U) or via the examination of spectral correlation coefficient (CC) as proposed in [16] (denoted as SPCA-CC). For an L-band image, its data covariance matrix Σ is L-by-L, from which the spectral correlation coefficient matrix $\mathbf{A}$ can be derived as

$$\mathbf{A}(i,j) = \frac{\Sigma(i,j)}{\sqrt{\Sigma(i,i)\Sigma(j,j)}} \tag{11.1}$$

$|\mathbf{A}|$ can be displayed as a gray-scale image. As illustrated in Fig. 11.4, a white pixel at location (i,j) means high correlation between the i-th and j-th band, and the highest value is 1 along the diagonal line. Obviously, the white blocks along the diagonal lines represent the adjacent bands being highly correlated, which should be grouped together. Thus, spectral bands in a hyperspectral image can be partitioned based on their correlation. Using the uniform partition, less correlated bands may be assigned to the same group.

Fig. 11.4 The spectral
correlation coefficient matrix
(224 × 224) of AVIRIS
Jasper Ridge data displayed
as an gray-scale image

Spectral partition can also be achieved by using mutual information in [17]. Since spectral correlation performs similarly as mutual information, resulting in similar band partitions, but with simpler computation, here we focus on SPCA with spectral correlation coefficient only.

Assume the original L bands are partitioned into p groups, i.e., $\sum_{i=1}^{p} L_i = L$ where L_i is the number of bands in the i-th group. There are three major advantages of SPCA:

(1) Band decorrelation may be more efficient since PCA is applied to highly correlated bands [16].
(2) Computational complexity is greatly reduced. The number of multiplications for calculating the data covariance matrix Σ is $N^2M^2L^2$; similarly, this number for Σ_i in the i-th group is $N^2M^2L_i^2$. Obviously, the number of multiplications in the SPCA is less than that of PCA due to the fact $\sum_{i=1}^{p} L_i^2 < L^2$. The computational complexity in the eigen-decomposition of Σ is $O(L^3)$, which is larger than that in the eigen-decomposition of Σ_i since $\sum_{i=1}^{p} L_i^3 < L^3$. Thus, SPCA can mitigate the computational burden of the original PCA through the reduction of matrix size [18].
(3) When SPCA is used for compression, the overhead related to the transformation matrix is reduced. Now the overhead bits are about $\sum_{i=1}^{p} 32L_i^2$, less than $32L^2$ in PCA.

3   SPCA+JPEG2000

When applying PCA+JPEG2000, the JPEG2000 encoder embeds the transform matrix into bitstream as an overhead. Assume there is an $L \times M \times N$ hyperspectral image with L bands and $M \times N$ pixels. The size of the transform matrix is $L \times L$, while the size of other vectors, e.g., mean vector of spectral bands, is negligible. Given the transform matrix is coded in 32-bit floating point value, the overhead occupies approximately $32L^2$ bits. In terms of bitrate, this overhead requires at least a bitrate of

Table 11.2 SNR in dB of PCA+JPEG2000 and DWT+JPEG2000 on the datasets with various spatial and spectral sizes

		R_{min}	$1.1R_{min}$	$1.5R_{min}$
$224 \times 512 \times 512$	DWT	13.85	14.09	14.95
	PCA	**5.45**	12.49	16.08
$224 \times 256 \times 256$	DWT	20.58	21.09	23.05
	PCA	**7.49**	16.24	23.96
$112 \times 256 \times 256$	DWT	21.38	21.91	23.78
	PCA	**7.97**	16.95	26.03
$224 \times 128 \times 128$	DWT	31.06	31.84	34.40
	PCA	**9.86**	21.67	33.28
$112 \times 128 \times 128$	DWT	32.46	33.21	35.78
	PCA	**11.85**	23.86	36.29
$56 \times 128 \times 128$	DWT	29.67	30.57	32.84
	PCA	**12.53**	15.25	27.16

$$R_{min} = \frac{32L}{MN} \text{bpppb}. \tag{11.2}$$

For PCA+JPEG2000, it requires $R > R_{min}$ to properly encode useful transform coefficients into bitstream. On the contrary, if $R < R_{min}$, no enough bits can be used for data encoding, resulting in poor compress performance. For instance, when the image size is $224 \times 64 \times 64$, it requires at least 1.75 bpppb for correct encoding; any compression ratio below 1.75 bpppb is simply insufficient for encoding actual data information. Table 11.2 shows the rate distortion performance when data size is changed. Obviously, PCA can provide better performance than DWT only when the image spatial size is relatively large enough.

Therefore, the size of the transformation matrix has to be dramatically reduced through SPCA to make the overhead negligible. To apply SPCA+JPEG2000 for hyperspectral image compression, the major steps of PCA+JPEG2000 are still needed. After a group of spectral bands are decorrelated by PCA, their principal components will be sent to JPEG2000 encoder for a three-stage process: spatial DWT, codeblock coding (CBC), and post-compression rate-distortion (PCRD) optimal truncation of codeblock bitstreams, to generate the final compressed bitstream. One approach is to process all three stages independently for each spectral group and then concatenate the bitstreams out of different encoders together, as illustrated in Fig. 11.5, where the PCRD optimizes the bitstream locally within each spectral group. The other approach is to process first two stages independently while applying PCRD globally within all codeblocks of all spectral groups, as shown in Fig. 11.6. The former approach is called local bit allocation (LBA) and the latter global bit allocation (GBA).

The AVIRIS Jasper Ridge dataset with different spatial and spectral sizes are spectrally partitioned into several band segments as listed in Table 11.3. The segments from CC have different sizes. It should be noted that the spectral correlation coefficient matrix **A** is recalculated when the image spatial size is different.

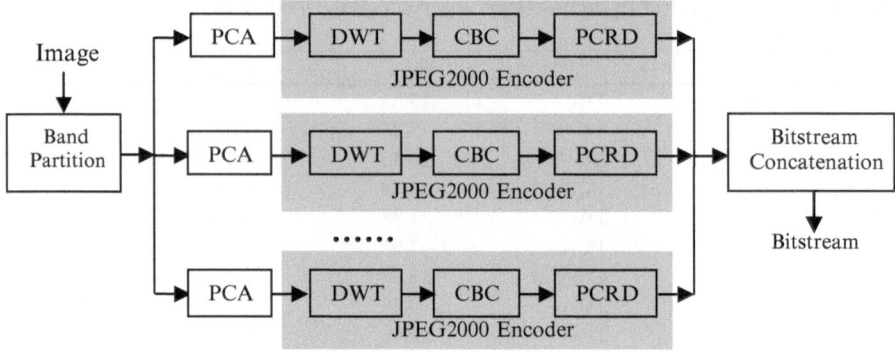

Fig. 11.5 The local bit allocation (*LBA*) procedure of SPCA+JPEG2000

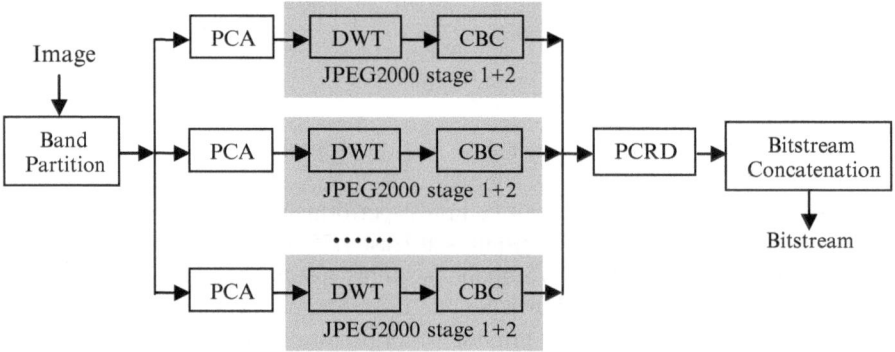

Fig. 11.6 The global bit allocation (*GBA*) procedure of SPCA+JPEG2000

Table 11.3 Band segments when the AVIRIS Jasper Ridge dataset is spatially and spectrally partitioned into subimages with different sizes

Spatial size	Spectral size	Number of Segments	SPCA-CC	Number of Segments SPCA-U
512 × 512	224	6	[1 38], [39 107], [108 113], [114 154], [155 166], [167 224]	[1 37], [38 74], [75 111], [112 148], [149 185], [186 224]
256 × 256	224	6	[1 38], [39 107], [108 113], [114 154], [155 166], [167 224]	[1 37], [38 74], [75 111], [112 148], [149 185], [186 224]
	112	3	[1 38], [39 107], [108 112]	[1 37], [38 74], [75 112]
128 × 128	224	7	[1 38], [39 82],[83 107], [108 113], [114 154], [155 166], [167 224]	[1 32], [33 64], [65 96], [97 128], [129 160], [161 192], [193 224]
	112	4	[1 38], [39 82], [83 107], [108 112]	[1 28], [29 56], [57 84], [85 112]
	56	3	[1 19], [20 38], [39 56]	[1 19], [20 38], [39 56]

Table 11.4 SNR in dB for various compression algorithms applied to AVIRIS Jasper Ridge dataset

Algorithms	R_{min}	$1.1R_{min}$	$1.5R_{min}$		R_{min}	$1.1R_{min}$	$1.5R_{min}$
$224 \times 512 \times 512, R_{min} = 0.0273$ bpppb				$224 \times 128 \times 128, R_{min} = 0.4375$ bpppb			
DWT	13.85	14.09	14.95	DWT	31.06	31.84	34.40
PCA	5.45	12.49	16.08	PCA	9.86	21.67	33.28
SPCA-U-LBA	15.01	15.29	16.34	SPCA-U-LBA	28.24	29.20	32.61
SPCA-U-GBA	15.60	15.97	17.52	SPCA-U-GBA	34.77	35.63	38.25
SPCA-CC-LBA	15.27	15.52	16.30	SPCA-CC-LBA	24.60	25.15	27.75
SPCA-CC-GBA	16.36	16.82	18.64	SPCA-CC-GBA	35.50	36.27	38.69
$224 \times 256 \times 256, R_{min} = 0.1094$ bpppb				$112 \times 128 \times 128, R_{min} = 0.2188$ bpppb			
DWT	20.58	21.09	23.05	DWT	32.46	33.21	35.78
PCA	7.49	16.24	23.96	PCA	11.85	23.86	36.29
SPCA-U-LBA	20.49	21.01	22.61	SPCA-U-LBA	33.75	34.81	38.01
SPCA-U-GBA	24.33	25.21	27.82	SPCA-U-GBA	34.93	35.94	39.16
SPCA-CC-LBA	19.07	19.33	20.34	SPCA-CC-LBA	33.94	35.08	38.12
SPCA-CC-GBA	25.54	26.45	28.99	SPCA-CC-GBA	35.75	36.77	39.91
$112 \times 256 \times 256, R_{min} = 0.0547$ bpppb				$56 \times 128 \times 128, R_{min} = 0.1094$ bpppb			
DWT	21.38	21.91	23.78	DWT	29.67	30.57	32.84
PCA	7.97	16.95	26.03	PCA	12.53	15.25	27.16
SPCA-U-LBA	22.63	23.48	26.65	SPCA-U-LBA	24.48	25.84	29.19
SPCA-U-GBA	24.37	25.52	28.60	SPCA-U-GBA	29.03	30.29	33.67
SPCA-CC-LBA	21.97	23.50	27.65	SPCA-CC-LBA	24.48	25.84	29.19
SPCA-CC-GBA	25.65	26.91	30.21	SPCA-CC-GBA	29.03	30.29	33.67

This is because the estimated correlation may be different when the enclosed pixels for estimation are changed.

The rate-distortion performance of SPCA+JPEG2000 is listed in Table 11.4. We can see that GBA-based SPCA clearly yielded better rate-distortion performance than LBA. In addition, SPCA-CC yielded similar rate-distortion results as SPCA-U.

To provide a more comprehensive view of performance, Figs. 11.7 and 11.8 show SNR for rates ranging from 0.1 to 1.0 bpppb for both the $224 \times 512 \times 512$ and $224 \times 128 \times 128$ subscene sizes. These sizes are chosen because, in Table 11.4, the non-segmented PCA transform achieves the best performance for the former, and the worst for the latter. As shown in Fig. 11.7, when PCA-based JPEG2000 coding works properly (i.e., the spatial size is relatively large with respect to the spectral size such that the PCA transform matrix occupies negligible rate overhead in the compressed bitstream), the performance of SPCA-CC-GBA is slightly below that of non-segmented PCA. However, as shown in Fig. 11.8, when the non-segmented PCA is overwhelmed with transform-matrix rate overhead (i.e., at rates close to R_{min} for datasets with spatial size relatively small with respect to spectral size), SPCA can significantly improve performance. In both cases, the CC-based partition outperforms the uniform partition.

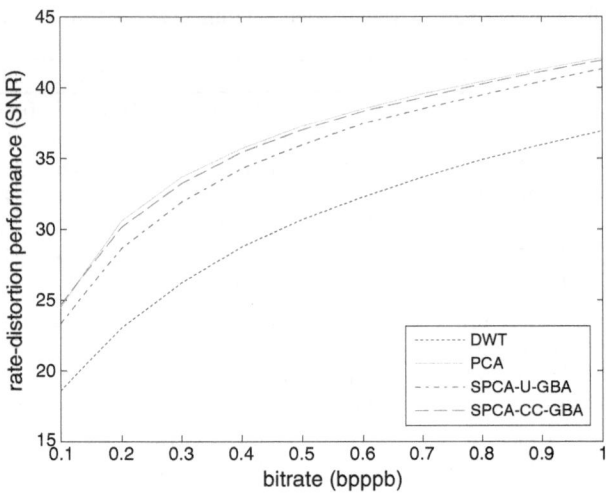

Fig. 11.7 Rate-distortion performance for Jasper Ridge dataset with size $224 \times 512 \times 512$

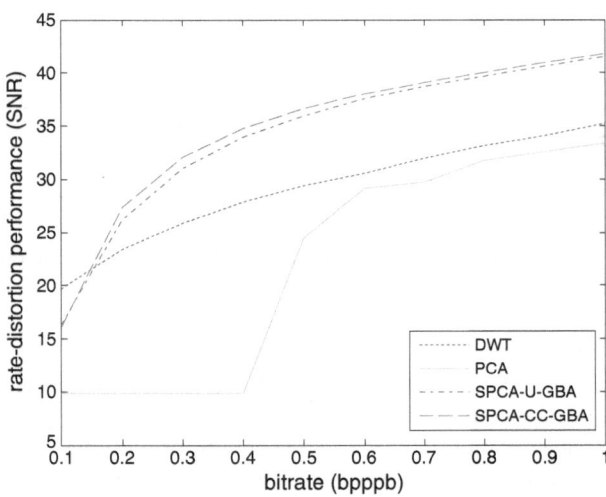

Fig. 11.8 Rate-distortion performance for Jasper Ridge dataset with size $224 \times 128 \times 128$

4 Data Analysis Performance of SPCA+JPEG2000

The performance of a lossy image compression algorithm is usually evaluated in terms of distortion and compression ratio. However, to evaluate the useful data information preserved during the compression process, we are particularly interested in application-specific distortions [19–21]. Depending on the nature of lost information, the impact on the following data analysis can be quite different. Thus, we will evaluate the performance of classification, detection, and spectral unmixing

Table 11.5 Spectral fidelity evaluated in average spectral angle

Size	Algorithms	0.2	0.4	0.6	0.8	1
224 × 512 × 512	DWT	3.15°	1.69°	1.14°	0.85°	0.67°
	PCA	1.37°	0.78°	0.56°	0.45°	0.37°
	SPCA-U-GBA	1.69°	0.90°	0.63°	0.50°	0.41°
	SPCA-CC-GBA	1.43°	0.80°	0.58°	0.46°	0.38°
224 × 128 × 128	DWT	2.42°	1.50°	1.11°	0.83°	0.66°
	PCA	7.07°	7.07°	1.32°	0.99°	0.81°
	SPCA-U-GBA	1.77°	0.76°	0.51°	0.40°	0.32°
	SPCA-CC-GBA	1.56°	0.69°	0.48°	0.38°	0.31°

using reconstructed data. Unsupervised algorithms are more useful due to the lack of prior information in practice.

Within the SPCA algorithms, we focus on GBA hereafter since its rate distortion performance is significantly better than LBA. The Jasper Ridge dataset of size 224 × 512 × 512 and 224 × 128 × 128 were used in the experiments. Five bitrates, i.e., 0.2, 0.4, 0.6, 0.8, 1.0 bpppb, were examined. For the 224 × 128 × 128 image, the $R_{\min}$ is 0.4375 bpppb, so PCA did not work at 0.2 and 0.4 bpppb.

4.1 Spectral Fidelity

Instead of data point fidelity assessed by rate-distortion performance, spectral fidelity can be evaluated by average spectral angle between original and reconstructed pixels [22]. It is important to evaluate spectral fidelity because many hyperspectral analysis algorithms utilize pixel spectral information. Let **r** and $\hat{\mathbf{r}}$ denote an original pixel vector and its reconstructed version after compression. Their spectral angle is defined as

$$\theta = \cos^{-1} \frac{\mathbf{r}^T \hat{\mathbf{r}}}{\|\mathbf{r}\|\|\hat{\mathbf{r}}\|} \tag{11.3}$$

A smaller angle corresponds to less spectral distortion.

As listed in Table 11.5, for the original image with relatively large spatial dimension, e.g., 224 × 512 × 512, PCA and SPCA performed similarly; for the image of size 224 × 128 × 128 after spatial shrinking, SPCA provided much smaller spectral angles. In both cases, the CC-based spectral partition yielded less spectral degradation than the uniform partition.

4.2 Unsupervised Classification

Unsupervised classification was conducted on the original and reconstructed image using independent component analysis (ICA). Specifically, the fastICA algorithm

Table 11.6 Classification performance evaluated in spatial correlation coefficient

Size	Algorithms	0.2	0.4	0.6	0.8	1
224 × 512 × 512	DWT	0.2996	0.3804	0.5357	0.5363	0.6245
	PCA	0.5563	0.5149	0.6591	0.7465	0.7281
	SPCA-U-GBA	0.4967	0.6615	0.8108	0.8100	0.8180
	SPCA-CC-GBA	0.5474	0.6238	0.6979	0.7563	0.7612
224 × 128 × 128	DWT	0.4024	0.4155	0.6259	0.6853	0.7695
	PCA	N/A	N/A	0.6035	0.6375	0.6747
	SPCA-U-GBA	0.4739	0.4842	0.6547	0.6726	0.8004
	SPCA-CC-GBA	0.3958	0.4707	0.6107	0.7023	0.7742

was adopted [23, 24]. Let $\mathbf{Z} = [\mathbf{z}_1 \mathbf{z}_2 \cdots \mathbf{z}_{MN}]$ be an $L \times MN$ data matrix with L-dimensional pixels. And let $\mathbf{w}$ be the desired projector and $\mathbf{y} = (y_1 y_2 \cdots y_{MN})$ be the projected data (after mean removal and data whitening). Denote $F(\bullet)$ as a function measuring independency. For instance, $F(y)$ can measure the kurtosis $\kappa(y)$ of the projected data, i.e.,

$$F(y) = \kappa(y) = E[(y)^4] - 3. \tag{11.4}$$

Then the task is to find an optimal $\mathbf{w}$ such that $\kappa(y)$ is maximal. This optimization problem can be formulated into the following objective function

$$J(\mathbf{w}) = \max_{\mathbf{w}} \{\kappa(y)\} = \max_{\mathbf{w}} \{\kappa(\mathbf{w}^T \mathbf{z})\}. \tag{11.5}$$

Taking the derivative with respect to $\mathbf{w}$ yields

$$\Delta \mathbf{w} = \frac{\partial \kappa}{\partial \mathbf{w}} = 4E(y^3 \mathbf{z}). \tag{11.6}$$

Then gradient-descent or fixed-point adaptation can be used to determine $\mathbf{w}$. After the first $\mathbf{w}$, denoted as $\mathbf{w}_1$, is found, it is used to transform the data for the first classification map. To find a second $\mathbf{w}$, denoted as $\mathbf{w}_2$, for another class, data matrix $\mathbf{Z}$ is projected onto the orthogonal subspace of $\mathbf{w}_1$ before searching $\mathbf{w}_2$. The algorithm continues until all the classes are classified.

The corresponding classification maps can be compared with spatial correlation coefficient; a larger average correlation coefficient means classification maps were closer. Let C and $\hat{C}$ denote classification maps using the original and reconstructed data, respectively. Their spatial correlation coefficient ρ is defined as

$$\rho = \frac{\sum_{x,y} (C(x,y) - \mu_C)(\hat{C}(x,y) - \mu_{\hat{C}})}{\sigma_C \sigma_{\hat{C}}} \tag{11.7}$$

where μ_C and $\mu_{\hat{C}}$ are the data mean of the two maps, and σ_C and $\sigma_{\hat{C}}$ are their corresponding standard deviation. In the experiment, ten classes were identified and compared. As shown in Table 11.6, both SPCA results were better than PCA result,

Table 11.7 Anomaly detection performance evaluated in the area covered by ROC

Size	Algorithms	0.2	0.4	0.6	0.8	1
224 × 512 × 512	DWT	0.999997	0.999998	0.999999	0.999999	0.999999
	PCA	0.999999	0.999999	0.999996	0.999989	0.999998
	SPCA-U-GBA	0.999998	0.999999	0.999999	0.999998	0.999998
	SPCA-CC-GBA	0.999999	0.999999	0.999999	0.999998	0.999998
224 × 128 × 128	DWT	0.980516	0.994166	0.996685	0.997357	0.998568
	PCA	0.5	0.5	0.990899	0.994811	0.998955
	SPCA-U-GBA	0.987356	0.998220	0.999515	0.999369	0.999688
	SPCA-CC-GBA	0.996060	0.996302	0.999661	0.999368	0.999790

which were also better than the DWT result. In this case, uniform partition generally provided slightly better performance than the CC-based partition.

4.3 Anomaly Detection

Anomaly detection is applied to the data before and after compression. The algorithm for anomaly detection is the well-known RX algorithm [25]:

$$\delta_{RXD}\left(\mathbf{r}\right) = \left(\mathbf{r} - \mathbf{\mu}\right)^T \mathbf{\Sigma}^{-1} \left(\mathbf{r} - \mathbf{\mu}\right) \tag{11.8}$$

where $\mathbf{\mu}$ is the sample mean vector. Thus, for a hyperspectral image with spatial size $M \times N$, the RX algorithm generates an $M \times N$ detection map. The receiver operating characteristic (ROC) curves are plotted, and the areas under the curves are computed. A large area means better performance. The maximum area is 1, and the minimum is 0.5.

As listed in Table 11.7, for the $224 \times 512 \times 512$ image, all the compression schemes worked well as the areas were close to 1, which is the ideal case. For the $224 \times 128 \times 128$ image, SPCA worked better than PCA and DWT; SPCA-CC-GBA provided slightly better performance than SPCA-U-GBA; PCA could not perform correctly at bitrates equal or below 0.4 bpppb, so the areas under ROC curves were 0.5.

4.4 Unsupervised Linear Unmixing

Let $\mathbf{E}$ be an $L \times p$ endmember signature matrix that is composed of $\left[\mathbf{e}_1, \mathbf{e}_2, \cdots, \mathbf{e}_p\right]$, where p is the number of endmembers in an image scene and $\mathbf{e}_i$ is an $L \times 1$ column vector representing the signature of the ith endmember material. Let $\alpha = \left(\alpha_1 \alpha_2 \cdots \alpha_p\right)^T$ be a $p \times 1$ abundance column vector, where α_i denotes the fraction of the ith signature presented in a pixel vector $\mathbf{r}$. With the assumption of

Table 11.8 Linear unmixing performance of SPCA+JPEG2000

Size	Algorithms		0.2	0.4	0.6	0.8	1
$224 \times 512 \times 512$	DWT	$\bar{\rho}$	0.97	1.00	1.00	1.00	1.00
		$\bar{\theta}$	6.34°	1.50°	0.82°	0.60°	0.45°
	PCA	$\bar{\rho}$	0.72	1.00	1.00	1.00	1.00
		$\bar{\theta}$	14.80°	0.57°	0.38°	0.33°	0.27°
	SPCA-U-GBA	$\bar{\rho}$	0.71	0.73	0.99	1.00	1.00
		$\bar{\theta}$	14.70°	14.63°	1.42°	0.31°	0.25°
	SPCA-CC-GBA	$\bar{\rho}$	1.00	1.00	1.00	1.00	1.00
		$\bar{\theta}$	1.26°	0.59°	0.38°	0.30°	0.22°
$224 \times 128 \times 128$	DWT	$\bar{\rho}$	0.60	0.99	0.85	1.00	1.00
		$\bar{\theta}$	14.89°	1.42°	2.03°	0.64°	0.52°
	PCA	$\bar{\rho}$	N/A	N/A	0.98	1.00	1.00
		$\bar{\theta}$	N/A	N/A	1.17°	0.60°	0.42°
	SPCA-U-GBA	$\bar{\rho}$	0.59	0.99	0.85	0.85	0.85
		$\bar{\theta}$	14.82°	0.81°	1.78°	1.70°	1.65°
	SPCA-CC-GBA	$\bar{\rho}$	0.59	0.99	0.85	0.85	0.85
		$\bar{\theta}$	14.43°	0.78°	1.76°	1.69°	1.64°

linear mixing, the spectral signature of a pixel vector **r** can be represented by the linear mixture model as

$$\mathbf{r} = \mathbf{E}\alpha + \mathbf{n} \tag{11.9}$$

where **n** is the noise or measurement error. There are two constraints for the abundance vector $\alpha = (\alpha_1 \alpha_2 \cdots \alpha_p)^T$: abundance sum-to-one constraint (ASC) and abundance nonnegativity constraint (ANC). The estimate from the least squares solution is to minimize the reconstruction residual

$$\min \ (\mathbf{r} - \mathbf{E}\alpha)^T (\mathbf{r} - \mathbf{E}\alpha). \tag{11.10}$$

There are no closed-form solutions to such a constrained optimization problem. But quadratic programming (QP) can be used to iteratively estimate the optimal solution, which minimizes the least squares estimation error and satisfies the two constraints, ASC and ANC, simultaneously [26]. The FCLS method requires a complete knowledge about the endmember signature matrix **E**. In order to handle a situation where no *a priori* information is available, an unsupervised FCLS (UFCLS) method can be used for endmember extraction [27, 28].

The corresponding endmembers are compared using spectral angles defined in Eq. 11.3, and a smaller average spectral angle $\bar{\theta}$ means better performance. The corresponding abundance maps are compared using spatial correlation coefficient defined in Eq. 11.7; the performance is claimed to be better if the average coefficient $\bar{\rho}$ is larger. As shown in Table 11.8, for the $224 \times 512 \times 512$ image, all the four schemes worked well; for the $224 \times 128 \times 128$ image, PCA did not perform correctly for 0.2 and 0.4 bpppb; SPCA-U-GBA and SPCA-CC-GBA were better than DWT and PCA, and SPCA-CC-GBA provided the best performance.

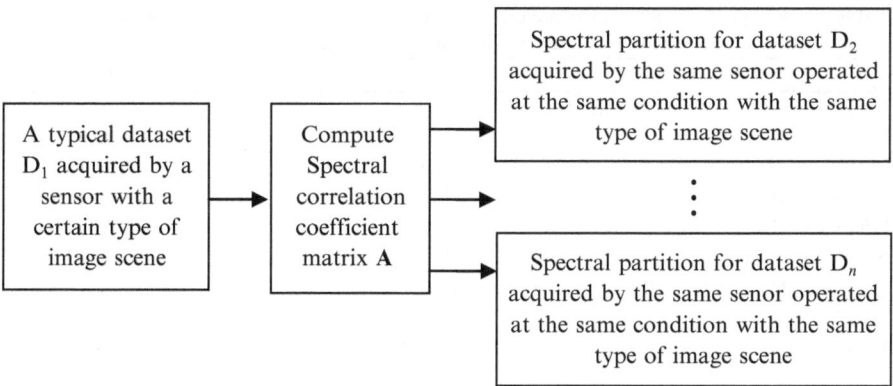

Fig. 11.9 The concept of the sensor-specific SPCA (SPCA-CC-S)

Table 11.9 Rate-distortion performance (SNR in dB) for Moffett data using spectral partitions based on correlation coefficients in Jasper Ridge data

Bitrate	R_{min}	$1.1R_{min}$	$1.5R_{min}$
Data size: 224 × 512 × 512 (R_{min} = 0.0273 bpppb)			
DWT	7.88	8.18	9.23
PCA	3.00	5.73	7.37
SPCA-U-GBA	8.71	9.32	11.20
SPCA-CC-GBA	8.76	9.38	11.58
SPCA-CC-S-GBA	8.63	9.26	11.54
Data size: 224 × 128 × 128 (R_{min} = 0.4375 bpppb)			
DWT	26.28	27.38	31.19
PCA	2.51	11.63	31.86
SPCA-U-GBA	33.53	34.77	38.28
SPCA-CC-GBA	34.89	36.03	39.11
SPCA-CC-S-GBA	34.89	36.03	39.11

5 Sensor-Specific Suboptimal Partitions for SPCA+JPEG2000

It is time-consuming if spectral partition is conducted for each specific dataset since data covariance coefficient matrix needs to be evaluated. It is helpful if a sensor-wide spectral partition is adopted for datasets collected by the same sensor. The resulting technique is denoted as "SPCA-CC-S". The concept is illustrated in Fig. 11.9.

In the experiment, the spectral partition for AVIRIS Jasper Ridge image scenes was applied to AVIRIS Moffett data. As presented in Table 11.9, the SNR values from such an SPCA-CC-S-GBA approach were slightly lower than the optimal SPCA-CC-GBA approach specific for Moffett data, but they were still higher than those from SPCA-U-GBA. Actually, for the image with small size (i.e., 224 × 128 × 128), the SNR values from the SPCA-CC-S-GBA approach were

Table 11.10 Spectral fidelity evaluated in average spectral angle for Moffett data using spectral partitions based on correlation coefficients in Jasper Ridge data

	$1.1R_{min}$	$1.5R_{min}$
Data size: $224 \times 512 \times 512$ ($R_{min} = 0.0273$ bpppb)		
DWT	13.46°	12.10°
PCA	18.94°	15.15°
SPCA-U-GBA	11.76°	9.51°
SPCA-CC-GBA	11.57°	8.97°
SPCA-CC-S-GBA	11.77°	9.04°
Data size: $224 \times 128 \times 128$ ($R_{min} = 0.4375$ bpppb)		
DWT	2.01°	1.32°
PCA	11.10°	1.23°
SPCA-U-GBA	0.88°	0.59°
SPCA-CC-GBA	0.76°	0.54°
SPCA-CC-S-GBA	0.76°	0.54°

Table 11.11 Classification performance for Moffett data using spectral partitions based on correlation coefficients in Jasper Ridge data

	$1.1R_{min}$	$1.5R_{min}$
Data size: $224 \times 512 \times 512$ ($R_{min} = 0.0273$ bpppb)		
DWT	0.61	0.62
PCA	0.53	0.54
SPCA-U-GBA	0.62	0.62
SPCA-CC-GBA	0.55	0.61
SPCA-CC-S-GBA	0.59	0.58
Data size: $224 \times 128 \times 128$ ($R_{min} = 0.4375$ bpppb)		
DWT	0.79	0.87
PCA	0.73	0.87
SPCA-U-GBA	0.90	0.89
SPCA-CC-GBA	0.89	0.89
SPCA-CC-S-GBA	0.89	0.89

similar to the optimal SPCA-CC-GBA approach. This proves the feasibility of a general sensor-specific spectral partition for the SPCA-based compression, which yields suboptimal compression performance but facilitates fast processing in practice.

The reconstructed AVIRIS Moffett data was also analyzed using the application-oriented evaluation approaches in Sect. 4. Here we evaluated two bitrates: $1.1R_{min}$ and $1.5R_{min}$. As shown in Table 11.10 for spectral fidelity, SPCA-CC-GBA and SPCA-CC-S-GBA yielded similar spectral angles, which were smaller than other methods. Table 11.11 lists the spatial correlation coefficients after applying fastICA to both original and reconstructed data. The three SPCA versions generated similar classification performance, which were better than DWT and PCA. Table 11.12 is

Table 11.12 Anomaly detection performance for Moffett data using spectral partitions based on correlation coefficients in Jasper Ridge data

	$1.1R_{min}$	$1.5R_{min}$
Data size: $224 \times 512 \times 512$ ($R_{min} = 0.0273$ bpppb)		
DWT	0.91	0.92
PCA	0.59	0.62
SPCA-U-GBA	0.91	0.93
SPCA-CC-GBA	0.90	0.93
SPCA-CC-S-GBA	0.90	0.93
Data size: $224 \times 128 \times 128$ ($R_{min} = 0.4375$ bpppb)		
DWT	0.62	0.63
PCA	0.51	0.60
SPCA-U-GBA	0.67	0.71
SPCA-CC-GBA	0.67	0.71
SPCA-CC-S-GBA	0.67	0.71

Table 11.13 Linear unmixing performance for Moffett data using spectral partitions based on correlation coefficients in Jasper Ridge data

		$1.1R_{min}$	$1.5R_{min}$
Data size: $224 \times 512 \times 512$ ($R_{min} = 0.0273$ bpppb)			
DWT	$\bar{\rho}$	0.33	0.36
	$\bar{\theta}$	$10.10°$	$8.92°$
PCA	$\bar{\rho}$	0.40	0.47
	$\bar{\theta}$	$11.12°$	$9.24°$
SPCA-U-GBA	$\bar{\rho}$	0.39	0.66
	$\bar{\theta}$	$11.07°$	$8.55°$
SPCA-CC-GBA	$\bar{\rho}$	0.39	0.67
	$\bar{\theta}$	$13.54°$	$9.58°$
SPCA-CC-S-GBA	$\bar{\rho}$	0.39	0.67
	$\bar{\theta}$	$13.30°$	$9.65°$
Data size: $224 \times 128 \times 128$ ($R_{min} = 0.4375$ bpppb)			
DWT	$\bar{\rho}$	0.58	0.41
	$\bar{\theta}$	$1.51°$	$2.63°$
PCA	$\bar{\rho}$	0.58	0.71
	$\bar{\theta}$	$7.34°$	$2.21°$
SPCA-U-GBA	$\bar{\rho}$	0.81	0.86
	$\bar{\theta}$	$2.16°$	$1.96°$
SPCA-CC-GBA	$\bar{\rho}$	0.75	0.87
	$\bar{\theta}$	$2.06°$	$1.95°$
SPCA-CC-S-GBA	$\bar{\rho}$	0.75	0.87
	$\bar{\theta}$	$2.06°$	$1.95°$

the anomaly detection result, which was evaluated by the area covered by the ROC curves. Again, the three SPCA versions had similar performance. Linear unmixing result is shown in Table 11.13. The three SPCA versions performed similarly on unsupervised linear unmixing.

6 Conclusion

There exists a minimum bitrate R_{min}, below which PCA+JPEG2000 cannot perform properly due to the overhead bits consumed by the large transformation matrix. When R_{min} is significantly large, we recommend the SPCA with CC-based band partition and global bit allocation, which can improve the performance dramatically. Even when PCA+JPEG2000 can perform properly, SPCA may yield comparable compression performance with lower computational complexity. The experimental results show that SPCA+JPEG2000 outperforms PCA+JPEG2000 in terms of preserving more useful data information. In addition, we propose a sensor-dependent suboptimal partition approach to accelerate the compression process without introducing much distortion.

References

1. Information Technology—JPEG 2000 Image Coding System—Part 1: *Core Coding System*, ISO/IEC 15444–1, 2000.
2. Information Technology—JPEG 2000 Image Coding System—Part 2: *Extensions*, ISO/IEC 15444–2, 2004.
3. J. E. Fowler and J. T. Rucker, "3D wavelet-based compression of hyperspectral imagery," in *Hyperspectral Data Exploitation: Theory and Applications*, C.-I. Chang, Ed., John Wiley & Sons, Inc., Hoboken, NJ, 2007.
4. B. Penna, T. Tillo, E. Magli, and G. Olmo, "Progressive 3-D coding of hyperspectral images based on JPEG 2000," *IEEE Geosciences and Remote Sensing Letters*, vol. 3, no. 1, pp. 125–129, 2006.
5. B. Penna, T. Tillo, E. Magli, and G. Olmo, "Transform coding techniques for lossy hyperspectral data compression," *IEEE Transactions on Geosciences and Remote Sensing*, vol. 45, no. 5, pp. 1408–1421, 2007.
6. Q. Du and J. E. Fowler, "Hyperspectral image compression using JPEG2000 and principal components analysis," *IEEE Geoscience and Remote Sensing Letters*, vol. 4, no. 2, pp. 201–205, 2007.
7. W. Zhu, *On the Performance of JPEG2000 and Principal Component Analysis in Hyperspectral Image Compression*, Master's Thesis, Mississippi State University, 2007.
8. B.-J. Kim, Z. Xiong, and W. A. Pearlman, "Low bit-rate scalable video coding with 3D set partitioning in hierarchical trees (3D SPIHT)," *IEEE Transactions on Circuits and Systems for Video Technology*, vol. 10, pp. 1374–1387, 2000.
9. X. Tang and W. A. Pearlman, "Scalable hyperspectral image coding," *Proceedings of IEEE International Conference on Acoustics, Speech, and Signal Processing*, vol. 2, pp. 401–404, 2005.
10. B. Penna, T. Tillo, E. Magli, and G. Olmo, "Hyperspectral image compression employing a model of anomalous pixels," *IEEE Geoscience and Remote Sensing Letters*, vol. 4, no. 4, pp. 664–668, 2007.
11. Q. Du, W. Zhu, and J. E. Fowler, "Anomaly-based JPEG2000 compression of hyperspectal imagery," *IEEE Geoscience and Remote Sensing Letters*, vol. 5, no.4, pp. 696–700, 2008.
12. W. Zhu, Q. Du, and J. E. Fowler, "Multi-temporal hyperspectral image compression," *IEEE Geoscience and Remote Sensing Letters*, vol. 8, no. 3, pp. 416–420, 2011.

13. F. Garcia-Vilchez and J. Serra-Sagrista, "Extending the CCSDS recommendation for image data compression for remote sensing scenarios," *IEEE Transactions on Geoscience and Remote Sensing*, vol. 47, no. 10, pp. 3431–3445, 2009.

14. H. Yang, Q. Du, W. Zhu, J. E. Fowler, and I. Banicescu, "Parallel data compression for hyperspectral imagery," *Proceedings of IEEE International Geoscience and Remote Sensing Symposium*, vol. 2, pp. 986–989, 2008.

15. Q. Du, W. Zhu, H. Yang, and J. E. Fowler, "Segmented principal component analysis for parallel compression of hyperspectral imagery," *IEEE Geoscience and Remote Sensing Letters*, vol. 6, no. 4, pp. 713–717, 2009.

16. X. Jia and J. A. Richards, "Segmented principal components transformation for efficient hyperspectral remote-sensing image display and classification," *IEEE Transactions on Geosciences and Remote Sensing*, vol. 37, no. 1, pp. 538–542, 1999.

17. V. Tsagaris, V. Anastassopoulos, and G. A. Lampropoulos, "Fusion of hyperspectral data using segmented PCT for color representation and classification," *IEEE Transactions on Geoscience and Remote Sensing*, vol. 43, no. 10, pp. 2365–2375, 2005.

18. Q. Du and J. E. Fowler, "Low-complexity principal component analysis for hyperspectral image compression," *International Journal of High Performance Computing Applications*, vol. 22, no. 4, pp. 438–448, 2008.

19. Q. Du, N. H. Younan, R. L. King, and V. P. Shah, "On the performance evaluation of pan-sharpening techniques," *IEEE Geoscience and Remote Sensing Letters*, vol. 4, no. 4, pp. 518–522, 2007.

20. G. Martin, V. Gonzalez-Ruiz, A. Plaza, J. P. Ortiz, and I. Garcia, "Impact of JPEG2000 compression on endmember extraction and unmixing of remotely sensed hyperspectral data," *Journal of Applied Remote Sensing*, vol. 4, Article ID 041796, 2010.

21. F. Garcia-Vilchez, J. Munoz-Mari, M. Zortea, I. Blanes, V. Gonzalez-Ruiz, G. Camps-Valls, A. Plaza, and J. Serra-Sagrista, "On the impact of lossy compression on hyperspectral image classification and unmixing," *IEEE Geoscience and Remote Sensing Letters*, vol. 8, no. 2, pp. 253–257, 2011.

22. F. A. Kruse, A. B. Lefkoff, J. W. Boardman, K. B. Heidebrecht, A. T. Shapiro, J. P. Barloon, and A. F. H. Goetz, "The spectral image processing system (SIPS) – Interactive visualization and analysis of imaging spectrometer data," *Remote Sensing of Environment*, vol. 44, no. 2–3, pp. 145–163, 1993.

23. A. Hyvärinen, "Fast and robust fixed-point algorithms for independent component analysis," *IEEE Transactions on Neural Network*, vol. 10, no. 3, pp. 626–634, 1999.

24. Q. Du, N. Raksuntorn, S. Cai, and R. J. Moorhead, "Color display for hyperspectral imagery," *IEEE Transactions on Geoscience and Remote Sensing*, vol. 46, no. 6, pp. 1858–1866, Jun. 2008.

25. I. S. Reed and X. Yu, "Adaptive multiple-band CFAR detection of an optical pattern with unknown spectral distribution," *IEEE Transactions on Acoustic, Speech and Signal Processing*, vol. 38, no. 10, pp. 1760–1770, 1990.

26. Q. Du, "Optimal linear unmixing for hyperspectral image analysis," *Proceedings of IEEE International Geoscience and Remote Sensing Symposium*, vol. 5, pp. 3219–3221, Anchorage, AK, Sep. 2004.

27. D. Heinz and C.-I Chang, "Fully constrained least squares linear mixture analysis for material quantification in hyperspectral imagery," *IEEE Transactions on Geoscience Remote Sensing*, vol. 39, no. 3, pp. 529–545, 2001.

28. Q. Du and C.-I Chang, "Linear mixture analysis-based compression for hyperspectal image analysis," *IEEE Transactions on Geoscience and Remote Sensing*, vol. 42, no. 4, pp. 875–891, 2004.

Chapter 12
Fast Precomputed Vector Quantization with Optimal Bit Allocation for Lossless Compression of Ultraspectral Sounder Data

Bormin Huang

Abstract The compression of three-dimensional ultraspectral sounder data is a challenging task given its unprecedented size. We develop a fast precomputed vector quantization (FPVQ) scheme with optimal bit allocation for lossless compression of ultraspectral sounder data. The scheme comprises of linear prediction, bit-depth partitioning, vector quantization, and optimal bit allocation. Linear prediction approach a Gaussian Distribution serves as a whitening tool to make the prediction residuals of each channel close to a Gaussian distribution. Then these residuals are partitioned based on bit depths. Each partition is further divided into several sub-partitions with various 2^k channels for vector quantization. Only the codebooks with 2^m codewords for 2^k-dimensional normalized Gaussian distributions are precomputed. A new algorithm is developed for optimal bit allocation among sub-partitions. Unlike previous algorithms [19, 20] that may yield a sub-optimal solution, the proposed algorithm guarantees to find the minimum of the cost function under the constraint of a given total bit rate. Numerical experiments performed on the NASA AIRS data show that the FPVQ scheme gives high compression ratios for lossless compression of ultraspectral sounder data.

1 Introduction

In the era of contemporary and future spaceborne ultraspectral sounders such as Atmospheric Infrared Sounder (AIRS) [1], Infrared Atmospheric Sounding Interferometer (IASI) [2], Geosynchronous Imaging Fourier Transform Spectrometer (GIFTS) [3] and Hyperspectral Environmental Suite (HES) [4], improved weather and climate prediction is expected. Given the large volume of 3D data that will be

B. Huang (✉)
Cooperative Institute for Meteorological Satellite Studies, Space Science and Engineering Center, University of Wisconsin–Madison, Madison, WI, USA
e-mail: bormin@ssec.wisc.edu

B. Huang (ed.), *Satellite Data Compression*, DOI 10.1007/978-1-4614-1183-3_12,   253
© Springer Science+Business Media, LLC 2011

generated each day by an ultraspectral sounder with thousands of infrared channels, the use of robust data compression techniques will be beneficial to data transfer and archiving. The main purpose of ultraspectral sounder data is to retrieve atmospheric temperature, moisture and trace gases profiles, surface temperature and emissivity, as well as cloud and aerosol optical properties. The physical retrieval of these geophysical parameters involves the inverse solution of the radiative transfer equation, which is a mathematically ill-posed problem, i.e. the solution is sensitive to the error or noise in the data [5]. Therefore, in order to avoid or near-lossless compression of ultraspectral sounder data to avoid potential degradation of geophysical parameters during retrieval, owing to lossy compression.

Vector Quantization (VQ) has been getting popular as a compression tool since the introduction of the Linde-Buzo-Gray (LBG) algorithm [6]. It has been previously applied to hyperspectral imager data compression [7–11]. To reduce the computational burden in ultraspectral sounder data compression, we develop the Fast Precomputed VQ (FPVQ) scheme that first converts the data into a Gaussian source via linear prediction, and then partitions the data based on their bit depths. Sub-partitions with various 2^k channels are created for each partition. Vector quantization using a set of precomputed 2^k-dimensional normalized Gaussian codebooks with 2^m codewords is then performed. Bit allocation for all sub-partitions is done via a new bit allocation scheme that reaches an optimal solution under the constraint of a given total bit rate. The FPVQ eliminates the time for online codebook generation and the precomputed codebooks are not required to be sent to decoder as side information.

The rest of the paper is arranged as follows. Section 2 describes the ultraspectral sounder data used in this study. Section 3 details the proposed compression scheme while Sect. 4 elaborates the results. Section 5 summarizes the paper.

2 Data

The ultraspectral sounder data could be generated from either a Michelson interferometer (e.g. CrIS, IASI and GIFTS) or a grating spectrometer (e.g. AIRS). Compression is performed on the standard ultraspectral sounder data set that is publicly available via anonymous ftp [12]. It consists of ten granules, five daytime and five nighttime, selected from the representative geographical regions of the Earth. Their locations, UTC times and local time adjustments are listed in Table 12.1. This standard ultraspectral sounder data set adopts the NASA AIRS digital counts made on March 2, 2004. The AIRS data includes 2,378 infrared channels in the 3.74–15.4 μm region of the spectrum. A day's worth of AIRS data is divided into 240 granules, each of 6 min durations. Each granule consists of 135 scan lines containing 90 cross-track footprints per scan line; thus there are a total of $135 \times 90 = 12,150$ footprints per granule. More information regarding the AIRS instrument may be acquired from the NASA AIRS website [13].

Table 12.1   Ten selected AIRS granules for hyperspectral sounder data compression studies

Granule 9	00:53:31 UTC	−12 H	(Pacific Ocean, daytime)
Granule 16	01:35:31 UTC	+2 H	(Europe, nighttime)
Granule 60	05:59:31 UTC	+7 H	(Asia, daytime)
Granule 82	08:11:31 UTC	−5 H	(North America, nighttime)
Granule 120	11:59:31 UTC	−10 H	(Antarctica, nighttime)
Granule 126	12:35:31 UTC	−0 H	(Africa, daytime)
Granule 129	12:53:31 UTC	−2 H	(Arctic, daytime)
Granule 151	15:05:31 UTC	+11 H	(Australia, nighttime)
Granule 182	18:11:31 UTC	+8 H	(Asia, nighttime)
Granule 193	19:17:31 UTC	−7 H	(North America, daytime)

The digital count data ranges from 12-bit to 14-bit for different channels. Each channel is saved using its own bit depth. To make the selected data more generic to other ultraspectral sounders, 271 bad channels identified in the supplied AIRS infrared channel properties file are excluded. It is assumed that they occur only in the AIRS sounder. Each resulting granule is saved as a binary file, arranged as 2,107 channels, 135 scan lines, and 90 pixels for each scan line. Figure 12.1 shows the AIRS digital counts at wavenumber 800.01 cm^{-1} for the ten selected granules on March 2, 2004. In these granules, coast lines are depicted by solid curves and multiple clouds at various altitudes are shown as different shades of colored pixels.

3   Compression Scheme

The proposed FPVQ scheme consists of the following five steps.

1. Linear Prediction: The purpose of this step is to reduce the data variance and make the data approach the Gaussian distribution. Popat and Zeger [23] proposed dispersive FIR filters to convert arbitrary data to appear Gaussian. Linear prediction appears to be a good whitening tool for ultraspectral sounder data. It employs a set of neighboring pixels to predict the current pixel [14–16]. For ultraspectral sounder data, the spectral correlation is generally much stronger than the spatial correlation [17]. Thus, it is natural to predict a channel as a linear combination of neighboring channels. The problem can be formulated as

$$\hat{\mathbf{X}}_i = \sum_{k=1}^{n_p} c_k \mathbf{X}_{i-k} \text{ or } \hat{\mathbf{X}}_i = X_p \mathbf{C} , \qquad (12.1)$$

where $\hat{\mathbf{X}}_i$ is the vector of the current channel representing a 2D spatial frame, $\mathbf{X}_p$ is the matrix consisting of n_p neighboring channels, and $\mathbf{C}$ is the vector of the prediction coefficients. The prediction coefficients is obtained from

$$\mathbf{C} = (X_p^T X_p)^{\dagger} (X_p^T \hat{\mathbf{X}}_i) , \qquad (12.2)$$

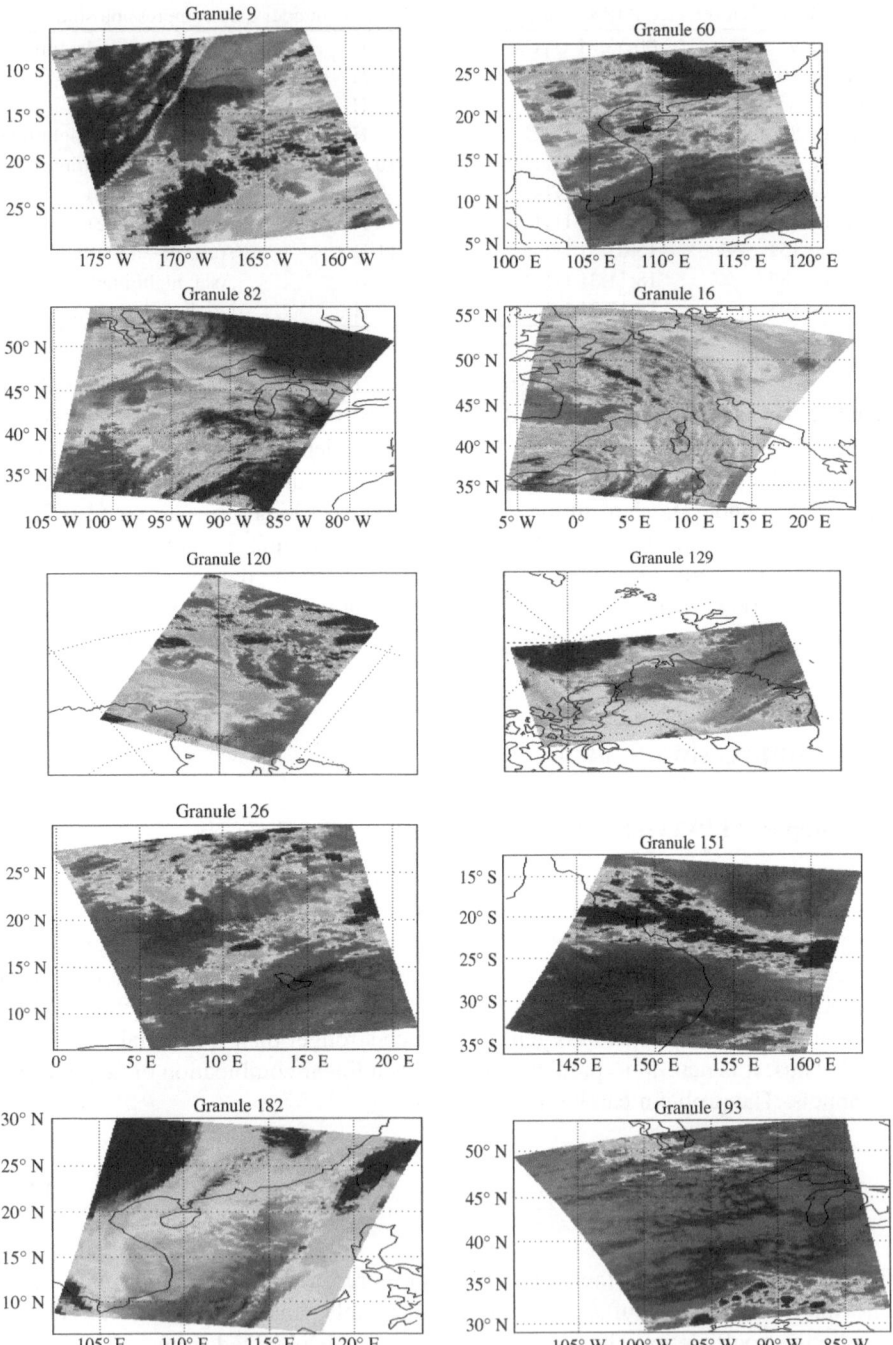

Fig. 12.1 AIRS digital counts at wavenumber 800.01 cm^{-1} for the ten selected granules on March 2, 2004

where the superscript † represents the pseudo-inverse that is robust against the case of the matrix being ill-conditioned [18]. The prediction error is the rounded difference between the original channel vector and its predicted counterpart.

2. Bit-depth Partitioning: To reduce the computational burden, channels with the same bit depth of prediction error are assigned to the same partition. Given n_d distinct bit depths, the channels are partitioned such that $\sum_{i=1}^{n_d} n_i = n_c$ where n_i is the number of channels in the i-th partition. The precomputed VQ codebooks are applied to each partition independently.

3. Vector Quantization with Precomputed Codebooks: Online VQ codebook generation using the well known Linde-Buzo-Gray (LBG) algorithm [6] is a costly operation. A precomputed VQ scheme is developed in order to avoid online codebook generation for ultraspectral sounder data compression. After the linear prediction, the prediction error of each channel is close to a Gaussian distribution with a different standard deviation. Only the codebooks with 2^m codewords for 2^k-dimensional normalized Gaussian distributions are precomputed via the LBG algorithm. It is known that any number of channels, n_i, in the i-th partition can be represented as a linear combination of 2^k as follows

$$n_i = \sum_{k=0}^{\lfloor \log_2^{n_i} \rfloor} d_{ik} 2^k, \; d_{ik} = 0 \text{ or } 1 \tag{12.3}$$

All the 2^k channels with $d_{ik} = 1$ form a sub-partition within the i-th bit-depth partition. The total number of the sub-partitions is

$$n_s = \sum_{i=1}^{n_d} n_{ib}, \tag{12.4}$$

where n_{ib} is the number of sub-partitions within the i-th bit-depth partition. It was reported that constraining the codebook size to be a power of two only slightly degrades the performance [20]. The actual, data-specific Gaussian codebook is the precomputed normalized Gaussian codebook scaled by the standard deviation spectrum, and the rounded quantization errors of the data within that sub-partition can be computed.

4. Optimal Bit Allocation: The number of bits used for representing the quantization errors within each sub-partition depends on its dimension and the codebook size. Several bit allocation algorithms [19, 20] based on marginal analysis [21] have been proposed in literature. These algorithms may not guarantee an optimal solution because they terminate as soon as the constraint of their respective minimization problems is met, and thus have no chance to move further along the hyperplane of the constraint to reach a minimum solution. Here we develop an improved bit allocation scheme that guarantees an optimal solution under the constraint. The minimization problem can be formulated as follows

$$f(b_{ij}^*) = \min_{b_{ij}} \sum_{i=1}^{n_d} \sum_{j=1}^{n_{ib}} L_{ij}(b_{ij}) \tag{12.5}$$

subject to

$$\sum_{i=1}^{n_d} \sum_{j=1}^{n_{ib}} b_{ij} = n_b, \tag{12.6}$$

where

$$L_{ij}(b_{ij}) = -n_{ij} \sum_{k=1}^{n_p(b_{ij})} p_k(b_{ij}) \, \log_2^{p_k(b_{ij})} + \frac{n_{ij}}{n_k} b_{ij} \tag{12.7}$$

is the expected total number of bits for the quantization errors in the *i-th* partition and the *j-th* sub-partition and for the quantization indices; n_{ij} the number of pixels within that sub-partition; n_k the number of channels in that sub-partition; b_{ij} the codebook size in bits for that sub-partition; n_b the total bits of all the codebooks; n_p is the number of distinct values of quantization errors, and p_k is the probability of the *k-th* distinct value. Both n_p and p_k depend on the codebook size b_{ij}. For lossless compression, measurement using the total bits for the quantization errors and the quantization indices appears superior to using the squared error measure.

The new optimal bit assignment algorithm for finding the solution to (12.5) with the constraint (12.6) consists of the following steps:

Step (1) Set $b_{ij} = 1, \quad \forall i,j$.
Step (2) Compute the marginal decrement $\Delta L_{ij}(b_{ij}) = L_{ij}(b_{ij} + 1) - L_{ij}(b_{ij}), \quad \forall i,j$.
Step (3) Find indices α, β for which $\Delta L_{\alpha\beta}(b_{\alpha\beta})$ is minimum.
Step (4) Set $b_{\alpha\beta} = b_{\alpha\beta} + 1$.
Step (5) Update $\Delta L_{\alpha\beta}(b_{\alpha\beta}) = L_{\alpha\beta}(b_{\alpha\beta} + 1) - L_{\alpha\beta}(b_{\alpha\beta})$.
Step (6) Repeat Steps 3–5 until $\sum_{i=1}^{n_d} \sum_{j=1}^{n_{ib}} b_{ij} = n_b$.
Step (7) Compute the next marginal decrement$\delta L_{ij} = L_{ij}(b_{ij} + 1) - L_{ij}(b_{ij}), \forall i,j$.
Step (8) Find $(\kappa, \lambda) = \underset{(i,j)}{\arg\min} \ \delta L_{ij}$ and $(v, \theta) = \underset{(i,j) \neq (\kappa,\lambda)}{\arg\max} \Delta L_{ij}(b_{ij})$.
Step (9) If $\delta L_{\kappa\lambda} < \Delta L_{v\theta}(b_{v\theta})$, set $b_{\kappa\lambda} = b_{\kappa\lambda} + 1$, $b_{v\theta} = b_{v\theta} - 1$, update $\delta L_{\kappa\lambda} = L_{\kappa\lambda}(b_{\kappa\lambda} + 1) - L_{\kappa\lambda}(b_{\kappa\lambda})$, and go to Step 8. else, STOP.

The idea of the marginal analysis in Steps 1–6 is similar to the algorithms developed by Riskin (with convexity assumption) [19] and Cuperman [20] for lossy compression. The operational rate-distortion functions derived from our training data with the precomputed Gaussian codebooks do not exhibit the convexity. An algorithm for nonconvex assumption was also proposed by Riskin [19], which requires time-consuming computation of all possible margins for all sub-divisions. It may not guarantee to satisfy the constraint because the number of deallocated bits is determined by the smallest slope of the rate-distortion function between the two possible rates. Our algorithm is much faster in the sense that it only needs to update the margin for the sub-partition that gives the minimum margin for both convex and nonconvex cases. After Step 6, the desired rate as the constraint in (12.6) is reached but the result may not be a minimum solution to (12.5). Steps 7–9 did not exist in the algorithms proposed by Riskin and Cuperman. They allow the comparison of neighboring bit allocations along the hyperplane of the constraint to reach to a (local) minimum of the cost function in (12.5). The proposed optimal bit algorithm is illustrated in Fig. 12.2 for a case of two partitions, where the green dots show the process of executing the first six steps and the red dots reflect the process of the last three steps. The black curve represents the constraint. As seen, the codebook sizes of the two partitions start with n1 = 1 and n2 = 1 and arrive at the green dot on the constraint curve after the tradition marginal analysis (Steps 1–6). The cost function values can be further decreased along the red dots on the constraint curve while executing Steps 7–9 of the proposed algorithm.

5. Entropy Coding: Arithmetic coding [22] is an entropy coding method that can represent symbols using variable number of bits. The basic idea of the arithmetic coding is to locate a proper interval $[I_l, I_u]$ for the given stream. An initial interval $[I_l^0 = 0, I_u^0 = 1]$ between 0 and 1 is iteratively reduced to a smaller subinterval $[I_l^i, I_u^i]$ based on the distribution model of the input source symbol. Assuming that the source symbols are numbered from 1 to n, and that symbol s has probability $\Pr(s)$, the interval reduction process for input source symbol x at iteration i can be computed according to (12.8) as follows.

$$I_l^{i+1} = I_l^i + (I_u^i - I_l^i) \sum_{s=1}^{x-1} \Pr(s)$$

$$I_u^{i+1} = I_u^i + (I_u^i - I_l^i) \sum_{s=1}^{x} \Pr(s) \qquad (12.8)$$

Symbols with a higher probability will reduce to bigger subintervals which require lesser number of bits to be represented while symbols with a lower probability will reduce to smaller subintervals which require more number of bits to be represented. This conforms to the general principle of compression. After the VQ stage, a context-based adaptive arithmetic coder [22] is used to encode the data quantization indices and quantization errors. In this way, the current interval is continued to be reduced to a smaller size until the end of stream. During the interval reduction process, the encoder can output the leading bits which are the same for the lower bound I_l^i and the upper bound I_u^i of the current interval. Then interval renormalization or rescaling can be applied to the interval bounds to save their precision bits. Using the leading bits and the same distribution model of symbols, the decoder can also iteratively locate the correct symbol and duplicate the same interval reduction process.

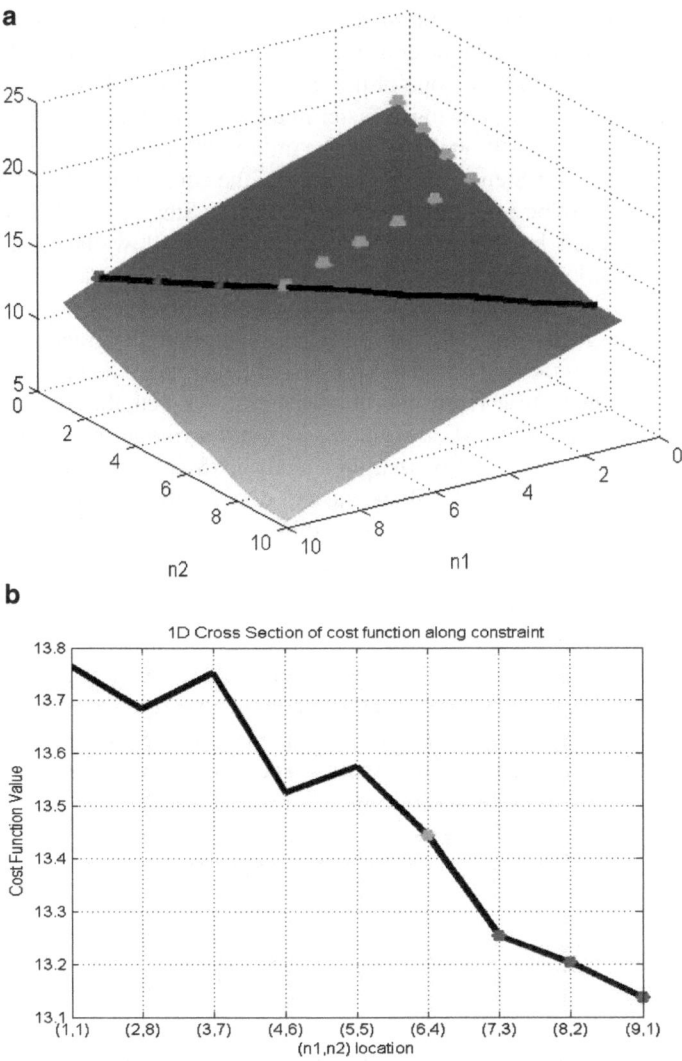

Fig. 12.2 (**a**) An example of the cost function in the new optimal bit allocation algorithm for two partitions with codebook sizes n1 and n2 bits, respectively (**b**) Cost function values along the constraint curve

4   Results

The aforementioned ten 3D NASA AIRS granules are studied for lossless compression of ultraspectral sounder data. Each granule has 12,150 spectra collected from the 2D spatial domain, consisting of 135 scan lines containing 90 cross-track footprints per

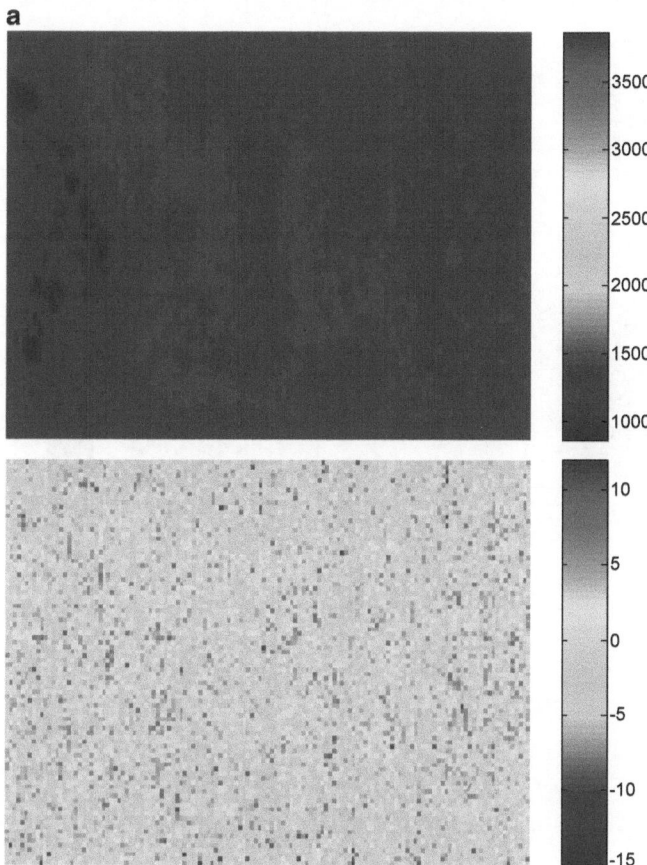

Fig. 12.3 (**a**) The original scene (*upper*) and the linear prediction residual scene (*lower*) for wavenumber 800.01 cm^{-1} in Granule 193 (**b**). Same as Fig. 3a except for wavenumber 1330.4 cm^{-1} (**c**) Same as Fig. 3a except for wavenumber 2197 cm^{-1}

scan line. Figure 12.3 shows the original 2D scene and the linear prediction residual scene for three selected channels in Granule 193. Thirty-two predictors are used in linear prediction. The randomness of the prediction residuals indicates that they are decorrelated quite well.

Figure 12.4 shows the data distributions after linear prediction for two selected channels in granule 82. As seen, the distributions appear close to Gaussian.

After the linear prediction, various channels may have different bit depths as illustrated in Fig. 12.5. The corresponding bit-depth partitioning result is shown in Fig. 12.6.

The results for FPVQ using a total bit budget of 50 are shown in Table 12.2. For comparison, we also show the compression ratios for JPEG2000 and for the

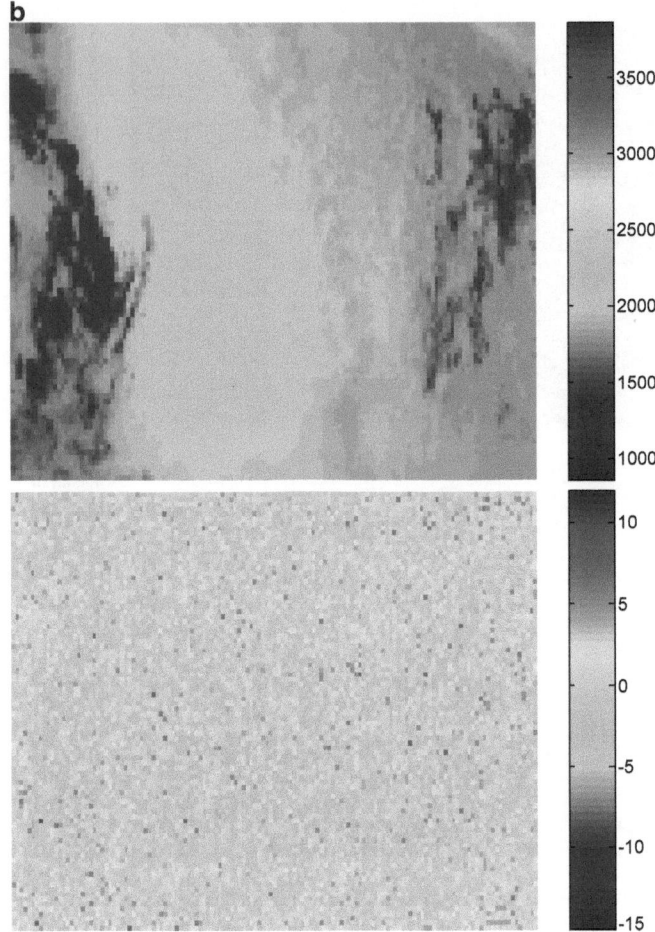

Fig. 12.3 (continued)

linear prediction (LP) followed by the entropy coding of the residual granule without VQ. As seen in Table 12.2, FPVQ produces significantly higher compress ratios than JPEG2000 and LP.

5 Summary

The Fast Precomputed VQ (FPVQ) scheme is proposed for lossless compression of ultraspectral sounder data. The VQ codebooks with 2^m codewords are precomputed for 2^k-dimensional normalized Gaussian distributions. The ultraspectral sounder data is converted to a Gaussian distribution via linear prediction in spectral

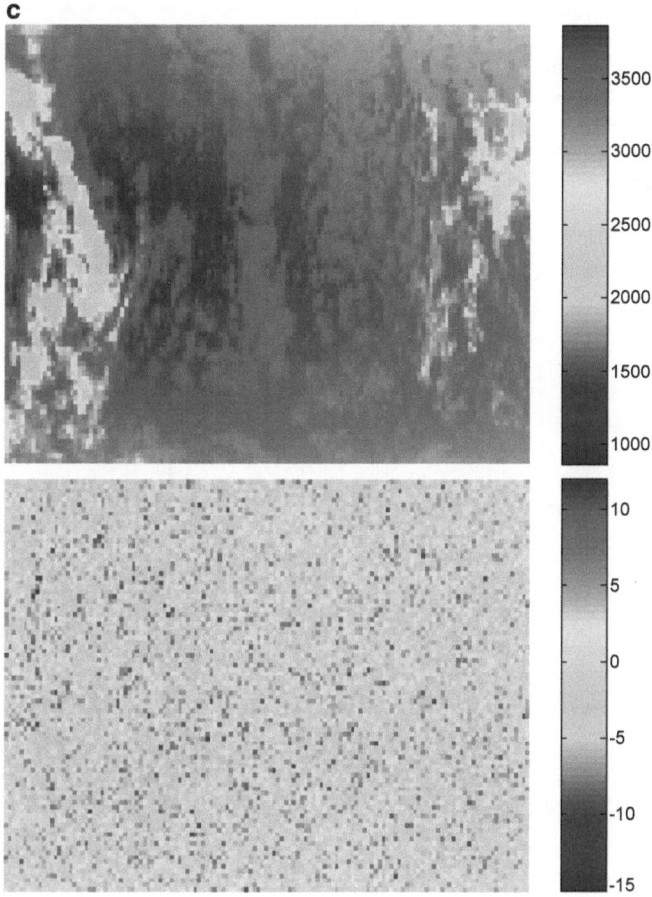

Fig. 12.3 (continued)

Table 12.2 Compression ratios of JPEG2000, LP and FPVQ for the ten tested AIRS granules

Granule No.	JPEG2000	LP	FPVQ
9	2.378	3.106	3.373
16	2.440	3.002	3.383
60	2.294	3.232	3.324
82	2.525	3.141	3.406
120	2.401	2.955	3.330
126	2.291	3.221	3.313
129	2.518	3.230	3.408
151	2.335	3.194	3.278
182	2.251	2.967	3.235
193	2.302	2.827	3.295
Average	2.374	3.087	3.334

Fig. 12.4 Data distribution
of linear prediction errors for
channels 244 (*upper*) and
1,237 (*lower*) in granule 82.
The fitted Gaussian
distributions and standard
deviations are also shown

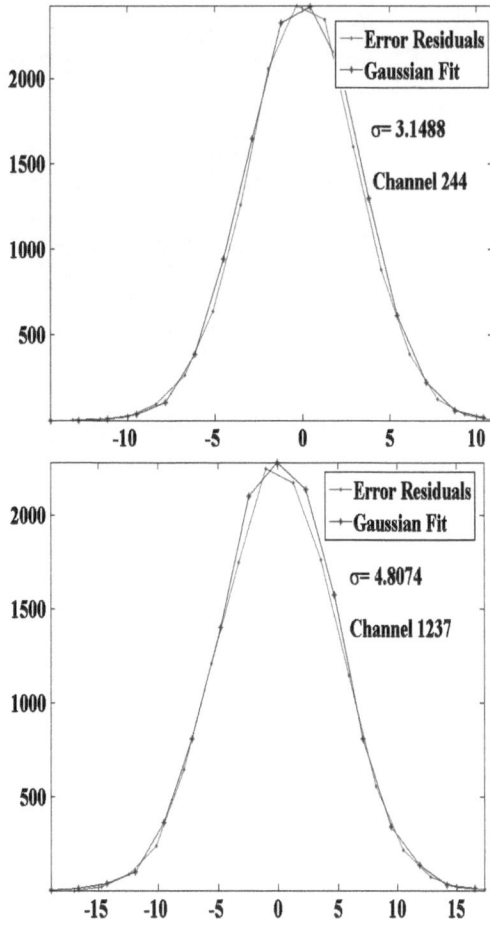

dimension. The data is first partitioned based on bit depths, followed by division of each partition into several sub-partitions with various 2^k channels. A novel marginal analysis scheme is developed for optimal bit allocation among sub-partitions. The FPVQ scheme is fast in the sense that we avoid the generation of time consuming online codebooks by using precomputed codebooks. Numerical results upon the ten NASA AIRS granules show that FPVQ yields significantly higher compression ratio than the other methods.

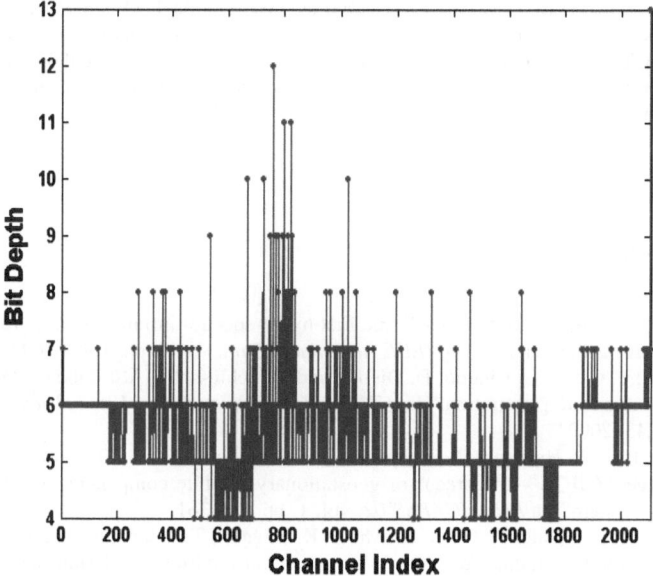

Fig. 12.5 Bit depth vs. original channel index after the linear prediction

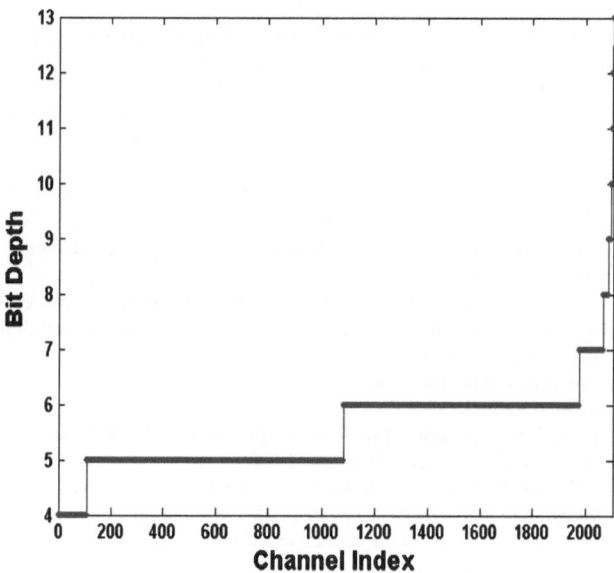

Fig. 12.6 Bit depth vs. new channel index after bit-depth partitioning

Acknowledgement This research was supported by National Oceanic and Atmospheric Administration's National Environmental Satellite, Data, and Information Service under grant NA07EC0676. The views, opinions, and findings contained in this report are those of the author(s) and should not be construed as an official National Oceanic and Atmospheric Administration or U.S. Government position, policy, or decision.

References

1. H. H. Aumann and L. Strow, "AIRS, the first hyper-spectral infrared sounder for operational weather forecasting," in *Proc. of IEEE Aerospace Conf.*, vol. 4, pp. 1683–1692, 2001.
2. T. Phulpin, F. Cayla, G. Chalon, D. Diebel, and D. Schlüssel, "IASI onboard Metop: Project status and scientific preparation," in *12th Int. TOVS Study Conf., Lorne, Victoria, Australia*, pp. 234–243, 2002.
3. W. L. Smith, F. W. Harrison, D. E. Hinton, H. E. Revercomb, G. E. Bingham, R. Petersen, and J. C. Dodge, "GIFTS – the precursor geostationary satellite component of the future Earth Observing System," in *Proc. IGARSS'02,* vol. 1, pp 357–361.
4. B. Huang, H.-L. Huang, H. Chen, A. Ahuja, K. Baggett, T. J. Schmit, and R. W. Heymann, "Data compression studies for NOAA hyperspectral environmental suite using 3D integer wavelet transforms with 3D set partitioning in hierarchical trees," in *SPIE Int. Symp. Remote Sensing Europe, 8–12 Sept. 2003, Barcelona, Spain, Proc. SPIE*, vol. 5238, pp. 255–265, 2003.
5. B. Huang, W. L. Smith, H.-L. Huang, H. M. Woolf, "Comparison of linear forms of the radiative transfer equation with analytic Jacobians," *Applied Optics*, 41 (21), 4209–4219.
6. Y. Linde, A. Buzo, and R. M. Gray, "An Algorithm for vector quantizer design," *IEEE Trans. Commun.*, vol. COM-28, pp. 84–95, Jan. 1980.
7. G. P. Abousleman, M. W. Marcellin, and B. R. Hunt, "Hyperspectral image compression using entropy-constrained predictive trellis coded quantization," *IEEE Trans. Image Processing*, vol. 6, no. 4, pp. 566–573, 1997.
8. G. R. Canta and G. Poggi, "Kronecker-product gain-shape vector quantization for multispectral and hyperspectral image coding," *IEEE Trans. Image Processing*, vol. 7, no. 5, pp. 668–678, 1998.
9. G. Gelli and G. Poggi, "Compression of multispectral images by spectral classification and transform coding," *IEEE Trans. Image Processing,* vol. 8, no. 4, pp. 476–489, 1999.
10. G. Motta, F. Rizzo, and J. A. Storer, "Compression of hyperspectral imagery," in *Proc. 2003 Data Comp. Conf.*, pp. 333–342, 2003.
11. F. Rizzo, B. Carpentieri, G. Motta, and J. A. Storer, "High performance compression of hyperspectral imagery with reduced search complexity in the compressed domain," in *Proc. 2004 Data Comp. Conf.*, pp. 479–488, 2004.
12. ftp://ftp.ssec.wisc.edu/pub/bormin/Count/.
13. http://www-airs.jpl.nasa.gov.
14. R. E. Roger, and M. C. Cavenor, "Lossless compression of AVIRIS images," *IEEE Trans. Image Processing*, vol. 5, no. 5, pp. 713–719, 1996.
15. X. Wu and N. Memon, "Context-based lossless interband compression – extending CALIC," *IEEE Trans. Image Processing*, vol. 9, pp. 994–1001, Jun. 2000.
16. D. Brunello, G. Calvagno, G. A. Mian, and R. Rinaldo, "Lossless compression of video using temporal information," *IEEE Tran. Image Processing*, vol. 12, no. 2, pp. 132–139, 2003.
17. B. Huang, A. Ahuja, H.-L. Huang, T. J. Schmit, and R. W. Heymann, "Lossless compression of 3D hyperspectral sounding data using context-based adaptive lossless image codec with Bias-Adjusted Reordering," *Opt. Eng.*, vol. 43, no. 9, pp. 2071–2079, 2004.
18. G. H. Golub and C. F. Van Loan, *Matrix Computations*, John Hopkins University Press, 1996.

19. E. A. Riskin, "Optimal bit allocation via the generalized BFOS algorithm," *IEEE Trans. Inform. Theory*, vol. 37, pp. 400–402, Mar. 1991.
20. V. Cuperman, "Joint bit allocation and dimensions optimization for vector transform quantization," *IEEE Trans. Inform. Theory*, vol. 39, pp. 302–305, Jan. 1993.
21. B. Fox, "Discrete optimization via marginal analysis," *Management Sci.*, vol. 13, no. 3, Nov. 1966.
22. I. H Witten., R. M. Neal, and J. Cleary, "Arithmetic coding for data compression," *Comm. ACM*, vol. 30, no. 6, pp. 520–540, 1987.
23. K. Popat and K. Zeger, "Robust quantization of memoryless sources using dispersive FIR filters," *IEEE Trans. Commun.*, vol. 40, pp. 1670–1674, Nov. 1992.

Chapter 13
Effects of Lossy Compression on Hyperspectral Classification

Chulhee Lee, Sangwook Lee, and Jonghwa Lee

1 Introduction

Rapid advancements in sensor technology have produced remotely sensed data with hundreds of spectral bands. As a result, there is now an increasing need for efficient compression algorithms for hyperspectral images. Modern sensors are able to generate a very large amount of data from satellite systems and compression is required to transmit and archive this hyperspectral data in most cases. Although lossless compression is preferable in some applications, its compression efficiency is around three [1–3]. On the other hand, lossy compression can achieve much higher compression rates at the expense of some information loss. Due to its increasing importance, many researchers have studied the compression of hyperspectral data and numerous methods have been proposed, including transform-based methods (2D and 3D), vector quantization [3–5], and predictive techniques [6]. Several authors have used principal component analysis to remove redundancy [7–9] and some researchers have used standard compression algorithms such as JPEG and JPEG 2000 for the compression of hyperspectral imagery [9–14]. The discrete wavelet transform has been applied to the compression of hyperspectral images [15, 16] and several authors have applied the SPIHT algorithm to the compression of hyperspectral imagery [17–23].

Hyperspectral images present opportunities as well as challenges. With detailed spectral information, it is possible to classify subtle differences among classes. However, as the data size increases sharply, transmitting and archiving hyperspectral data is a challenge. In most cases, it is impossible to exchange or store the hyperspectral images as raw data, and as a result, compression is required. However, when hyperspectral images are compressed with conventional image compression algorithms, which have been developed to minimize mean-squared

C. Lee (✉) • S. Lee • J. Lee
Yonsei University, South Korea
e-mail: chulhee@yonsei.ac.kr

B. Huang (ed.), *Satellite Data Compression*, DOI 10.1007/978-1-4614-1183-3_13,
© Springer Science+Business Media, LLC 2011

Fig. 13.1 Illustration of
discriminant features which
are small in energy

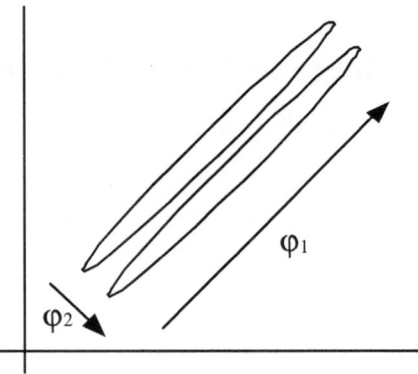

errors between the original and reconstructed data, discriminant features of the
original data may not be well preserved, even though the overall mean squared error
is small. Figure 13.1 shows an illustration of such a case. Although most energies
are distributed along φ_1, all the information necessary to distinguish between the
classes is distributed along φ_2.

In order to preserve the discriminant information for pattern classification
problems, care should be taken when applying compression algorithms. In this
chapter, we examine the discriminant characteristics of high dimensional data and
investigate how hyperspectral images can be compressed with minimal loss of these
discriminant characteristics.

2   Compression and Classification Accuracy

Principal component analysis (PCA) is optimal for signal representation in the
sense that it provides the smallest mean-squared error. However, quite often those
features defined by PCA are not optimal with regard to classification. Preserving the
discriminating power of the hyperspectral images is important if quantitative
analyses are later performed to compress the data.

Figure 13.2 shows an AVIRIS (Airborne Visible/Infrared Imaging Spectrometer) image containing 220 spectral bands [24]. Figure 13.3 shows a SNR comparison of 2D SPIHT, 3D SPIHT and PCA-based compression. Although the PCA-
based compression method produces the best SNR performance, its classification
accuracy is not satisfactory (as can be seen in Fig. 13.4). Higher classification
accuracy at lower bit rates is mainly due to the nature of the AVIRIS data, which
was taken in agricultural areas. The classification accuracy of the original data was
about 89.5%. However, the classification accuracy of the PCA-based compression
method was about 88.6% at 1 bit per pixel per band (bpppb) even though the SNR was
higher than 45.59 dB. Table 13.1 shows the eigenvalues along with their proportions
and accumulations. The first three eigenimages provided about 98.9% of the total
energy and the first six eigenimages provided 99.5%. It appears that using more than

Fig. 13.2 AVIRIS image with selected classes

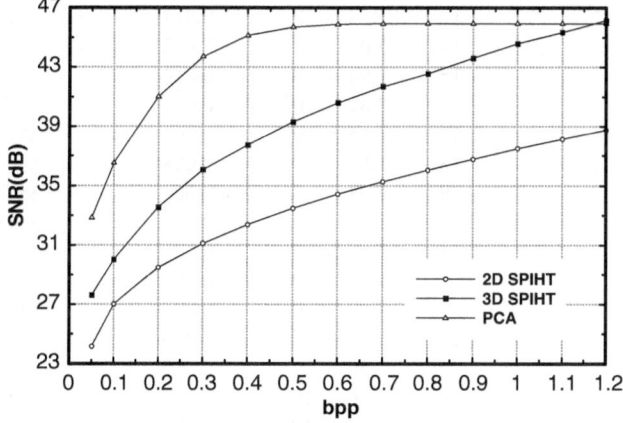

Fig. 13.3 Performance comparison of three compression methods (2D SPIHT, 3D SPIHT, PCA) in terms of SNRs

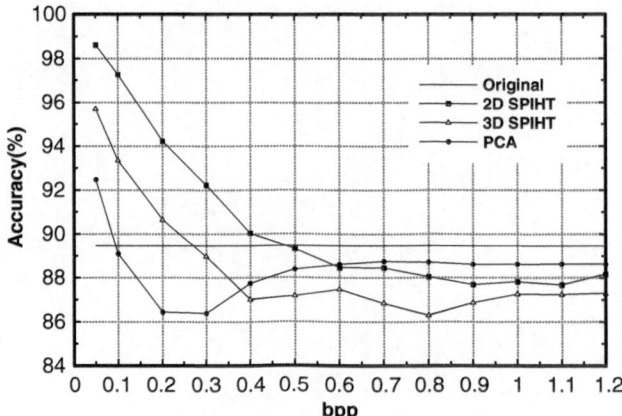

Fig. 13.4 Performance comparison of three compression methods (2D SPIHT, 3D SPIHT, PCA) in terms of classification accuracy

Table 13.1 Eigenvalues of hyperspectral images along with proportions and accumulations

	Eigenvalues	Proportion (%)	Accumulation (%)
1	1.67×10^7	67.686	67.686
2	6.84×10^6	27.601	95.287
3	8.93×10^5	3.601	98.889
4	6.82×10^4	0.275	99.164
5	5.28×10^4	0.213	99.377
6	2.59×10^4	0.105	99.482
7	2.40×10^4	0.097	99.578
8	1.28×10^4	0.051	99.630
9	7.72×10^3	0.031	99.661
10	5.62×10^3	0.023	99.684
11	4.90×10^3	0.020	99.704
12	4.25×10^3	0.017	99.721
13	3.54×10^3	0.014	99.735
14	3.02×10^3	0.012	99.747
15	2.94×10^3	0.012	99.759
16	2.82×10^3	0.011	99.770
17	2.15×10^3	0.009	99.779
18	1.88×10^3	0.008	99.787
19	1.77×10^3	0.007	99.794
20	1.60×10^3	0.006	99.800
21	1.53×10^3	0.006	99.806
22	1.52×10^3	0.006	99.813
23	1.46×10^3	0.006	99.818
24	1.43×10^3	0.006	99.824
25	1.40×10^3	0.006	99.830
26	1.36×10^3	0.005	99.835
27	1.33×10^3	0.005	99.841
28	1.29×10^3	0.005	99.846
29	1.25×10^3	0.005	99.851
30	1.19×10^3	0.005	99.856

7–10 eigenimages may not improve the mean squared error. All these results indicate that minimizing mean squared errors may not necessarily preserve the discriminant information required to distinguish between classes. Since most compression algorithms have been developed to minimize mean squared errors, they should show similar performance to the PCA-based compression method.

3   Feature Extraction and Compression

A number of feature extraction algorithms have been proposed for pattern classification [11, 25–29]. In canonical analysis [26], a within-class scatter matrix Σ_w and a between-class scatter matrix Σ_b can be used to formulate a criterion function and a vector d can be selected to maximize:

$$\frac{d^t \Sigma_b d}{d^t \Sigma_w d} \tag{13.1}$$

where the within-class scatter matrix and between-class scatter matrix are computed as follows:

$$\Sigma_w = \sum_i P(\omega_i) \Sigma_i$$

(within-class scatter matrix),

$$\Sigma_b = \sum_i P(\omega_i)(M_i - M_0)(M_i - M_0)^t$$

(between-class scatter matrix),

$$M_0 = \sum_i P(\omega_i) M_i.$$

Here M_i, Σ_i, and $P(\omega_i)$ are the mean vector, the covariance matrix, and the prior probability of class ω_i, respectively. In canonical analysis, the effectiveness of feature vectors for classification is quantified by (13.1).

In the decision boundary feature extraction method [27], feature vectors are directly extracted from decision boundaries. The decision boundary feature matrix Σ_{DBFM} can be defined as:

$$\Sigma_{DBFM} = \frac{1}{K} \int_S N(X) N^t(X) p(X) \, dX$$

where $N(X)$ represents the unit normal vector to the decision boundary at the point X and $p(X)$ is a probability density function, $K = \int_S p(X) \, dX$, and S is the decision boundary, and the integral function is performed over the entire decision boundary.

Table 13.2 Angles between the eigenvectors and the feature vectors produced by the decision boundary feature extraction method

	ψ_1	ψ_2	ψ_3	ψ_4	ψ_5	ψ_6	ψ_7	ψ_8	ψ_9	ψ_{10}
φ_1	90.7	90.5	87.2	93.7	96.6	89.0	90.1	90.5	88.6	88.1
φ_2	89.3	88.5	89.2	88.3	84.9	86.8	92.2	91.2	89.1	88.6
φ_3	90.5	90.4	86.0	87.9	92.0	87.2	92.5	92.1	88.0	87.2
φ_4	90.4	91.0	92.6	92.3	92.0	93.0	87.5	88.0	91.8	93.5
φ_5	91.9	91.3	87.3	90.3	94.9	95.7	89.0	90.4	89.2	89.0
φ_6	89.6	88.5	87.0	88.7	86.0	87.2	92.0	91.4	88.1	86.0
φ_7	90.8	90.3	86.9	90.9	96.9	88.7	91.2	91.5	87.7	86.3
φ_8	91.1	90.8	89.9	88.3	98.4	92.3	90.9	91.7	88.7	89.8
φ_9	91.0	90.9	88.5	90.9	95.8	88.1	90.9	90.9	88.1	88.7
φ_{10}	91.0	90.6	89.7	90.5	94.4	90.5	89.9	90.2	90.0	90.0

The eigenvectors of the decision boundary feature matrix of a pattern recognition problem corresponding to the non-zero eigenvalues are the necessary feature vectors to achieve the same classification accuracy as in the original space [15]. In decision boundary feature extraction, the effectiveness of feature vectors for classification is represented by the corresponding eigenvalue.

As can be seen in the two examples, a feature extraction method produces a set of feature vectors, which can be used to compute new features. For example, let $\{\varphi_i\}$ be a new feature vector set produced by a feature extraction algorithm, and let $\{\varphi_i\}$ be a basis of the N-dimensional Euclidean space. It is possible to make an independent feature vector set into an orthonormal basis by using the Gram-Schmidt procedure [30]. Let $\{\psi_i\}$ be the set of eigenvectors produced by principal component analysis, where $\{\psi_i\}$ is orthogonal. Then, the inner product between $\{\varphi_i\}$ and ψ_i indicates how well $\{\varphi_i\}$ can be represented by the eigenvectors of principal component analysis. Table 13.2 shows the angles between the eigenvectors and the feature vectors produced by the decision boundary feature extraction method. As can be seen, the eigenvectors are almost perpendicular to the feature vectors in most cases. These results indicate that the important feature vectors may not be well preserved in most conventional compression algorithms since they are not large in signal energy. Therefore, to preserve the discriminant information during compression, this discriminant information needs to be enhanced in some way prior to compression.

4   Preserving Discriminant Features

4.1   Pre-enhancing Discriminant Features

One possible way of preserving discriminant information is to increase the energy of the discriminant features found by feature extraction. Also, several methods have been proposed to enhance discriminant information [31–34]. In [32], enhancing the discriminant information was proposed before compression was applied to the hyperspectral images. As explained in the previous section, effectiveness can be quantized in most feature extraction algorithms. In general, feature extraction methods produce a new feature vector set $\{\varphi_i\}$, where class separability is better represented. In particular, the subset of $\{\varphi_i\}$ retains the most discriminating power. Thus, in [32, 34], the dominant discriminant feature vectors were enhanced and then a conventional compression algorithm was used, such as the 3D SPIHT algorithm or JPEG 2000 [8, 22].

Let $\{\varphi_i\}$ be a new feature vector set produced by a feature extraction algorithm, and let $\{\varphi_i\}$ be a basis. Then, X can be expressed as follows:

$$X = \sum_{i=1}^{N} \alpha_i \varphi_i \tag{13.2}$$

To enhance the discriminant features, the coefficients of feature vectors that are dominant in discriminant power can be enhanced as follows:

$$X' = \sum_{i=1}^{N} w_i \alpha_i \varphi_i \tag{13.3}$$

where w_i represents a weight that reflects the discriminating power of the corresponding feature vector. Then, a conventional compression algorithm can be applied to this pre-enhanced data (X'). During the decoding procedure, the weights should be considered to obtain the reconstructed data:

$$\hat{X} = \sum_{i=1}^{N} \frac{1}{w_i} \beta_i \varphi_i \tag{13.4}$$

where $\hat{X}$ represents reconstructed data. In this case, both $\{\varphi_i\}$ and $\{w_i\}$ need to be transmitted or stored.

There are a number of possible weight functions for (13.3). In [32, 34], the following weight functions were proposed:

Weight function 1. $w_i = \sqrt{\lambda_i} + 1$ (λ_i: eigenvalue of the decision boundary feature matrix)
Weight function 2. $w_i =$ a stair function (width $= 5$ bands)

Weight function 3. $w_i = \begin{cases} K & 0 \le i \le L < N \\ 1 & otherwise \end{cases}$ (K, L: constants)

Figure 13.5 illustrates some of the weighting functions.

In hyperspectral data, the covariance matrix of the original dimension may not be invertible, even if the covariance matrix is estimated from a large number of training data. In order to address this problem, two methods have been proposed: grouping the spectral bands [31] and combining the adjacent bands [32, 34]. In the grouping method, the spectral bands are divided into a number of groups and a covariance matrix is estimated from each group. With the reduced dimensions, the

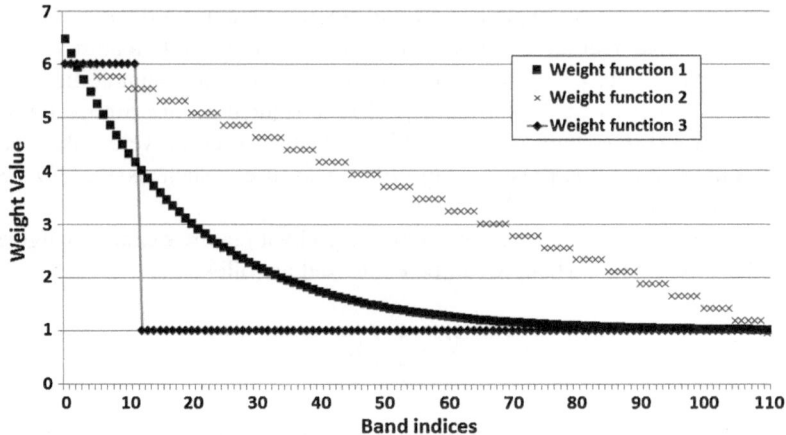

Fig. 13.5 Illustration of weight functions

covariance matrices can be invertible. Then, feature extraction is performed in each group. A problem with this approach is that by dividing the spectral bands, the correlation information between the groups is lost. As a result, feature extraction may be sub-optimal.

In the combining method, the dimensionality is first reduced by combining the adjacent bands and feature extraction is then performed in this reduced dimension [32, 34]. For example, every two adjacent bands can be combined as follows:

$$Y = AX$$

where

$$A = \begin{bmatrix} 1 & 1 & 0 & 0 & 0 & 0 & 0 & ... & 0 & 0 \\ 0 & 0 & 1 & 1 & 0 & 0 & 0 & ... & 0 & 0 \\ 0 & 0 & 0 & 0 & 1 & 1 & 0 & ... & 0 & 0 \\ . & . & . & . & . & . & . & ... & . & . \\ 0 & 0 & 0 & 0 & 0 & 0 & 0 & ... & 1 & 1 \end{bmatrix}.$$

Let $\{\psi_j\}$ be a set of feature vectors in the reduced dimension. The corresponding feature vector in the original space can be obtained by repeating every element as follows:

$$\psi_i^{\text{expand } by\, 2} = \frac{1}{\sqrt{2}} [\psi_{i,1}, \psi_{i,1}, \psi_{i,2}, \psi_{i,2}, \psi_{i,3}, \psi_{i,3}, ..., \psi_{i,110}, \psi_{i,110}]^T,$$

where $\psi_i = [\psi_{i,1}, \psi_{i,2}, ... \psi_{i,110}]^T$ represents a feature vector in the reduced dimension and $1/\sqrt{2}$ is a normalization constant. Using the Gram-Schmidt procedure [35], $\{\psi_i^{\text{expand } by2}\}$ can be expanded to an orthonormal basis in the original space, which can then be used as a feature vector in the original space.

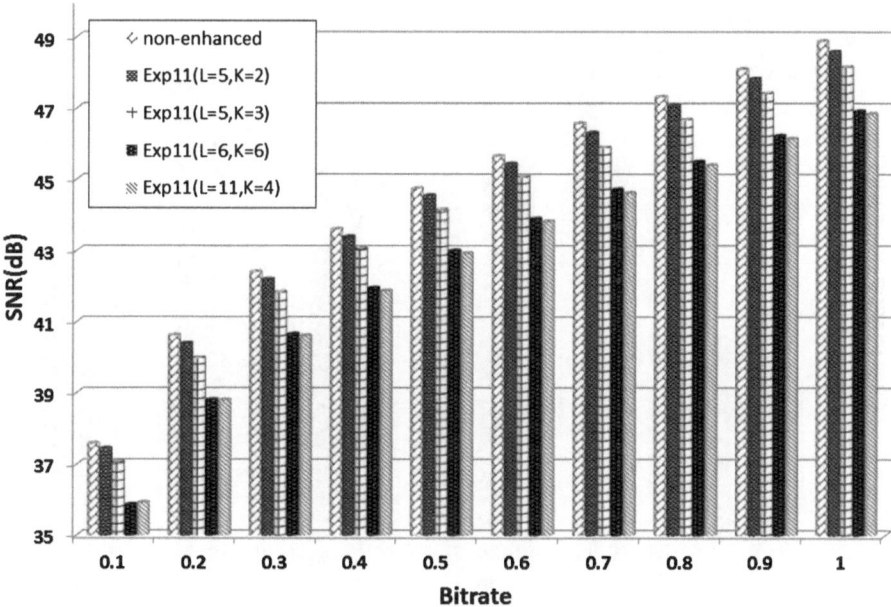

Fig. 13.6 Performance comparison when the weight function 3 was used (SNR)

Figure 13.6 shows the SNR comparison when the weight function 3 was used. As expected, the pre-enhanced images showed lower SNR performance. On the other hand, Figs. 13.7–13.9 show classification accuracy comparison at 0.1 bpppb, 0.4 bpppb and 0.8 bpppb. The pre-enhanced images showed noticeably improved classification accuracies.

4.2 Enhancing Discriminant Informative Spectral Bands

In feature extraction, a new feature can be computed by taking the inner product between an observation and a feature vector:

$$y_i = X \bullet \varphi_i = \sum_{j=1}^{N} x_j \varphi_{i,j}$$

where y_i represents the new feature, X represents the observation, y_i represents a feature vector, x_j represents the j-th spectral band of the observation, and $\varphi_{i,j}$ is the j-th element of the feature vector. Thus, it is possible to measure the importance of each spectral band by examining the elements of the feature vector. If the feature vector has a large element, the corresponding spectral band will play a key role in classification since the spectral band will be heavily reflected in computing the new feature. Therefore, by examining the elements of the feature vectors, it is

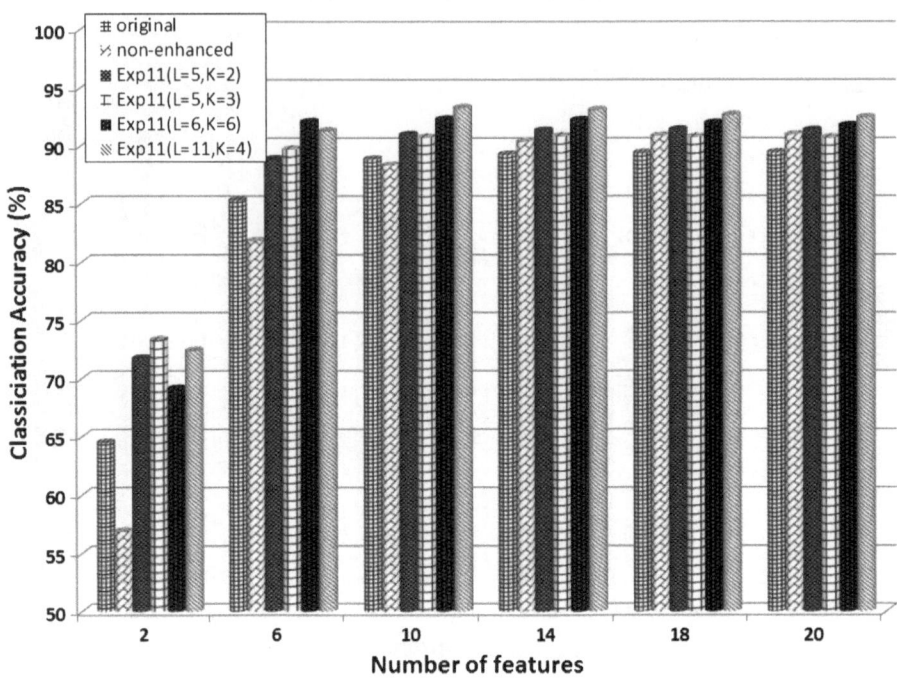

Fig. 13.7 Comparison of classification accuracies of test data with 100 training samples at 0.1 bpppb when the weight function 3 was used

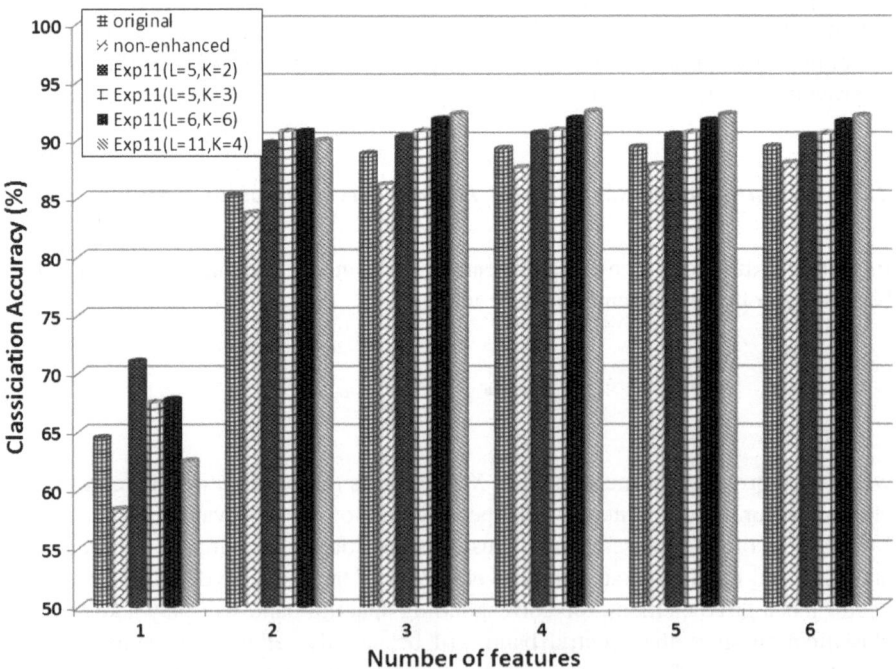

Fig. 13.8 Comparison of classification accuracies of test data with 100 training samples at 0.4 bpppb when the weight function 3 was used

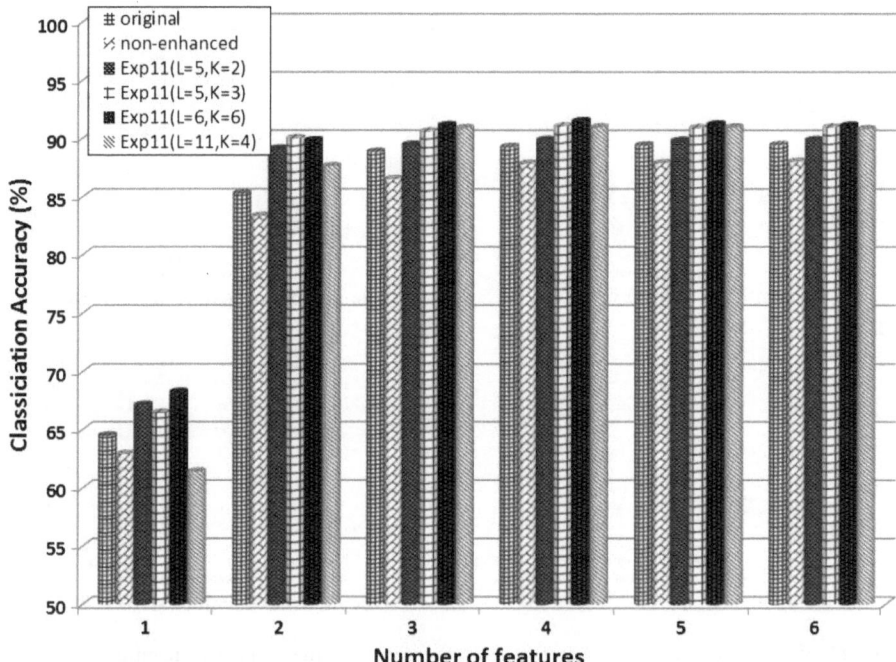

Fig. 13.9 Comparison of classification accuracies of test data with 100 training samples at 0.8 bpppb when the weight function 3 was used

possible to determine the discriminant usefulness of each spectral band. For example, it is possible to calculate the absolute mean vector of the feature vectors (φ_{AS}) from k dominant feature vectors as follows:

$$\varphi_{AS} = [\varphi_{AS}^1, \varphi_{AS}^2, \ldots, \varphi_{AS}^L]^T = \frac{1}{k} \sum_{i=1}^{k} |\varphi_i| \qquad (13.5)$$

where φ_i represents a feature vector ($L \times 1$), L represents the number of channels and k represents the number of selected feature vectors [33, 34]. Figure 13.10 shows the absolute mean vector for different numbers of feature vectors. Although the numbers are different, similar patterns were observed. Generally, a large element of the absolute mean vector indicates that the corresponding spectral band is important for classification.

Although 3D compression methods are more efficient in encoding hyperspectral images, 2D compression methods can be more suitable if random access to spectral band images is desirable [7, 36]. If 2D compression methods are used, it is possible to allocate more bits to the spectral bands, which is a useful way of discriminating

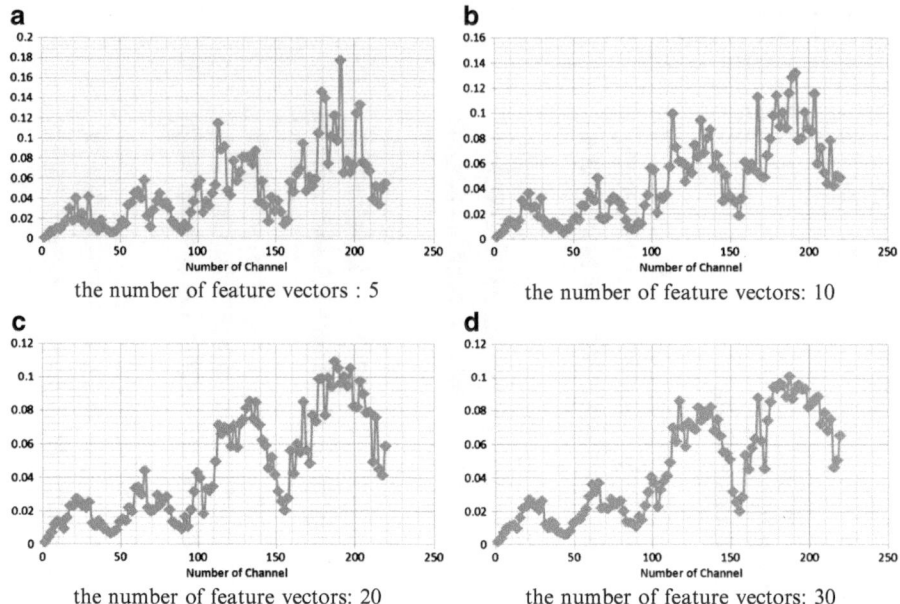

Fig. 13.10 The absolute mean vector of feature vectors. (**a**) the number of feature vectors : 5. (**b**) the number of feature vectors: 10. (**c**) the number of feature vectors: 20. (**d**) the number of feature vectors: 30

among classes. First, a portion of the total bits must be retained to enhance the discriminant information as follows [33]:

$$N_{discriminant} = \alpha N_{total} \ (0 \le \alpha \le 1) \tag{13.6}$$

where N_{total} represents the total number of bits to be used for compression, $N_{discriminant}$ represents the number of bits used to enhance the discriminant information, and α represents a coefficient. These discriminant bits can be assigned to the j-th spectral band image as follows:

$$N^j{}_{discriminant} = N_{discriminant} \frac{\varphi^j{}_{AS}}{\sum\limits_{j=1}^{L} \varphi^j{}_{AS}} \tag{13.7}$$

where $\varphi_{AS} = [\varphi^1{}_{AS}, \varphi^2{}_{AS}, \dots \varphi^L{}_{AS}]^T, N^j{}_{discriminant}$ represents the number of bits assigned to the j-th spectral band image and φ^j_{AS} is the j-th element of vector φ_{AS}. The remaining bits $((1 - \alpha) N_{total})$ can be used to minimize the mean squared error. For example, the remaining bits can be equally divided and assigned to each spectral band image.

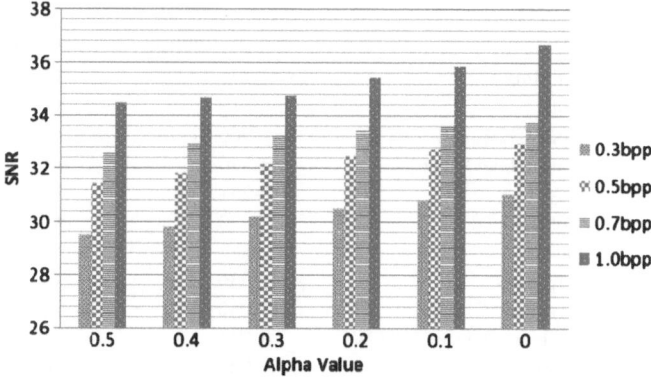

Fig. 13.11   SNR comparison

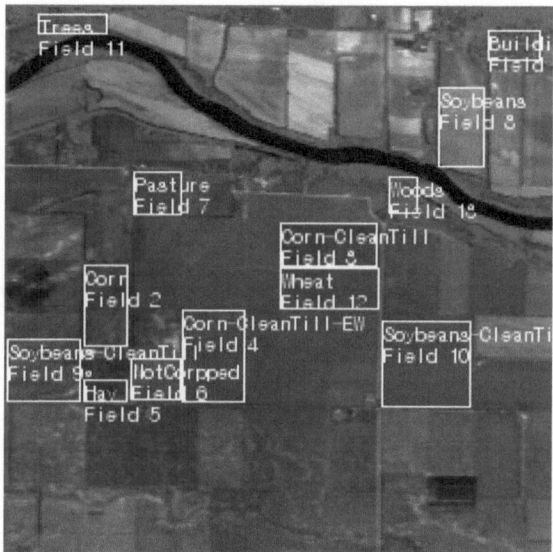

Fig. 13.12   The selected classes

Figure 13.11 shows SNR comparison for different values of α and bit rates, and α=0 indicates that all bits are equally assigned to each spectral band images. Figure 13.12 shows some selected classes and Table 13.3 provides the class description. Figure 13.13 shows classification accuracy for the different values of α and bit rates. Although the SNR of the algorithm with discriminant bit allocation [33, 34] was slightly lower, it provided better classification performance. Table 13.4 shows the SNRs and classification accuracy in terms of bit rate, the number of feature images and the value of α.

Table 13.3 Class description

Class index	Class species	No. samples	No. training samples
1	Buildings	375	200
2	Corn	819	200
3	Corn-Clean Till	966	200
4	Corn-Clean Till-EW	1,320	200
5	Hay	231	200
6	NotCorpped	480	200
7	Pasture	483	200
8	Soybeans	836	200
9	Soybeans-Clean Till	1,050	200
10	Soybeans-Clean Till-EW	1,722	200
11	Trees	330	200
12	Wheat	940	200
13	Woods	252	200

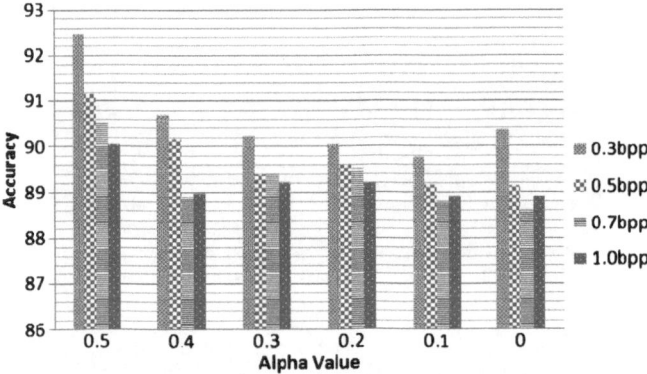

Fig. 13.13 Comparison of classification accuracy

5 Conclusions

Compression is an important consideration when dealing with hyperspectral images. Although many compression algorithms provide good compression performance in terms of the mean squared error, discriminant information, which may not be necessarily large in energy, might not be well preserved. In this chapter, we investigated this problem and presented several ways of enhancing the discriminant features of hyperspectral images.

Table 13.4 SNRs and classification accuracy

	bpp	No. feature image	Alpha value					
			0.5	0.4	0.3	0.2	0.1	0
SNR	0.3	5	29.48	29.86	30.2	30.5	30.78	31.04
		10	29.45	29.81	30.16	30.47	30.77	31.04
		30	29.35	29.75	30.12	30.45	30.76	31.04
	0.5	5	31.45	31.83	32.17	32.45	32.7	32.92
		10	31.39	31.79	32.14	32.44	32.69	32.92
		30	31.31	31.74	32.11	32.42	32.66	32.92
	0.7	5	32.6	32.92	33.19	33.41	33.57	33.71
		10	32.57	32.9	33.2	33.41	33.56	33.71
		30	32.54	32.89	33.17	33.4	33.57	33.71
	1	5	34.44	34.66	34.82	35.43	35.89	36.67
		10	34.43	34.66	34.72	35.41	35.85	36.67
		30	34.38	34.61	34.6	35.41	35.86	36.67
Accuracy	0.3	5	91.87	90.68	90.25	90.19	89.71	90.36
		10	92.47	90.69	90.23	90.05	89.76	90.36
		30	92.12	90.07	90.24	90.58	89.92	90.36
	0.5	5	91.2	90.04	89.43	89.4	88.92	89.12
		10	91.14	90.16	89.37	89.58	89.14	89.12
		30	91.15	89.87	89.62	89.79	89.45	89.12
	0.7	5	90.42	88.96	89.22	89.45	88.91	88.58
		10	90.5	88.88	89.41	89.51	88.78	88.58
		30	90.74	89.08	89.31	89.28	88.53	88.58
	1	5	90.09	88.85	89.15	89.15	88.87	88.89
		10	90.06	88.98	89.21	89.19	88.92	88.89
		30	90.29	89.2	89.02	88.91	88.84	88.89

References

1. B. Aiazzi, P. Alba, L. Alparone, and S. Baronti, "Lossless compression of multi/hyper-spectral imagery based on a 3-D fuzzy prediction," *IEEE Trans. Geosci. Remote Sens.,* vol. 37, no. 5, pp. 2287–2294, 1999.
2. E. Magli, G. Olmo, and E. Quacchio, "Optimized onboard lossless and near-lossless compression of hyperspectral data using CALIC," *IEEE Geosci. Remote Sens. Lett.,* vol. 1, no. 1, pp. 21–25, 2004.
3. M. J. Ryan, and J. F. Arnold, "Lossy compression of hyperspectral data using vector quantization," *Remote Sensing of Environment,* vol. 61, no. 3, pp. 419–436, 1997.
4. M. J. Ryan, and J. F. Arnold, "The lossless compression of AVIRIS images by vector quantization," *IEEE Trans. Geosci. Remote Sens.,* vol. 35, no. 3, pp. 546–550, 1997.
5. Q. Shen-En, "Hyperspectral data compression using a fast vector quantization algorithm," *IEEE Trans. Geosci. Remote Sens.,* vol. 42, no. 8, pp. 1791–1798, 2004.
6. G. Gelli, and G. Poggi, "Compression of hyperspectral images by spectral classification and transform coding," *IEEE Trans. Image Process.,* vol. 8, pp. 476–489, 1999.
7. A. Kaarna, "Integer PCA and wavelet transforms for hyperspectral image compression," in Proc. IEEE IGARSS'01, 2001, pp. 1853–1855.
8. S. Lim, K. Sohn, and C. Lee, "Principal component analysis for compression of hyperspectral images," in Proc. IEEE IGARSS'01, 2001, pp. 97–99.

9. D. Qian, and J. E. Fowler, "Hyperspectral Image Compression Using JPEG2000 and Principal Component Analysis," *IEEE Geosc. Remote Sens. Lett.,* vol. 4, no. 2, pp. 201–205, 2007.
10. H. S. Lee, N. H. Younan, and R. L. King, "Hyperspectral image cube compression combining JPEG-2000 and spectral decorrelation," in IGARSS '02., 2002, pp. 3317–3319.
11. M. D. Pal, C. M. Brislawn, and S. R. Brumby, "Feature extraction from hyperspectral images compressed using the JPEG-2000 standard," in Image Analysis and Interpretation, 2002. Proceedings. Fifth IEEE Southwest Symposium on, 2002, pp. 168–172.
12. J. T. Rucker, J. E. Fowler, and N. H. Younan, "JPEG2000 coding strategies for hyperspectral data," in IEEE IGARSS '05, 2005, pp. 4 pp.
13. B. Penna, T. Tillo, E. Magli, and G. Olmo, "Progressive 3-D coding of hyperspectral images based on JPEG 2000," *Geosci. Remote Sens. Lett.,* vol. 3, no. 1, pp. 125–129, 2006.
14. B. Aiazzi, P. S. Alba, L. Alparone, and S. Baronti, "Reversible compression of hyperspectral imagery based on an enhanced inter-band JPEG prediction," in Proc. IEEE IGARSS'97, 1997, pp. 1990–1992.
15. L. M. Bruce, C. H. Koger, and L. Jiang, "Dimensionality reduction of hyperspectral data using discrete wavelet transform feature extraction," *IEEE Trans. Geosci. Remote Sens.,* vol. 40, no. 10, pp. 2331–2338, 2002.
16. S. Kaewpijit, J. Le Moigne, and T. El-Ghazawi, "Automatic reduction of hyperspectral imagery using wavelet spectral analysis," *IEEE Trans. Geosci. Remote Sens.,* vol. 41, no. 4, pp. 863–871, 2003.
17. P. L. Dragotti, G. Poggi, and A. R. P. Ragozini, "Compression of hyperspectral images by three-dimensional SPIHT algorithm," *IEEE Trans. Geosci. Remote Sens.,* vol. 38, pp. 416–428, 2000.
18. B.-J. Kim, and Z. Xiong, "Low bit-rate scalable video coding with 3-D set partition in hierarchical trees(3-D SPIHT)," *IEEE Trans. Circuits Syst. Video Technol.,* vol. 10, no. 8, pp. 1374–1387, 2000.
19. E. Christophe, C. Mailhes, and P. Duhamel, "Hyperspectral Image Compression: Adapting SPIHT and EZW to Anisotropic 3-D Wavelet Coding," *IEEE Trans. Image Process.,* vol. 17, no. 12, pp. 2334–2346, 2008.
20. T. Xiaoli, C. Sungdae, and W. A. Pearlman, "Comparison of 3D set partitioning methods in hyperspectral image compression featuring an improved 3D-SPIHT," in Data Compression Conference, 2003. Proceedings. DCC 2003, 2003, pp. 449.
21. X. Tang, S. Cho, and W. A. Pearlman, "3D set partitioning coding methods in hyperspectral image compression," in IEEE ICIP'03, 2003, pp. II-239–42 vol.3.
22. A. Said, and W. A. Pearlman, "A new fast and efficient image codec based on set partitioning in hierarchical trees," *IEEE Trans. Circuits Syst. Video Technol.,* vol. 6, pp. 243–250, 1996.
23. S. Lim, K. Sohn, and C. Lee, "Compression for hyperspectral images using three dimensional wavelet transform," in Proc. IEEE IGARSS'01, 2001, pp. 109–111
24. R. O. Green, M. L. Eastwood, C. M. Sarture, T. G. Chrien, M. Aronsson, B. J. Chippendale, J. A. Faust, B. E. Pavri, C. J. Chovit, M. Solis, M. R. Olah, and O. Williams, "Imaging Spectroscopy and the Airborne Visible/Infrared Imaging Spectrometer (AVIRIS)," *Remote Sensing of Environment,* vol. 65, no. 3, pp. 227–248, 1998.
25. Fukunaga, *Introduction to Statistical Pattern Recognition*, New York: Academic Press, 1990.
26. J. A. Richards, *Remote Sensing Digital Image Analysis*, Berlin, Germany: Springer-Verlag, 1993.
27. C. Lee, and D. A. Landgrebe, "Feature extraction based on the decision boundaries," *IEEE Trans. Pattern Anal. Machine Intell.,* vol. 15, pp. 388–400, 2002.
28. S. Kumar, J. Ghosh, and M. M. Crawford, "Best-bases feature extraction algorithms for classification of hyperspectral data," *IEEE Trans. Geosci. Remote Sens.,* vol. 39, no. 7, pp. 1368–1379, 2001.
29. A. Cheriyadat, and L. M. Bruce, "Why principal component analysis is not an appropriate feature extraction method for hyperspectral data," in IGARSS '03., 2003, pp. 3420–3422 vol.6.

30. G. Strang, *Linear Algebra and Its Applications*, Tokyo, Japan: Harcourt Brace Jovanovich, Inc., 1998.
31. C. Lee, and E. Choi, "Compression of hyperspectral images with enhanced discriminant features," in IEEE Workshop on Advances in Techniques for Analysis of Remotely Sensed Data, 2003, pp. 76–79.
32. C. Lee, E. Choi, J. Choe, and T. Jeong, "Dimension Reduction and Pre-emphasis for Compression of Hyperspectral Images," in Proc. LNCS (ICIAR), 2004, pp. 446–453.
33. S. Lee, J. Lee, and C. Lee, "Bit allocation for 2D compression of hyperspectral images for classification," in Proc. SPIE 7455, 2009, pp. 745507.
34. C. Lee, E. Choi, T. Jeong, S. Lee, and J. Lee, "Compression of hyperspectral images with discriminant features enhanced," *Journal of Applied Remote Sensing, Satellite Data Compression Special*, 2010.
35. C. Lee, and D. Landgrebe, "Analyzing high dimensional hyperspectral data," *IEEE Trans. Geosci. Remote Sens.*, vol. 31, no. 4, pp. 792–800, 2002.
36. J. A. Saghri, A. G. Tescher, and J. T. Reagan, "Practical transform coding of hyperspectral imagery," *IEEE Signal Process. Mag.*, vol. 12, pp. 32–43, 1995.

Chapter 14
Projection Pursuit-Based Dimensionality Reduction for Hyperspectral Analysis

Haleh Safavi, Chein-I Chang, and Antonio J. Plaza

Abstract Dimensionality Reduction (DR) has found many applications in hyperspectral image processing. This book chapter investigates Projection Pursuit (PP)-based Dimensionality Reduction, (PP-DR) which includes both Principal Components Analysis (PCA) and Independent Component Analysis (ICA) as special cases. Three approaches are developed for PP-DR. One is to use a Projection Index (PI) to produce projection vectors to generate Projection Index Components (PICs). Since PP generally uses random initial conditions to produce PICs, when the same PP is performed in different times or by different users at the same time, the resulting PICs are generally different in terms of components and appearing orders. To resolve this issue, a second approach is called PI-based PRioritized PP (PI-PRPP) which uses a PI as a criterion to prioritize PICs. A third approach proposed as an alternative to PI-PRPP is called Initialization-Driven PP (ID-PIPP) which specifies an appropriate set of initial conditions that allows PP to produce the same PICs as well as in the same order regardless of how PP is run. As shown by experimental results, the three PP-DR techniques can perform not only DR but also separate various targets in different PICs so as to achieve unsupervised target detection.

H. Safavi (✉)
Remote Sensing Signal and Image Processing Laboratory, Department of Computer Science and Electrical Engineering, University of Maryland, Baltimore, MD, USA
e-mail: haleh1@umbc.edu

C.-I Chang
Remote Sensing Signal and Image Processing Laboratory, Department of Computer Science and Electrical Engineering, University of Maryland, Baltimore, MD, USA

Department of Electrical Engineering, National Chung Hsing University, Taichung, Taiwan
e-mail: cchang@umbc.edu

A.J. Plaza
Department of Technology of Computers and Communications, University of Extremadura, Escuela Politecnica de Caceres, Caceres, SPAIN
e-mail: aplaza@unex.es

B. Huang (ed.), *Satellite Data Compression*, DOI 10.1007/978-1-4614-1183-3_14,
© Springer Science+Business Media, LLC 2011

1 Introduction

One of great challenging issues in hyperspectral analysis is how to deal with enormous data volumes acquired by hyperspectral imaging sensors' hundreds of contiguous spectral bands. A general approach is to perform Dimensionality Reduction (DR) as a pre-processing step to represent the original data in a manageable low dimensional data space prior to data processing. The Principal Components Analysis (PCA) is probably the most commonly used DR technique that reduces data dimensionality by representing the original data via a small set of Principal Components (PCs) in accordance with data variances specified by eigenvalues of the data sample covariance matrix. However, the PCA can only capture information characterized by second-order statistics as demonstrated in [1] where small targets may not be preserved in PCs. In order to retain the information preserved by statistically independent statistics, Independent Component Analysis (ICA) [2] was suggested for DR in [1]. This evidence was further confirmed by [3] where High-Order Statistics (HOS) were able to detect subtle targets such as anomalies, small targets. Since the independent statistics can be measured by mutual information [2], it will be interesting to see if PCA (i.e., 2nd order statistics), HOS and ICA can be integrated into a general setting so that each of these three cases can be considered as a special circumstance of this framework. However, this is easier said than done because several issues need to be addressed. First of all, both PCA and ICA have different ways to generate their projection vectors. For the PCA it first calculate the characteristic polynomial to find eigenvalues from which their associated eigenvectors can be generated as projection vectors to produce Principal Components (PCs) ranked by sample data variances. On the other hand, unlike the PCA, the ICA does not have a similar characteristic polynomial that allows it to find solutions from which projection vectors can be generated. Instead, it must rely on numerical algorithms to find these projection vectors. In doing so it makes use of random initial vectors to generate projection vectors in a random order so that the projection vector-produced Independent Components (ICs) also appear randomly. As a result of using different sets of random initial conditions, the generated ICs not only appear in a random order, but also are different even if they appear in the same order. Consequently, the results produced by different runs using different sets of random initial conditions or by different users will be also different. This same issue is also encountered in the ISODATA (K-means) clustering method in [4]. Therefore, the first challenging issue is how to rank components such as ICs in an appropriate order like PCs ranked by data variances. Recently, this issue has been addressed for ICA in [1] and for HOS-based components in [3] where a concept of using priority scores to rank components was developed to prioritize components according to the significance of the information contained in each component measured by a specific criterion. The goal of this chapter is to extend the ICA-based DR in [1] and the work in [3] by unifying these approaches in context of a more general framework, Projection Pursuit (PP) [5]. In the mean time it also generalizes the prioritization criteria in [3, 4] to the Projection Index (PI) used by PP where the components generated by the PP using a specific PI are referred to as Projection Index Components (PICs). For example, the PI used to rank PCA-generated components is reduced to data variance with PICs being PCs, while the PI used to generate components by the FastICA in [2] turns out to be the neg-entropy and PICs become ICs.

Three approaches are proposed in this chapter to implement the PP with the PI specified by a particular prioritization criterion. The first approach is commonly used in the literature which makes use of the PI to produce components, referred to as PIPP. With this interpretation when the PI is specified by data variance, the resulting PP becomes PCA. On the other hand, if the PI is specified by mutual information to measure statistical independence, the resulting PP turns out to be ICA. While the first approach is focused on component generation, the second and third approaches can be considered as component prioritization. More specifically, the second approach, referred to as PI-PRioritized PP (PI-PR PP) utilizes the PI as a criterion to prioritize PICs produced by PIPP. In other words, due to the use of random initial conditions the PIPP-generated PICs generally appear in a random order. Using a PI-PRPP allows users to rank and prioritize PICs regardless of what random initial condition is used. Despite that the PI-PRPP resolves the issue of random order in which PICs appear, it does not necessarily imply that PICs ranked by the same order are identical. To further remedy this problem, the third approach, referred to as Initialization-Driven PP (ID-PIPP) is proposed to specify an appropriate set of initial conditions for PIPP so that as long as the PIPP uses the same initial condition, it always produces identical PICs which are ranked by the same order. By means of the 2nd and 3rd approaches the PP-DR can be accomplished by retaining a small number of components whose priorities are ranked by PI-PRPP using a specific prioritization criterion or the first few components produced by ID-PIPP. In order to evaluate the three different versions of the PP, PIPP, PI-PRPP and ID-PIPP, real hyperspectral image experiments are conducted for performance analysis.

2   Projection Pursuit-Based Component Analysis for Dimensionality Reduction

Dimensionality reduction is an important preprocessing technique that represents multi-dimensional data in a lower dimensional data space without significant loss of desired data information. A common approach which has been widely used in many applications such as data compression is the PCA which represents data to be processed in a new data space whose dimensions are specified by eigenvectors in descending order of eigenvalues. Another example is the ICA which represents data to be processed in a new data space whose dimensions are specified by a set of statistically independent projection vectors. This section presents a Projection Index (PI)-based dimensionality reduction technique, referred to as PI-based Project Pursuit (PIPP) which uses a PI as a criterion to find directions of interestingness

of data to be processed and then represents the data in the data space specified by
these new interesting directions. Within the context of PIPP the PCA and ICA can
be considered as special cases of PIPP in the sense that PCA uses data variance as a
PI to produce eigenvectors while the ICA uses mutual information as a PI to
produce statistically independent projection vectors.

The term of "Projection Pursuit (PP)" was first coined by Friedman and
Tukey [5] which was used as a technique for exploratory analysis of multivariate
data. The idea is to project a high dimensional data set into a low dimensional data
space while retaining the information of interest. It designs a PI to explore projections
of interestingness. Assume that there are N data points $\{\mathbf{X}_n\}_{n-1}^N$ each with dimension-
ality K and $\mathbf{X} = [\mathbf{r}_1\mathbf{r}_2 \cdots \mathbf{r}_N]$ is a $K \times N$ data matrix and $\mathbf{a}$ is a K-dimensional column
vector which serves as a desired projection. Then $\mathbf{a}^T\mathbf{X}$ represents an N-dimensional
row vector that is the orthogonal projections of all sample data points mapped onto
the direction $\mathbf{a}$. Now if we let $H(\cdot)$ is a function measuring the degree of the
interestingness of the projection $\mathbf{a}^T\mathbf{X}$ for a fixed data matrix $\mathbf{X}$, a Projection Index
(PI) is a real-valued function of $\mathbf{a}, I(\mathbf{a}) : R^K \to R$ defined by

$$I(\mathbf{a}) = H(\mathbf{a}^T\mathbf{X}) \tag{14.1}$$

The PI can be easily extended to multiple directions, $\{\mathbf{a}_j\}_{j-1}^J$. In this case, $\mathbf{A} =$
$[\mathbf{a}_1\mathbf{a}_2 \cdots \mathbf{a}_J]$ is a $K \times J$ projection direction matrix and the corresponding projection
index is also a real-valued function, $I(\mathbf{A}) : R^{K \times J} \to R$ is given by

$$I(\mathbf{A}) = H(\mathbf{A}^T\mathbf{X}) \tag{14.2}$$

The choice of the $H(\cdot)$ in (14.1) and (14.2) is application-dependent. Its purpose
is to reveal interesting structures within data sets such as clustering. However,
finding an optimal projection matrix $\mathbf{A}$ in (14.2) is not a simple matter [6]. In this
chapter, we focus on PIs which are specified by statistics of high orders such as
skewness, kurtosis, etc. [7].

Assume that the ith projection index-projected component can be described by a
random variable ζ_i with values taken by the gray level value of the nth pixel denoted
by z_n^i. In what follows, we present a general form for the kth-order orders of
statistics: kth moment by solving the following eigen-problem [7, 8]

$$\left(E\left[\mathbf{r}_i\left(\mathbf{r}_i^T\mathbf{w}\right)^{k-2}\mathbf{r}_i^T\right] - \lambda'\mathbf{I}\right)\mathbf{w} = 0 \tag{14.3}$$

It should be noted that when $k = 2, 3, 4$ in (14.3) is then reduced to variance,
skewness and kurtosis respectively.

An algorithm for finding a sequence of projection vectors to solve (14.3) can be
described as follows [7, 8]

Projection-Index Projection Pursuit (PIPP)

1. Initially, assume that $\mathbf{X} = [\mathbf{r}_1\mathbf{r}_2 \cdots \mathbf{r}_N]$ is data matrix and a PI is specified.

2. Find the first projection vector $\mathbf{w}_1^*$ by maximizing the PI.
3. Using the found $\mathbf{w}_1^*$, generate the first projection image $\mathbf{Z}^1 = \left(\mathbf{w}_1^*\right)^T \mathbf{X} = \left\{ \mathbf{z}_i^1 | \mathbf{z}_i^1 = \left(\mathbf{w}_1^*\right)^T \mathbf{r}_i \right\}$ which can be used to detect the first endmember.
4. Apply the orthogonal subspace projector (OSP) specified by $P_{\mathbf{w}_1}^\perp = \mathbf{I} - \mathbf{w}_1 (\mathbf{w}_1^T \mathbf{w}_1)^{-1} \mathbf{w}_1^T$ to the data set $\mathbf{X}$ to produce the first OSP-projected data set denoted by $\mathbf{X}^1$, $\mathbf{X}^1 = P_{\mathbf{w}_1}^\perp \mathbf{X}$.
5. Use the data set $\mathbf{X}^1$ and find the second projection vector $\mathbf{w}_2^*$ by maximizing the same PI again.
6. Apply $P_{\mathbf{w}_2}^\perp = \mathbf{I} - \mathbf{w}_2 (\mathbf{w}_2^T \mathbf{w}_2)^{-1} \mathbf{w}_2^T$ to the data set $\mathbf{X}^1$ to produce the second OSP-projected data set denoted by $\mathbf{X}^2$, $\mathbf{X}^2 = P_{\mathbf{w}_2}^\perp \mathbf{X}^1$ which can be used to produce the third projection vector $\mathbf{w}_3^*$ by maximizing the same PI again. Or equivalently, we define a matrix projection matrix $\mathbf{W}^2 = [\mathbf{w}_1 \mathbf{w}_2]$ and apply $P_{\mathbf{W}^2}^\perp = \mathbf{I} - \mathbf{W}^2 \left((\mathbf{W}^2)^T \mathbf{W}^2 \right)^{-1} (\mathbf{W}^2)^T$ to the data set $\mathbf{X}$ to obtain $\mathbf{X}^2 = P_{\mathbf{W}^2}^\perp \mathbf{X}$.
7. Repeat the procedure of steps 5 and 6 over and over again to produce $\mathbf{w}_3^*, \cdots, \mathbf{w}_k^*$ until a stopping criterion is met. It should be noted that a stopping criterion can be either a predetermined number of projection vectors required to be generated or a predetermined threshold for the difference between two consecutive projection vectors.

3 Projection Index-Based Prioritized PP

According to the PIPP described in Sect. 2 a vector is randomly generated as an initial condition to produce projection vectors that are used to generate PICs. Accordingly, different initial condition may produce different projection vectors and so are their generated PICs. In other words, if the PIPP is performed in different times by different sets of random initial vectors or different users who will use different sets of random vectors to run the PIPP, the resulting final PICs will also be different. In order to correct this problem, this section presents a PI-based Prioritized PP (PI-PRPP) which uses a PI as a prioritization criterion to rank PIPP-generated PICs so that all PICs will be prioritized in accordance with the priorities measured by the PI. In this case, the PICs will be always ranked and prioritized by the PI in the same order regardless of what initial vectors are used to produce projection vectors. It should be noted that there is a major distinction between PIPP and PI-PRPP. While the PIPP uses a PI as criterion to produce a desired projection vector for each of PICs, the PI-PRPP uses a PI to prioritize PICs and this PI may not the same PI used by the PIPP to generate PICs. Therefore, the PI used in both PP and PI-PRPP is not necessarily the same and can be different. As a matter of fact, on many occasions, different PIs can be used in various applications. In what follows, we describe various criteria that can be used to define PI. These criteria are statistics-based measures that go beyond the 2nd order statistics.

Projection Index (PI)-Based Criteria

1. Sample mean of 3rd order statistics: skewness for ζ_j.

$$\text{PI}_{\text{skewness}}(\text{PIC}_j) = \left[\kappa_j^3\right]^2 \tag{14.4}$$

where $\kappa_j^3 = E\left[\zeta_j^3\right] = (1/MN)\sum_{n=1}^{MN}\left(z_n^j\right)^3$ is the sample mean of the 3rd order of statistics in the PIC_j.

2. Sample mean of 4th order statistics: kurtosis for ζ_i.

$$\text{PI}_{\text{kurtosis}}(\text{PIC}_j) = \left[\kappa_j^4\right]^2 \tag{14.5}$$

where $\kappa_j^4 = E\left[\zeta_j^4\right] = (1/MN)\sum_{n=1}^{MN}\left(z_n^j\right)^4$ is the sample mean of the 4th order of statistics in the PIC_j.

3. Sample mean of k-th order statistics: k-th moments for ζ_j.

$$\text{PI}_{\text{k-moment}}(\text{PIC}_j) = \left[\kappa_j^k\right]^2 \tag{14.6}$$

where $\kappa_j^k = E\left[\zeta_j^k\right] = (1/MN)\sum_{n=1}^{MN}\left(z_n^j\right)^k$ is the sample mean of the k-th moment of statistics in the PIC_j.

4. Negentropy: combination of 3rd and 4th orders of statistics for ζ_j.

$$\text{PI}_{\text{negentropy}}(\text{PIC}_j) = (1/12)\left[\kappa_j^3\right]^2 + (1/48)\left[\kappa_j^4 - 3\right]^2 \tag{14.7}$$

It should be note that (14.7) is taken from (5.35) in Hyvarinen and Oja [2, p. 115], which is used to measure the negentropy by high-order statistics.

5. Entropy

$$\text{PI}_{\text{entropy}}(\text{PIC}_j) = -\sum_{j=1}^{MN} p_{ji} \log p_j \tag{14.8}$$

where $p_j = \left(p_{j1}, p_{j2}, \cdots, p_{jMN}\right)^T$ is the probability distribution derived from the image histogram of PIC_i.

6. Information Divergence (ID)

$$\text{PI}_{\text{ID}}(\text{PIC}_j) = \sum_{j=1}^{MN} p_{ji} \log\left(p_{ji}/q_i\right) \tag{14.9}$$

where $p_j = \left(p_{j1}, p_{j2}, \cdots, p_{jMN}\right)^T$ is the probability distribution derived from the image histogram of PIC_i and $\mathbf{q}_j = \left(q_{j1}, q_{j2}, \cdots, q_{jMN}\right)^T$ is the Gaussian probability distribution with the mean and variance calculated from PIC_i.

4 Initialization-Driven PIPP

The PI-PRPP in Sect. 3 intended to remedy the issue that PICs can appear in a random order due to the use of randomly generated initial vectors. The PI-PRPP allows users to prioritize PICs according to information significance measured by a specific PI. Despite the fact that the PICs ranked by PI-PRPP may appear in the same order independent of different sets of random initial conditions they are not necessarily identical because the slight discrepancy in two corresponding PICs at the same appearing order may be caused by randomness introduced by their used initial conditions. Although such a variation may be minor compared to different appearing orders of PICs without prioritization, the inconsistency may still cause difficulty in data analysis. Therefore, this section further develops a new approach, called Initialization-Driven PP (ID-PIPP) which custom-designs an initialization algorithm to produce a specific set of initial conditions for PIPP so that the same initial condition is used all along whenever PIPP is implemented. Therefore, the ID-PIPP-generated PICs are always identical. When a particular initial algorithm, say X, is used to produce an initial set of vectors for the ID-PIPP to converge to projection vectors to produce PICs, the resulting PIPP is referred to as X-PIPP.

One such initialization algorithm described above that can be used for the ID-PIPP is the Automatic Target Generation Process (ATGP) previously developed in [10]. It makes use of an orthogonal subspace projector defined in [11] by

$$P_{\mathbf{U}}^{\perp} = \mathbf{I} - \mathbf{U}\mathbf{U}^{\#} \tag{14.10}$$

where $\mathbf{U}^{\#} = \left(\mathbf{U}^{T}\mathbf{U}\right)^{-1}\mathbf{U}^{T}$ is the pseudo inverse of the $\mathbf{U}$, repeatedly to find target pixel vectors of interest from the data without prior knowledge regardless of what types of pixels are these targets. Details of implementing the ATGP are provided in the following steps.

Automatic Target Generation Process (ATGP)

1. Initial condition: Let p be the number of target pixels needed to be generated. Select an initial target pixel vector of interest denoted by $\mathbf{t}_0$. In order to initialize the ATGP without knowing $\mathbf{t}_0$, we select a target pixel vector with the maximum length as the initial target $\mathbf{t}_0$, namely, $\mathbf{t}_0 = \arg\{\max_{\mathbf{r}}\mathbf{r}^{T}\mathbf{r}\}$, which has the highest intensity, i.e., the brightest pixel vector in the image scene. Set $n = 1$ and$\mathbf{U}_0 = [\mathbf{t}_0]$. (It is worth noting that this selection may not be necessarily the best selection. However, according to our experiments it was found that the brightest pixel vector was always extracted later on, if it was not used as an initial target pixel vector in the initialization).

2. At nth iteration, apply $P_{t_0}^{\perp}$ via (14.10) to all image pixels $\mathbf{r}$ in the image and find the nth target $\mathbf{t}_n$ generated at the n-th stage which has the maximum orthogonal projection as follows.

$$\mathbf{t}_n = \arg\left\{\max_{\mathbf{r}}\left[\left(P_{[\mathbf{U}_{n-1}\mathbf{t}_n]}^{\perp}\mathbf{r}\right)^T\left(P_{[\mathbf{U}_{n-1}\mathbf{t}_n]}^{\perp}\mathbf{r}\right)\right]\right\} \qquad (14.11)$$

where $\mathbf{U}_{n-1} = [\mathbf{t}_1\mathbf{t}_2\cdots\mathbf{t}_{n-1}]$ is the target matrix generated at the $(n-1)$st stage.
3. Stopping rule: If $n < p - 1$, let $\mathbf{U}_n = [\mathbf{U}_{n-1}\mathbf{t}_n] = [\mathbf{t}_1\mathbf{t}_2\cdots\mathbf{t}_n]$ be the n-th target matrix, go to step 2. Otherwise, continue.
4. At this stage, the ATGP is terminated. At this point, the target matrix is $\mathbf{U}_{p-1}$, which contains $p-1$ target pixel vectors as its column vectors, which do not include the initial target pixel vector $\mathbf{t}_0$.

As a result of the ATGP, the final set of target pixel vectors produced by the ATGP at step 4 is the final target set which comprises p target pixel vectors, $\{\mathbf{t}_0, \mathbf{t}_1, \mathbf{t}_2, \cdots, \mathbf{t}_{p-1}\} = \{\mathbf{t}_0\} \cup \{\mathbf{t}_1, \mathbf{t}_2, \cdots, \mathbf{t}_{p-1}\}$ which were found by repeatedly using (14.11). It should be noted that the stopping rule used in the above ATGP was set by a pre-determined number of targets that should be generated. Of course this stopping rule can be replaced by any other rule such as the one used in [10]. Finally, a PIPP using the ATGP as its initialization algorithm is called ATGP-PIPP.

5 Real Hyperspectral Image Experiments

The image scene to be studied for experiments is a real image scene collected by HYperspectral Digital Imagery Collection Experiments (HYDICE) sensor shown in Fig. 14.1a, which has a size of 64×64 pixel vectors with 15 panels in the scene and the ground truth map in Fig. 14.1b. It was acquired by 210 spectral bands with a spectral coverage from 0.4 μm to 2.5 μm. Low signal/high noise bands: bands 1–3 and bands 202–210; and water vapor absorption bands: bands 101–112 and bands 137–153 were removed. So, a total of 169 bands were used in experiments. The spatial resolution is 1.56 m and spectral resolution is 10 nm.

Within the scene in Fig. 14.1a, there is a large grass field background, and a forest on the left edge. Each element in this matrix is a square panel and denoted by p_{ij} with rows indexed by i and columns indexed by $j = 1, 2, 3$. For each row $i = 1, 2, \cdots, 5$, there are three panels p_{i1}, p_{i2}, p_{i3}, painted by the same paint but with three different sizes. The sizes of the panels in the first, second and third columns are 3m × 3m and 2m × 2m and 1m × 1m respectively. Since the size of the panels in the third column is 1m × 1m, they cannot be seen visually from Fig. 14.1a due to the fact that its size is less than the 1.56 m pixel resolution. For

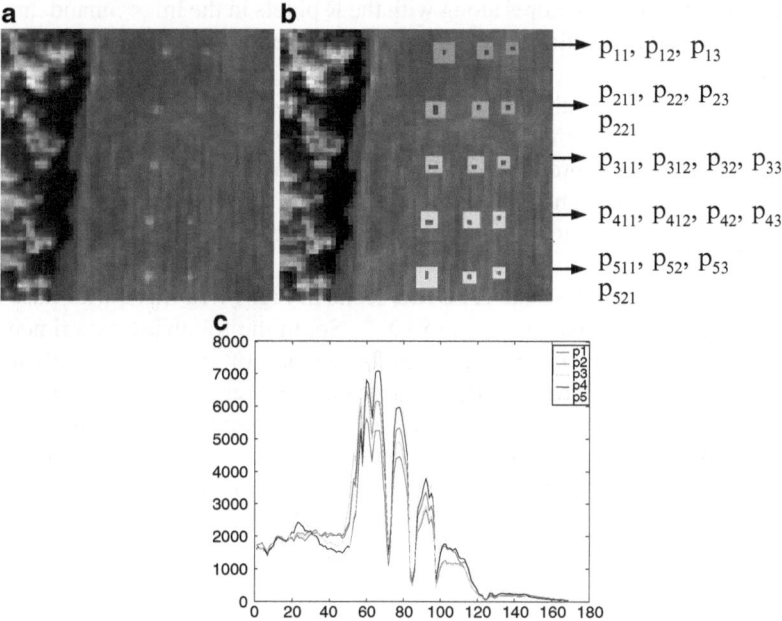

Fig. 14.1 (**a**) A HYDICE panel scene which contains 15 panels; (**b**) Ground truth map of spatial locations of the 15 panels; (**c**) Spectral signatures of $\mathbf{p}_1$, $\mathbf{p}_2$, $\mathbf{p}_3$, $\mathbf{p}_4$ and $\mathbf{p}_5$

each column $j = 1, 2, 3$, the five panels, $\mathbf{p}_{1j}$, $\mathbf{p}_{2j}$, $\mathbf{p}_{3j}$, $\mathbf{p}_{4j}$, $\mathbf{p}_{5j}$ have the same size but with five different paints. However, it should be noted that the panels in rows 2 and 3 were made by the same material with two different paints. Similarly, it is also the case for panels in rows 4 and 5. Nevertheless, they were still considered as different panels but our experiments will demonstrate that detecting panels in row 5 (row 4) may also have effect on detection of panels in row 2 (row 3). The 1.56 m-spatial resolution of the image scene suggests that most of the 15 panels are one pixel in size except that $\mathbf{p}_{21}$, $\mathbf{p}_{31}$, $\mathbf{p}_{41}$, $\mathbf{p}_{51}$ which are two-pixel panels, denoted by $\mathbf{p}_{211}$, $\mathbf{p}_{221}$, $\mathbf{p}_{311}$, $\mathbf{p}_{312}$, $\mathbf{p}_{411}$, $\mathbf{p}_{412}$, $\mathbf{p}_{511}$, $\mathbf{p}_{521}$. Since the size of the panels in the third column is 1m × 1m, they cannot be seen visually from Fig. 14.1a due to the fact that its size is less than the 1.56 m pixel resolution. Figure 14.1b shows the precise spatial locations of these 15 panels where red pixels (R pixels) are the panel center pixels and the pixels in yellow (Y pixels) are panel pixels mixed with the background. Figure 14.1c plots the five panel spectral signatures $\mathbf{p}_i$ for $i = 1, 2, \cdots, 5$ obtained by averaging R pixels in the 3m × 3mand 2m × 2m panels in row i in Fig. 14.1b. It should be noted the R pixels in the 1m × 1m panels are not included because they are not pure pixels, mainly due to that fact that the spatial resolution of the R pixels in the 1m × 1m panels is 1 m smaller than the pixel resolution is

1.56 m. These panel signatures along with the R pixels in the3m $\times$ 3mand 2m $\times$ 2m panels were used as required prior target knowledge for the following comparative studies.

In order to perform dimensionality reduction we must know how many PICs needed to be retained, denoted by p after PIPP. Over the past years this knowledge has been obtained by preserving a certain level of energy percentage based on accumulative sum of eignevlaues. Unfortunately, it has been shown in [6, 12] that this was ineffective. Instead, a new concept, called Virtual Dimensionality (VD) was proposed to address this issue and has shown success and promise in [1, 3] where the VD estimated for the HYDICE scene in Fig. 14.1a was 9 with false alarm probability P_F greater than or equal to 10^{-4}. So, in the following experiments, the value of p was set to $p = 9$ where three versions of PIPP were evaluated for applications in dimensionality reduction and endmember extraction. Due to limited space including all experimental results are nearly impossible. In this case, only representatives are included in this chapter, which are PI = skewness (3rd order statistics), kurtosis (4th order statistics) and negentropy (infinite order statistics, i.e., statistical independence).

5.1   PIPP with Random Initial Conditions

In order to demonstrate inconsistent results from using two different sets of random initial vectors by the PIPP Figs. 14.2–14.4 show nine PICs resulting from performing PIPP with PI = skewness, kurtosis and negentropy respectively where it should be noted that the PIPP with PI = negentropy was carried out by the FastICA developed by Hyvarinen and Oja in [9] to produce PICs.

As we can see from these figures, due to the use of random initial conditions the appearing orders of interesting PICs generally are not the same in each run. In particular, some PICs which did not appear in one run actually appeared in another run. In addition, some of the PICs that contained little information showed up among the first a few PICs.

Now, these nine components obtained in Figs. 14.2–14.4 can then be used for endmember extraction performed by the well-known algorithm developed by Winter, referred to as N-FINDR [13] and the results of nine endmember extracted by the N-FINDR are shown in Figs. 14.5–14.7 where it is clear to see that endmembers were extracted in different orders by the PIPP with the same PI using two different sets of random initial vectors.

According to the results in Figs. 14.5–14.7, the best performance was given by the PIPP using PI = negentropy where all five panel signatures were extracted as endmembers.

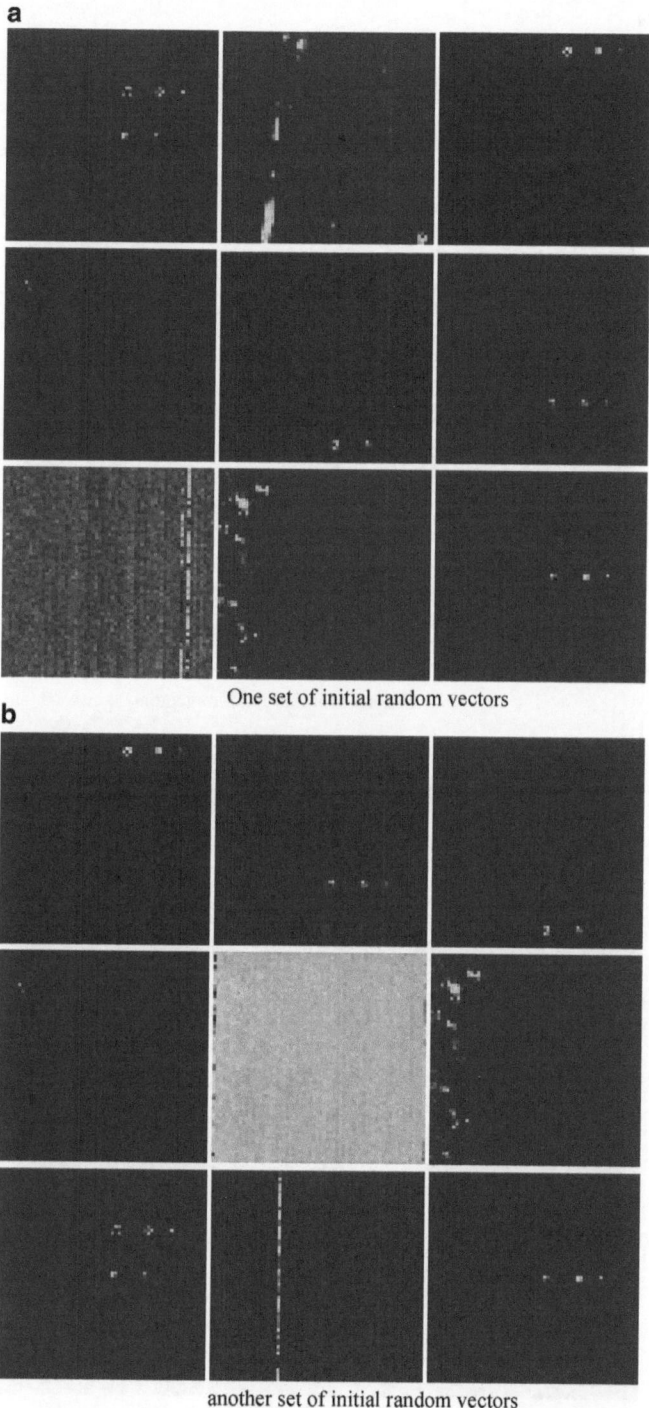

Fig. 14.2 First nine PICs extracted by PI-PP with PI = skewnes using random initial conditions.
(**a**) One set of initial random vectors. (**b**) Another set of initial random vectors

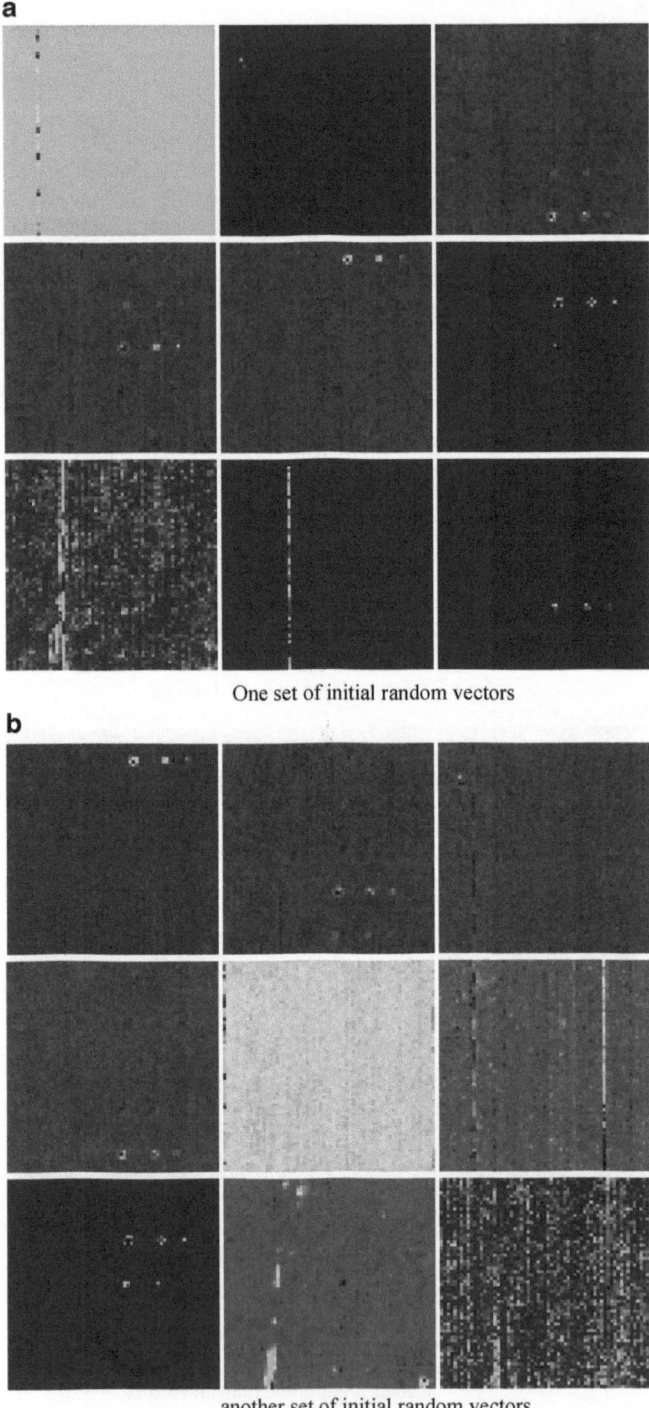

Fig. 14.3 First nine PICs extracted by PI-PP with PI = kurtosis using random initial conditions.
(**a**) One set of initial random vectors. (**b**) Another set of initial random vectors

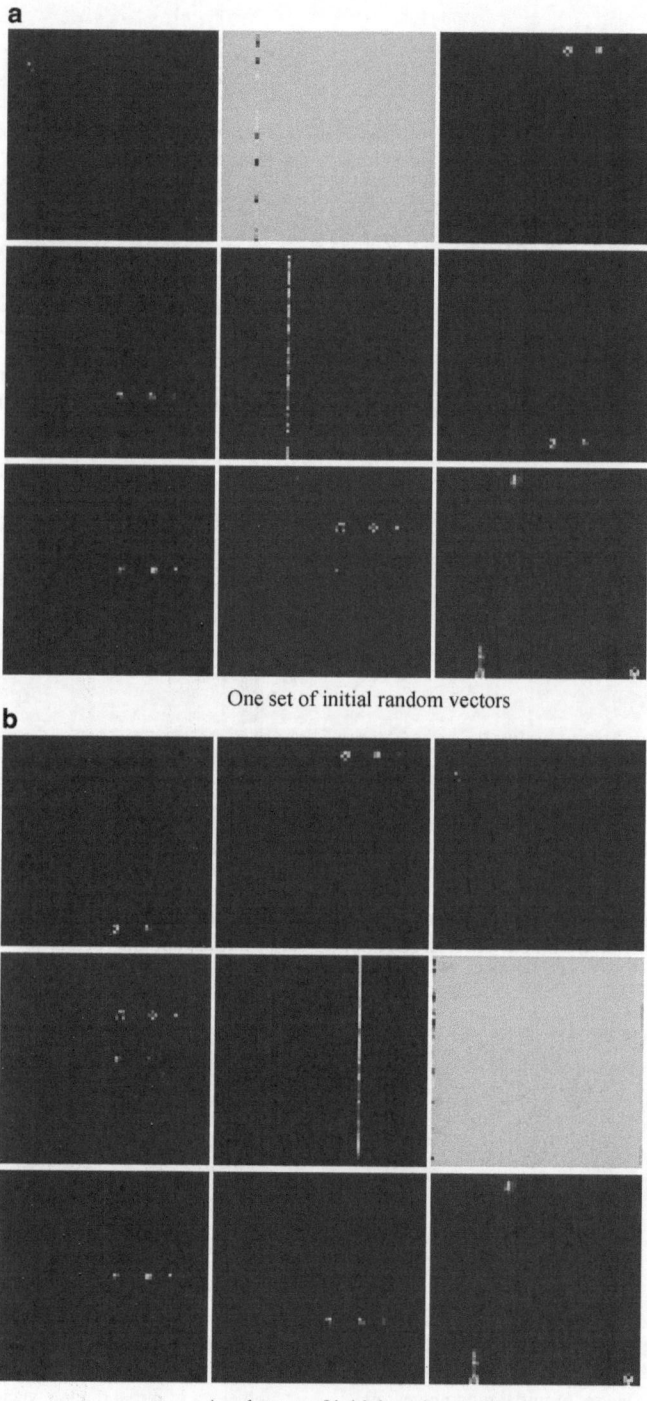

One set of initial random vectors

Another set of initial random vectors

Fig. 14.4 First nine PICs extracted by PI-PP with PI = negentropy PI using random initial conditions. (**a**) One set of initial random vectors. (**b**) Another set of initial random vectors

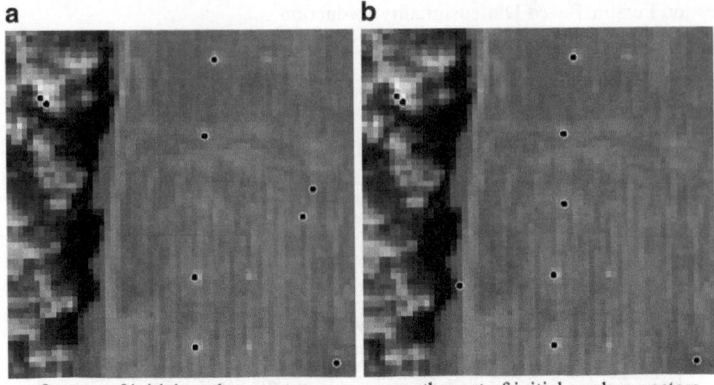

One set of initial random vectors   another set of initial random vectors

Fig. 14.5 Nine endmembers extracted by PIPP with PI = skewnes using random initial conditions. (**a**) One set of initial random vectors. (**b**) another set of initial random vectors

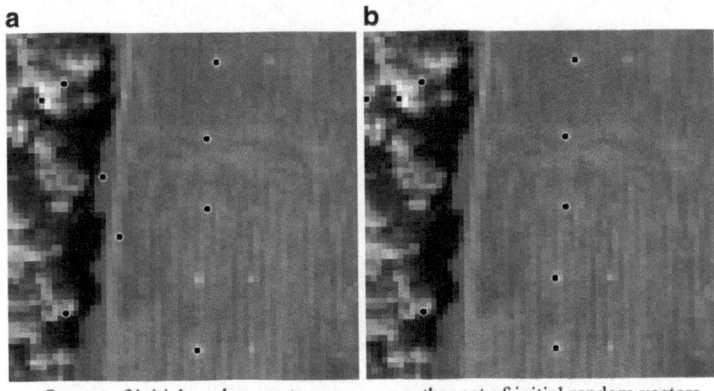

One set of initial random vectors   another set of initial random vectors

Fig. 14.6 Nine endmembers extracted by PIPP with PI = kurtosis using random initial conditions (**a**) One set of initial random vectors. (**b**) another set of initial random vectors

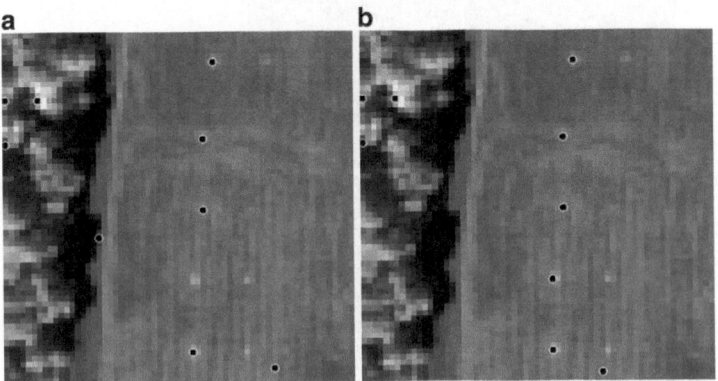

One set of initial random vectors   another set of initial random vectors

Fig. 14.7 Nine endmembers extracted by PIPP with PI = negentropy using random initial conditions. (**a**) One set of initial random vectors. (**b**) another set of initial random vectors

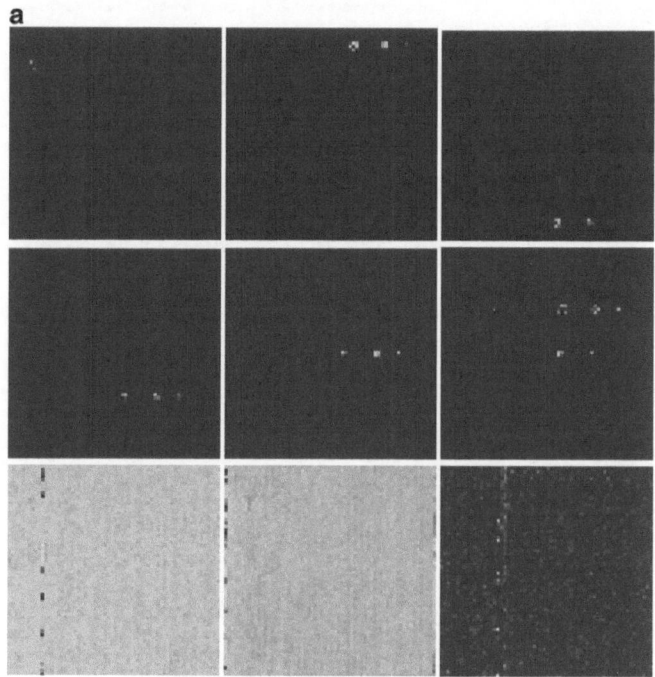

9 PICs produced by PI-PRPP using PI = skewness.

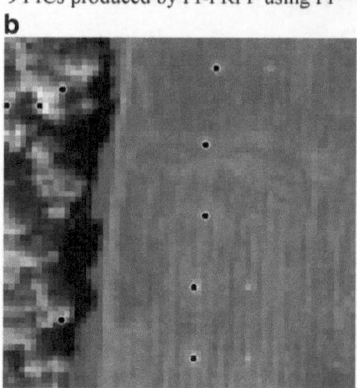

9 endmembers extracted by N-FINDR using the 9 PICs in (a)

Fig. 14.8 PI-PRPP with PI = skewness. (**a**) 9 PICs produced by PI-PRPP using PI = skewness. (**b**) Nine endmembers extracted by N-FINDR using the nine PICs in (**a**)

5.2 PI-PRPP

As noted, the first nine PIPP-generated components in each of Figs. 14.2–14.4 were different not only in order but also in information contained in components. Accordingly, the endmember extraction results were also different in Figs. 14.5–14.7. The PI-PRPP was developed to remedy this problem. In order to make comparison with the results in Figs. 14.2–14.4, the PI-PRPP used the same three PIs, skewness, kurtosis and negentroy to produce PICs which were further prioritized by the same PI = negentropy. Figures 14.8a, 14.9a and 14.10a show the

a

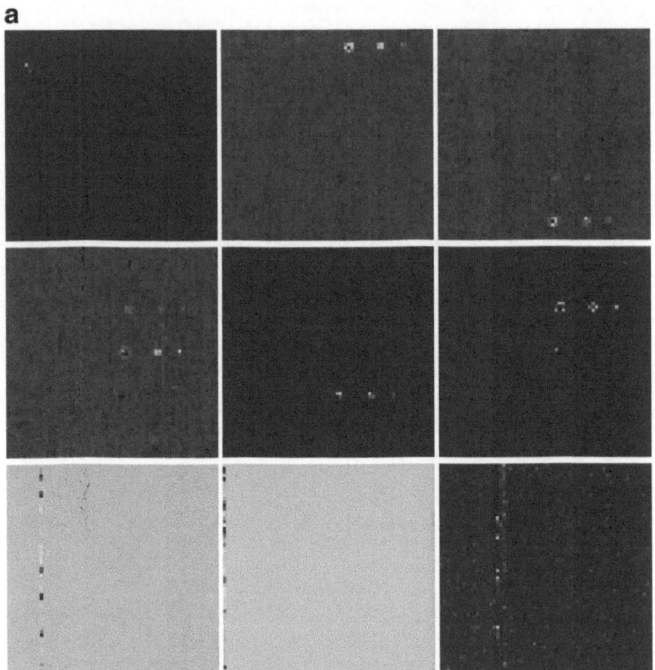

9 PICs produced by PI-PRPP using PI = kurtosis.

b

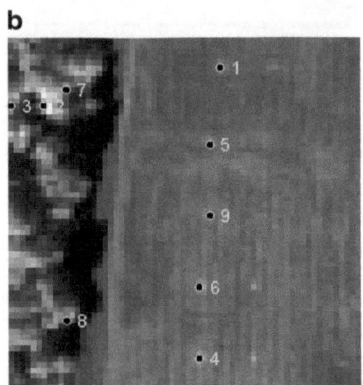

9 endmembers extracted by N-FINDR using the 9 PICs in (a).

Fig. 14.9 PI-PRPP with PI = kurtosis. (a) Nine PICs produced by PI-PRPP using PI = kurtosis. (b) Nine endmembers extracted by N-FINDR using the nine PICs in (a)

nine negentropy-prioritized PICs produced by PI = skewness, kurtosis and negentropy respectively. Figures 14.8b, 14.9b and 14.10b also show the endmember extraction results by applying the N-FINDR to the nine prioritized PICs as image cubes where five panel pixels corresponding to all the five endmembers were successfully extracted compared to only four endmembers

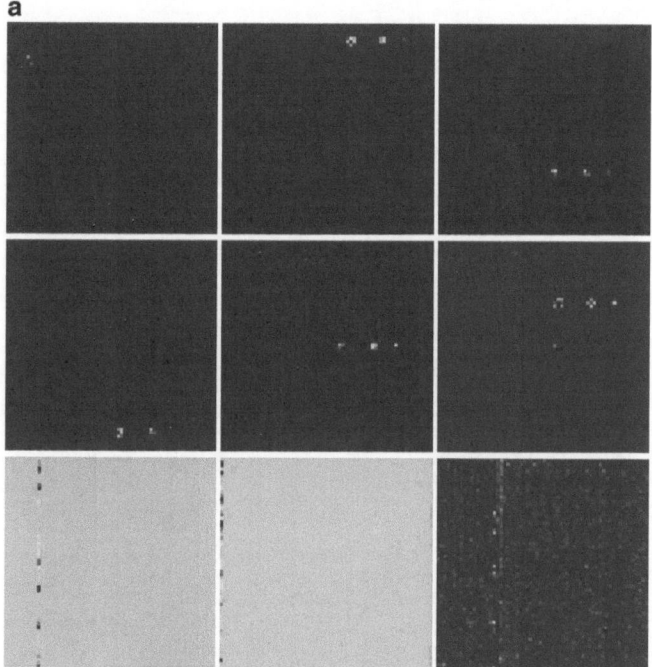

9 PICs produced by PI-PRPP using PI = negentropy.

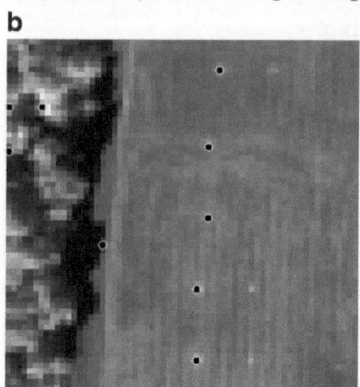

9 endmembers extracted by N-FINDR using the 9 PICs in (a).

Fig. 14.10 PI-PRPP with PI = negentropy. (a) Nine PICs produced by PI-PRPP using PI = negentropy. (b) Nine endmembers extracted by N-FINDR using the nine PICs in (a)

extracted by the PIPP in one run shown in Figs. 14.5a, 14.6a and 14.7a. These experiments clearly demonstrated advantages of using the PR-PIPP over the PIPP.

It should be noted that similar results were also obtained for nine PICs prioritized by PI = skewness and kurtosis. Thus, their results are not included here.

a

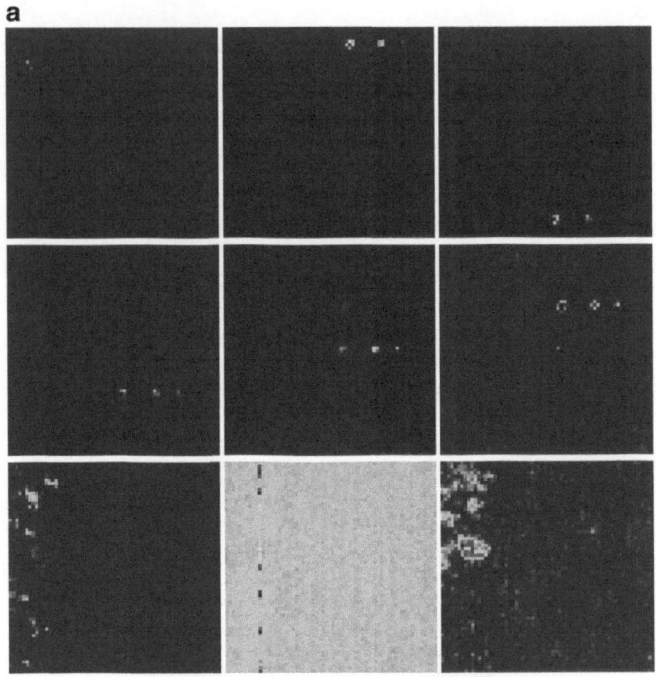

9 PICs produced by ID-PIPP using PI = skewness.

b

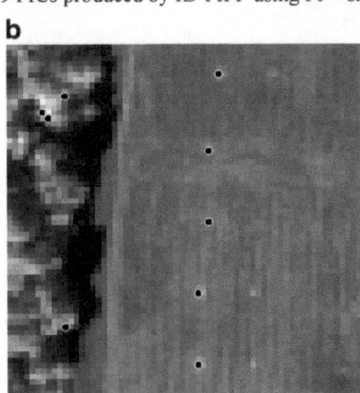

9 endmembers extracted by N-FINDR using the 9 PICs in (a).

Fig. 14.11 ATGP-PIPP with PI = skewness. (**a**) Nine PICs produced by ID-PIPP using PI = skewness. (**b**) Nine endmembers extracted by N-FINDR using the nine PICs in (**a**)

5.3 ID-PIPP

In the experiments of the PIPP and PI-PRPP, the initial conditions were generated by a random generator. This section investigates the ID-PIPP and compares its performance against the PIPP and PI-PRPP. The ATGP was the initialization algorithm used to produce a set of initial vectors to initialize the PIPP. Figures 14.11–14.13 show the results of 9 PICs produced by the ATGP-PIPP and

a

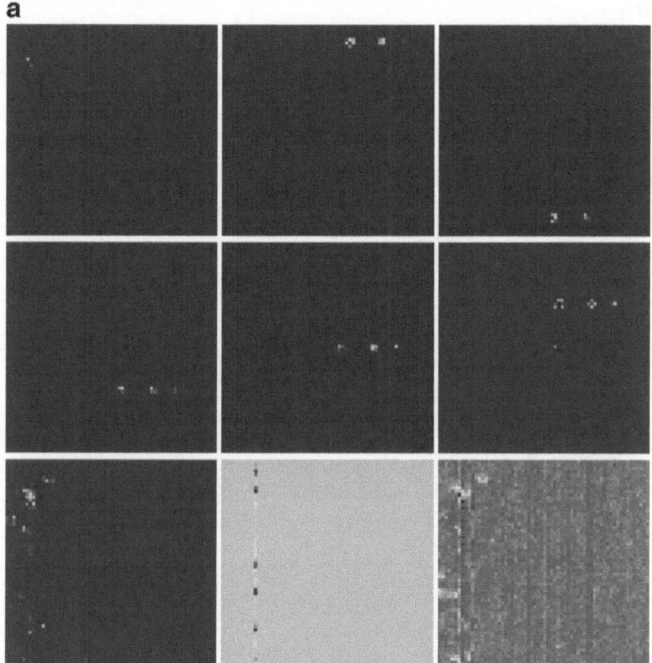

9 PICs produced by ATGP-PIPP using PI = kurtosis

b

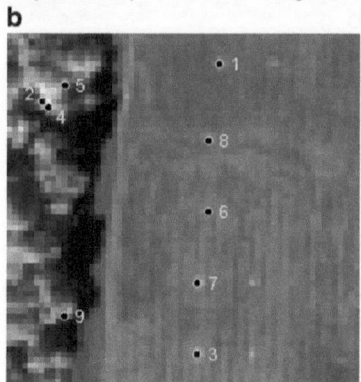

9 endmembers extracted by N-FINDR using the 9 PICs in (a).

Fig. 14.12 ATGP-PIPP with PI = kurtosis. (**a**) Nine PICs produced by ATGP-PIPP using PI = kurtosis. (**b**) Nine endmembers extracted by N-FINDR using the Nine PICs in (**a**)

endmember extraction by the N-FINDR using the nine ATGP-PIPP-generated PICs using PI = skewness, kurtosis and negentropy respectively where like the PI-PRPP there were five panel pixels corresponding to five endmembers which were successfully extracted.

Finally, in order to complete our comparative study and analysis, we also included experiments performed by the PCA which is the 2nd order statistics.

a

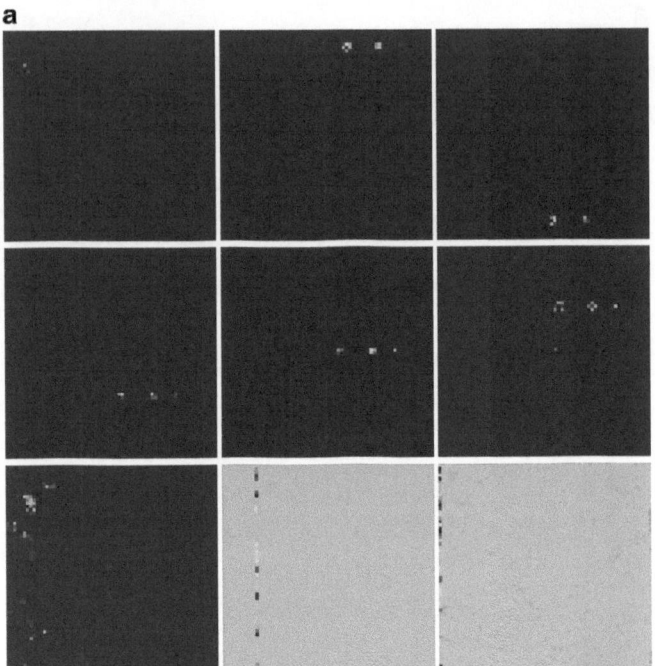

9 PICs produced by ATGP-PIPP using PI = negentropy

b

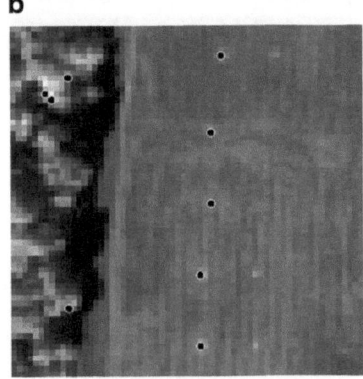

9 endmembers extracted by N-FINDR using the 9 PICs in (a).

Fig. 14.13 ATGP-PIPP with PI = negentropy. (**a**) Nine PICs produced by ATGP-PIPP using PI = negentropy. (**b**) Nine endmembers extracted by N-FINDR using the nine PICs in (**a**)

Figure 14.14a shows the nine PCs produced by the PCA. These nine PCs were then formed as an image cube to be processed by the N-FINDR to extract nine endmembers shown in Fig. 14.14b where only three panel pixels in rows 1, 3 and 5 corresponding to three endmembers were extracted. These experiments provided simple evidence that the PCA was ineffective in preserving endmember information in its PCs.

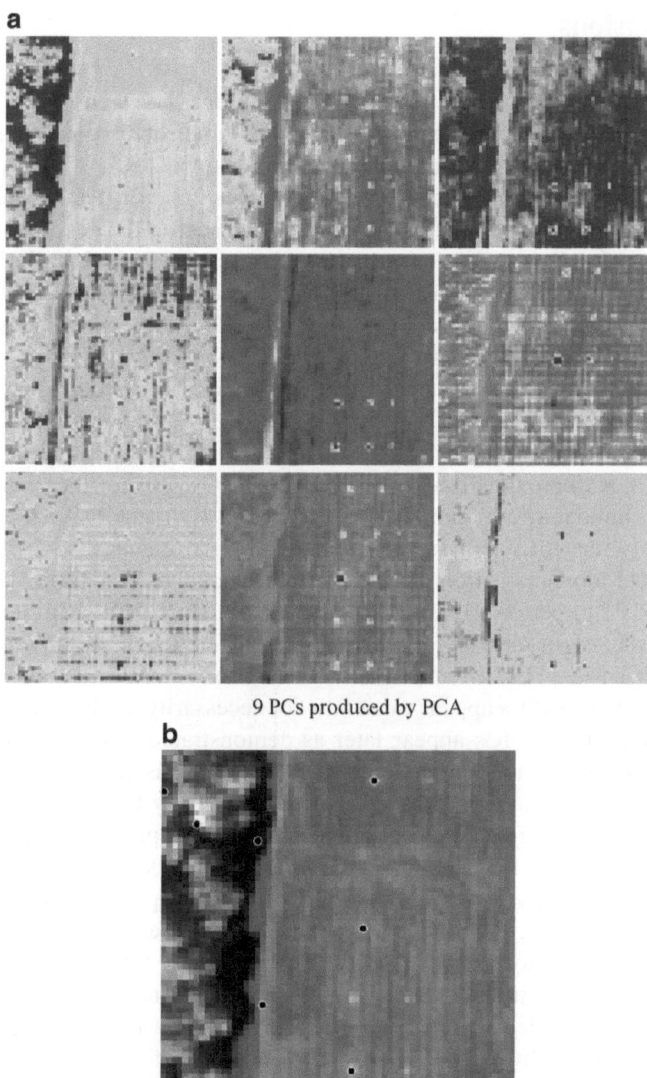

9 PCs produced by PCA

9 endmembers extracted by N-FINDR using the 9 PCs in (a).

Fig. 14.14 PCA results. (**a**) Nine PCs produced by PCA. (**b**) Nine endmembers extracted by N-FINDR using the nine PCs in (**a**)

As a final concluding remark, the same experiments can also be conducted for other data sets such as various synthetic image-based scenarios [14] and another HYDICE scene in [15]. Similar results and conclusions can be also drawn from these experiments. Therefore, in order to avoid duplicates, these results are not presented here.

6   Conclusions

Dimensionality Reduction (DR) is a general pre-processing technique to reduce the vast amount of data volumes provided by multi-dimensional data while retaining most significant data information in a lower dimensional space. One type of multi-dimensional data involving enormous data volumes is hyperspectral imagery. This chapter presents a new approach to DR for hyperspectral data exploitation, called Projection- Index-based Projection Pursuit-based (PIPP) DR technique which includes the commonly used PCA and a recently developed ICA as its special cases. Three version of PIPP are developed for DR, Projection Index-Based PP (PIPP), PI-Prioritized PP (PI-PRPP) and Initialization-Driven PIPP (ID-PIPP). The PIPP uses a selected PI as a criterion to produce projection vectors that are used to specify components, referred to as Projection Index Components (PICs) for data representation. For example, when PI is used as sample data variance, the PIPP is reduced to PCA where the PICs are considered as PCs. On the other hand, when the PI is used as mutual information, the PIPP becomes ICA and PICs are actually ICs. Unfortunately, the PIPP still suffers from two major drawbacks which prevent it from practical implementation. This may be reasons that very little work done on the use of PIPP to perform DR. One of most serious problems with implementing the PIPP is the use of random initial conditions which result in different appearing orders of PICs when two different sets of random initial conditions are used. Under these circumstances PICs appear earlier do not necessarily imply that they are more significant than those PICs appear later as demonstrated in our experiments (see Figs. 14.2–14.4). Another problem is that when DR is performed, there is no appropriate guideline to be used to determine how many PICs should be selected. This issue is very crucial and closely related to the first problem addressed above. In order to cope with the second problem, a recently developed concept, called Virtual Dimensionality (VD) can be used for this purpose. However, this only solves half a problem. As noted, because the PIPP uses a random generator to generate initial conditions, a PIC generated earlier by the PIPP does not necessarily have more useful information that the one generated later by the PIPP. Once the number of PIC, p is determined by the VD, we must ensure that all desired PICs appear as the first p PICs. To this end, two versions of the PIPP are developed to address this issue. One is referred to as PI-based Prioritized PP (PI-PRPP) which uses a PI to rank appearing order of PICs in accordance with the information contained in PICs prioritized by the PI. It should be noted that the PI used for prioritization is not necessarily the same one used by the PIPP as also illustrated in Figs. 14.8–14.10. Although the PI-PRPP prioritizes PICs by a specific PI, it does not necessarily imply that the PI-PRPP generated PICs with same priorities are identical due to the randomness caused by random initial conditions. Therefore, a second version of the PIPP is further developed to mitigate this inconsistency issue. It is called Initialization-Driven PP (ID-PIPP) which uses a custom-designed initialization algorithm to produce a specific set of initial conditions for the PIPP. Since the initial conditions are always the same, the final PP-generated PICs are always consistent. Finally, experiments are conducted to demonstrate the utility of these three PP-based DR techniques and results show their potentials in various applications.

References

1. Wang, J. and Chang, C.-I, "Independent component analysis-based dimensionality reduction with applications in hyperspectral image analysis," IEEE Trans. on Geoscience and Remote Sensing, 44(6), 1586–1600 (2006).
2. Hyvarinen, A., Karhunen, J. and Oja, E., [Independent Component Analysis], John Wiley & Sons (2001).
3. Ren, H., Du, Q., Wang, J., Chang, C.-I and Jensen, J., "Automatic target recognition hyperspectral imagery using high order statistics," IEEE Trans. on Aerospace and Electronic Systems, 1372–1385 (2006).
4. Duda, R. O. and Hart, P. E., Pattern Classification and Scene analysis, John Wiley & Sons, 1973.
5. Friedman, J. H., and Tukey, J. W., "A projection pursuit algorithm for exploratory data analysis," IEEE Transactions on Computers, c-23(9), 881–889 (1974).
6. Chang, C.-I, Hyperspectral Imaging: Techniques for Spectral Detection and Classification, Kluwer Academic/Plenum Publishers, New York (2003).
7. Ren, H., Du, Q., Wang, J., Chang, C.-I and Jensen, J., "Automatic traget recognition hyperspectral imagery using high order statistics," IEEE Trans. on Aerospace and Electronic Systems, 42(4), 1372–1385 (2006).
8. Chu, S., Ren, H. and Chang, C.-I, "High order statistics-based approaches to endmember extraction for hyperspectral imagery," to be presented in SPIE 6966, (2008).
9. Hyvarinen, A. and Oja, E., "A fast fixed-point for independent component analysis," Neural Comp., 9(7), 1483–1492 (1997).
10. Ren, H. and Chang, C.-I, "Automatic spectral target recognition in hyperspectral imagery," IEEE Trans. on Aerospace and Electronic Sys., 39(4), 1232–1249 (2003).
11. Harsanyi, J. and Chang, C.-I, "Hyperspectral image classification and dimensionality reduction: an orthogonal subspace projection approach," IEEE Trans. on Geoscience and Remote Sensing, 32(4), 779–785 (1994).
12. Chang, C.-I, and Du, Q., "Estimation of number of spectrally distinct signal sources in hyperspectral imagery," IEEE Trans. on Geoscience and Remote Sensing, vol. 42, no. 3, pp. 608–619, March 2004.
13. Winter, M.E., "N-finder: an algorithm for fast autonomous spectral endmember determination in hyperspectral data," Image Spectrometry V, Proc. SPIE 3753, pp. 266–277, 1999.
14. Chang, Y.-C., Ren, H., Chang, C.-I and Rand, B., "How to design synthetic images to validate and evaluate hyperspectral imaging algorithms," SPIE Conference on Algorithms and Technologies for Multispectral, Hyperspectral, and Ultraspectral Imagery XIV, March 16–20, Orlando, Florida, 2008.
15. Safavi, H., and Chang, C.-I , "Projection pursuit-based dimensionality reduction," SPIE Conference on Algorithms and Technologies for Multispectral, Hyperspectral, and Ultraspectral Imagery XIV, March 16–20, Orlando, Florida, 2008.